Carbon Farming

Nancy Loria • Rattan Lal

Carbon Farming

Science and Practice

Springer

Nancy Loria
CFAES Rattan Lal Center for Carbon
Management and Sequestration
The Ohio State University
Columbus, OH, USA

Rattan Lal
CFAES Rattan Lal Center for Carbon
Management and Sequestration
The Ohio State University
Columbus, OH, USA

ISBN 978-3-032-00841-1 ISBN 978-3-032-00842-8 (eBook)
https://doi.org/10.1007/978-3-032-00842-8

This Springer imprint is published by the registered company Springer Nature Switzerland AG
The registered company address is: Gewerbestrasse 11, 6330 Cham, Switzerland

Preface

Since the beginning of the industrial era in 1750, Earth has undergone dramatic environmental changes. What started as gradual changes have now escalated into a global crisis—climate change. Initial warnings were met with skepticism, but over time, the surge in extreme weather events, biodiversity loss, and rising food insecurity made the consequences undeniably evident. By the 1990s, global awareness of the crisis gained traction. The Kyoto Protocol of 1997 marked a first step toward international cooperation on greenhouse gas (GHG) emission reductions, followed by the Paris Agreement in 2015, which aimed to limit global temperature rise to well below 2 °C, ideally 1.5 °C. However, global warming has already surpassed the 1.5 °C threshold and continues to rise at a rate of 0.15–0.20 °C per decade. If this trend persists, global temperatures could reach 2 °C by 2050. At the same time, land degradation is accelerating, with UNESCO warning that up to 90 percent of the Earth's land surface could be degraded by mid-century. The urgency of the situation is escalating, and land-based carbon (C) sinks now represent a critical strategy for slowing this trajectory.

Meeting mid-century net zero emissions targets has become a defining global climate objective. However, stabilizing the climate and avoiding irreversible ecological damage will require achieving net negative emissions, where more C is removed from the atmosphere than is released. To date, climate mitigation efforts have largely focused on sectors such as energy, industry, and transport, while the agricultural sector has remained underrepresented, despite its vast and underutilized potential. Among the most powerful tools for removal of C lies beneath our feet: soil. Capable of storing up to three times more C than the atmosphere, soils represent a critical natural climate solution. When managed effectively, soils not only sequester C but also support ecological restoration, improve biodiversity, and enhance land productivity. With over one-third of the world's land under agriculture and nearly half the global population relying on agrifood systems for their livelihoods, the agricultural sector holds a unique and strategic position to lead climate solutions through C farming.

C farming offers a solution that directly addresses the interconnections between soil health, climate mitigation, and sustainability. By enhancing soil organic C

(SOC) through nature-based practices such as reduced tillage, cover cropping, organic amendments, agroforestry, and integrated livestock systems, C farming regenerates soil function, increases productivity, and reduces GHG emissions. These practices not only draw down atmospheric C but also restore water cycles, improve nutrient availability, and build resilience to droughts and floods—core pillars of sustainable agriculture. Across regions, farmers, particularly those with large landholdings, have long adopted such practices out of necessity. Informal experimentation with techniques to enhance yields, manage pests and diseases, and reduce input costs has led to widespread adoption of many C-enhancing practices. However, this grassroots innovation remains largely undocumented and disconnected from formal C programs. Neighboring farmers frequently adopt such practices by observing visible improvements in yield or soil conditions and through informal knowledge exchange. While this peer-to-peer learning supports the spread of C farming practices, it often occurs without a clear understanding of the underlying biophysical processes or long-term climate benefits. As a result, implementation can be inconsistent, and opportunities for optimizing C sequestration are sometimes missed. To unlock the full potential of C farming, it is essential to connect local knowledge with scientific evidence, and field practices with policy and markets.

C Farming: Science and Practice has been written to serve that purpose. The book provides a comprehensive, science-based resource that explores the scientific principles behind C sequestration, field-level practices, and enabling frameworks such as C markets and digital monitoring tools. It aims to strengthen the understanding of how soil C management contributes to climate solutions, agricultural resilience, and long-term sustainability. This book is an attempt to connect the dots: to provide a clear, science-based understanding of C farming practices; to highlight their climate, ecological, and socio-economic benefits; and to explore how they can be scaled, measured, and supported through policy, technology, and markets. It brings together current scientific findings, practical methods, and global experiences to support everyone involved in shaping the future of sustainable agriculture. The goal is to demystify C farming by explaining not just the "how," but also the "why"—why these practices work, why they matter, and why now is the time to act.

C farming supports the achievement of the United Nations Sustainable Development Goals (SDGs) by 2030. By improving soil fertility, increasing crop productivity, and stabilizing farm incomes, it makes a direct contribution to SDG 1 (No Poverty) and SDG 2 (Zero Hunger). Water-efficient practices and improved soil structure support SDG 6 (Clean Water and Sanitation), while enhanced C sequestration and climate resilience on agricultural lands drive progress toward SDG 13 (Climate Action). Additionally, nature-based practices such as agroforestry, cover cropping, and composting promote biodiversity and help reverse land degradation, supporting SDG 15 (Life on Land). Together, these benefits position C farming as a strategic approach to aligning agriculture with global sustainability goals. This book also contributes to global initiatives such as the 4p1000, the Adaptation of African Agriculture (AAA) launched at COP22 in Marrakesh in 2016, the Coalition of Action 4 Soil Health (CA4SH) launched during the UN Food Systems Summit in 2021, and the Global Future Council on Soils launched by the World Economic

Forum in 2025, by offering a deeper understanding of how increasing SOC can enhance food security and climate resilience.

The book is intended for a diverse audience, including farmers, researchers, practitioners, policymakers, students, and development professionals. It serves as both an academic reference and a practical guide, designed to deepen understanding, inform decisions, and inspire action.

C farming is more than a climate solution, not merely a set of techniques—it is a systems-based approach that integrates agronomy with environmental stewardship to regenerating landscapes, supporting livelihoods, and addressing the climate crisis. By transforming agriculture into a net sink of C, the sector can play a pivotal role in achieving climate neutrality while restoring ecosystems and advancing sustainability.

This book aims to contribute meaningfully to that transformation by building scientific understanding, supporting policy integration, and accelerating on-the-ground action. The implementation of C farming can be framed within a trinity of acts: the Clean Air Act, the Clean Water Act, and the urgently needed Healthy Soil Act. While the first two acts are already in place, the Healthy Soil Act is essential to institutionalize soil health and unlock the full potential of soils as a long-term C sink. The knowledge and frameworks presented in this book provide the scientific foundation and practical insights necessary to inform such legislation.

The hope is that this book will inform, connect, and empower stakeholders across sectors in the pursuit of a sustainable, low-carbon, and climate-resilient future.

Columbus, OH, USA

Nancy Loria
Rattan Lal

Acknowledgments

The preparation of this book began under a project that was partially supported through funding from Microsoft. The authors gratefully acknowledge the School of Environment and Natural Resources at The Ohio State University for providing laboratory and desk space that facilitated the research and writing process. This institutional support was instrumental in shaping the content and direction of the book.

Competing Interests The authors have no competing interests to declare that are relevant to the content of this manuscript.

Contents

List of Figures

List of Tables

About the Authors

Nancy Loria, Ph.D., is a researcher at the CFAES Rattan Lal Center for Carbon Management and Sequestration at The Ohio State University. She holds a Ph.D. in Environmental Science and a master's in (Forestry) Environmental Management, both awarded with gold medals from Dr. Y S Parmar University of Horticulture and Forestry, Solan, India. Her doctoral research examined the impacts of agricultural practices on natural resources and the vulnerability of farming communities to environmental changes. She has also contributed to the National Mission for Sustaining the Himalayan Ecosystem during her tenure at the ICAR–Indian Institute of Soil and Water Conservation, Research Centre, Chandigarh, India. Her research interests include carbon farming, soil health, soil carbon sequestration, sustainable agriculture, and agroforestry. She has worked across diverse agroecological zones, including the Indian Himalayan region and North America, evaluating land management practices that enhance soil health and support climate-resilient agriculture. While pursuing her research interests, she has authored multiple peer-reviewed research articles and book chapters. She continues to advance her research in carbon farming, with a focus on developing innovative solutions for soil carbon management.

Rattan Lal, Ph.D., is a Distinguished University Professor and Director of the CFAES Rattan Lal Center for Carbon Management and Sequestration at The Ohio State University. He also holds adjunct/honorary professorships at multiple institutions, including Punjab Agricultural University (Ludhiana, India), the Indian Agricultural Research Institute (New Delhi), Amity University (Noida, India), Sher-e-Kashmir University of Agricultural Sciences and Technology (Srinagar, India), and the National University of Uzbekistan (Tashkent), and serves as a Visiting Professor at the Pontifical Catholic University of Valparaíso, Chile. He is Chair in Soil Science and Goodwill Ambassador for Sustainable Development Issues of the Inter-American Institute for Cooperation on Agriculture, San José, Costa Rica, and Global Ambassador of Regenagri, London, United Kingdom. He was President of the World Association of the Soil and Water Conservation, the International Soil Tillage Research Organization, the Soil Science Society of America (2006–2008), and the International Union of Soil Sciences (2017–2018). He has researched soil-centric and regenerative agricultural management for food and climate security, soil carbon sequestration, sustainable/eco-intensification, input use efficiency of agroecosystems, sustainable management of soils, and soil health. His career spans over six decades and six continents. With an h-index of 204 and 177,072 citations (April, 2025), he has authored over 1,200 journal articles, mentored 425 researchers, and was ranked #1 globally by Research.com in 2023 and 2024. He is laureate of GCHERA World Agriculture Prize (2018), Glinka World Soil Prize of FAO (2018), Japan Prize (2019), U.S. Awasthi IFFCO Prize of India (2019), World Food Prize (2020), Arrell Global Food Innovation Award of Canada (2020), Padma Shri Award of India (2021), Mahatma Gandhi Environmental Protection Award (2022), and Gulbenkian Prize for Humanity (2024).

Abbreviations

ACR	American Carbon Registry
AFOLU	Agriculture, Forestry, and Other Land Use
AFS	Agroforestry Systems
AI	Artificial Intelligence
BC	Biogenic Carbonates
BMPs	Best Management Practices
C	Carbon
CA	Conservation Agriculture
CAR	Climate Action Reserve
CARB	California Air Resources Board
CCP	Core Carbon Principles
CDM	Clean Development Mechanism
CEC	Cation Exchange Capacity
CFCs	Chlorofluorocarbons
CFI	Carbon Farming Initiative
CH_4	Methane
cLHS	Conditioned Latin Hypercube Sampling
CO_2	Carbon Dioxide
CO_2eq	Carbon Dioxide Equivalent
COP	Conference of the Parties
CRP	Conservation Reserve Program
CS	Carbon Sequestration
CSA	Climate-Smart Agriculture
DIC	Dissolved Inorganic carbon
DOC	Dissolved Organic Carbon
dton	Dry ton
EPA	Environmental Protection Agency
ESs	Ecosystem Services
ET	Emissions Trading
EUETS	European Union Emissions Trading System
FAO	Food and Agriculture Organization

GC	Geogenic Carbonates
Gg	Gigagram = 10^9g
GHG	Greenhouse Gas
GPS	Global Positioning System
GS	Gold Standard
GSP	Global Soil Partnership
GWP	Global Warming Potential
Ha	Hectares
ICVCM	Integrity Council for the Voluntary Carbon Market
INM	Integrated Nutrient Management
IoT	Internet of Things
IPCC	Intergovernmental Panel on Climate Change
IPM	Integrated Pest Management
IUV	Indirect Use Values
JI	Joint Implementation
K	Potassium
LCA	Life Cycle Assessment
M Ha	Million Hectares
MaOM	Mineral-Associated Organic Matter
MBC	Microbial Biomass Carbon
Mg	Megagram =10^6g
M Ha	Million Hectares
MRT	Mean Residence Time
MRV	Measurement, Reporting, and Verification
MWD	Mean Weight Diameter
N_2O	Nitrous Oxide
NDCs	Nationally Determined Contributions
NT	No-Tillage
O_3	Ozone
OC	Oxidative Carbon
OM	Organic Matter
P	Phosphorus
PC	Pedogenic Carbonates
PES	Payments for Ecosystem Services
Pg	Petagram =10^{15}g
PLF	Precision Livestock Farming
POC	Particulate Organic Carbon
POM	Particulate Organic Matter
POXC	Permanganate Oxidizable Carbon
PPM	Parts Per Million
R^2	Coefficient of Determination
RA	Regenerative Agriculture
RMSE	Root Mean Square Error
RS	Remote Sensing
S	Sulfur

SA	Sustainable Agriculture
SAN	Sustainable Agriculture Network
SAPs	Sustainable Agriculture Practices
SBTi	Science Based Targets Initiative
SDGs	Sustainable Development Goals
SEP	Soil Enrichment Protocol
SIC	Soil Inorganic Carbon
SLM	Sustainable Land Management
SOC	Soil Organic Carbon
SOM	Soil Organic Matter
TC	Total Carbon
Tg	Teragram = 10^{12}g
UAV	Unmanned Aerial Vehicles
UNFCCC	United Nations Framework Convention on Climate Change
USDA	United States Department of Agriculture
VCM	Voluntary Carbon Market
VCMI	Voluntary Carbon Markets Integrity Initiative
VCS	Verified Carbon Standard
WCI	Western Climate Initiative
WFP	Water Footprint
WRP	Wetlands Reserve Program

Chapter 1
Introduction

Abstract This chapter lays the foundation for carbon (C) farming, building from sustainable agriculture to introduce it as a climate solution. It provides an integrated overview of its scientific basis, global efforts, and the growing urgency for its adoption. The chapter highlights that, with soils holding far more C than the atmosphere, even modest improvements in soil management can make a significant difference in reducing greenhouse gas emissions and enhancing resilience. It also discusses international initiatives such as the United Nations' Sustainable Development Goals, COP28, and the 4 per 1000 initiative that support these efforts, positioning C farming as a practical and scalable approach for a sustainable future.

1.1 Introduction

The global climate is undergoing profound changes, with far-reaching consequences. Progressive surface and atmospheric temperatures, erratic precipitation patterns, and rising sea levels are reshaping the planet. Concurrently, extreme weather events like heatwaves, wildfires, and drought-flood syndrome are becoming distressingly frequent. These events, coupled with the spread of disease-carrying vectors, pose immediate risks to lives and livelihoods. Moreover, the shrinking of land areas due to rising sea levels exacerbates the vulnerability of coastal communities. In regions of high altitude, species are being forced to migrate, disrupting ecosystems and biodiversity. Agriculture, a cornerstone of global sustenance, faces unprecedented challenges with frequent crop losses, leading to food insecurity and poverty. In the face of these interconnected crises, the urgency for immediate and collective action to combat climate change is undeniable. Climate crisis is primarily driven by the escalation of greenhouse gas (GHG) emissions, namely carbon dioxide (CO_2), methane (CH_4), nitrous oxide (N_2O), ozone (O_3), and Chlorofluorocarbons (CFCs), which have surged over the past two centuries (Calvin et al. 2023). Human activities predominantly fuel this increasing emissions trend, despite the influence of both natural and human-induced factors on Earth's climate (Hayhoe et al. 2018; Calvin et al. 2023). Human activities, notably alterations in land use, the

N. Loria, R. Lal, *Carbon Farming*, https://doi.org/10.1007/978-3-032-00842-8_1

combustion of fossil fuels, excessive exploitation of natural resources, inefficient energy consumption, inadequate waste management, and unsustainable agricultural intensification, have all incessantly contributed to the rise in GHG emissions (IPCC 2019; WMO 2024).

Climate change is now widely recognized as one of the most critical challenges of the twenty-first Century. Its effects are pervasive and acutely felt across various sectors including energy, transportation, health, forests, oceans, and agriculture (Marvel et al. 2023). While all sectors grapple with its impacts, agriculture emerges as particularly vulnerable due to its profound reliance on climatic conditions, underscoring the urgency for targeted adaptation measures. Worldwide agri-food systems are responsible for generating 31 percent of the GHG emissions (Crippa et al. 2021). Between 2010 and 2019, the Agriculture, Forestry, and Other Land Use (AFOLU) sector contributed around 13–21% of the overall anthropogenic GHG emissions worldwide (Pathak et al. 2023). Approximately half of the global anthropogenic CH_4 emissions and three-quarters of the overall N_2O emissions arise from agricultural activities (Tubiello et al. 2022). These emissions primarily result from biological processes intrinsic to agricultural systems: enteric fermentation leads to the release of CH_4, which is subsequently expelled into the atmosphere through belching; manure management can produce CH_4 as livestock waste decomposes under specific conditions; flooded rice cultivation creates anaerobic conditions in the soil, fostering CH_4 production by specialized microbes; synthetic pesticides release N_2O through processes like nitrification and denitrification; and the burning of crop residues in certain practices releases CO_2 and other pollutants into the atmosphere. While GHG emissions are a significant driver of climate change, it's important to note that they cannot single-handedly reverse its impacts. However, they do play a vital role in enhancing the agricultural sector's ability to adapt to changing climatic conditions.

Altering weather patterns, erratic rainfall, and the surge in extreme events like droughts, floods, and heatwaves disrupt the delicate balance of agricultural systems. This disruption can further exacerbate climate change through a feedback loop if not addressed promptly (Fig. 1.1). The escalating climate crisis is straining agriculture, leading to crop failures, livestock challenges, and a significant decline in overall farm productivity. This inability to adapt not only jeopardizes the livelihoods of those dependent on agriculture but also fuels global hunger, as changing climatic conditions disrupt the food production backbone. In the next three decades, a substantial increase in global food demand is anticipated, and is projected to rise significantly by approximately 35–56% (van Dijk et al. 2021). In parallel, the population vulnerable to hunger is expected to experience a shift, ranging from a potential reduction of 91% to a possible increase of 8% during the period 2010–2050 (van Dijk et al. 2021). Without prompt and decisive action to implement adaptive measures and sustainable practices, this cycle of vulnerability will persist, underscoring the urgent need for a concerted global effort to mitigate climate change and secure a resilient and sustainable future for agriculture and food systems worldwide.

International efforts, such as the United Nations' Sustainable Development Goals (SDGs) have identified addressing food security as a critical component in the battle against climate change. These goals, with 'No Poverty' and 'Zero Hunger'

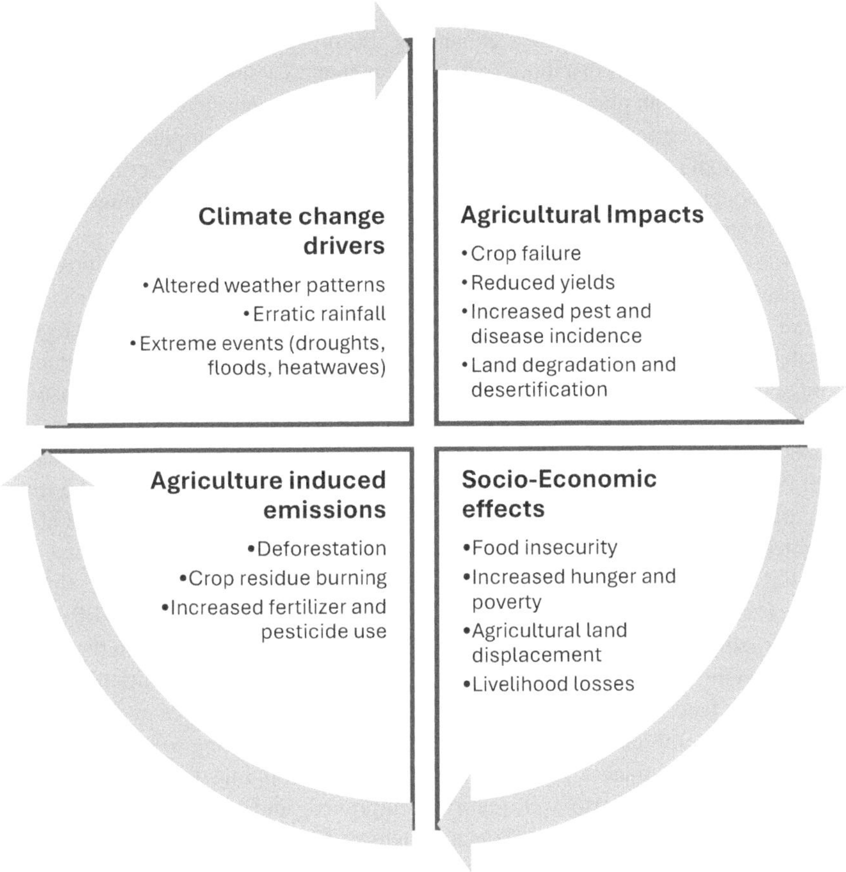

Fig. 1.1 Climate change impacts and feedback loop in agriculture

at the forefront, emphasize the necessity of ensuring a growing global population has access to not only sufficient but also nutritious food, all while protecting the environment from the adverse impacts of climate change. Effectively tackling the challenges faced by agriculture involves a two-fold approach: mitigating GHG emissions originating from agricultural activities and preventing them in the first place. Central to this effort are the prudent management of crops and soil, as a significant portion of emissions stems from their misuse. This can be effectively managed by enhancing the capacity of crops to store carbon (C), both in their biomass and the underlying soils. This dual-pronged strategy holds tremendous potential, with the capacity to reduce agriculture's negative impact on climate change by an impressive 5.5–6.0 gigatons (Gt = 1billion ton = Pg = 10^{15} g) of CO_2 equivalent per year (Carlisle et al. 2019). Even modest adjustments in agricultural practices, such as fine-tuning crop management techniques, optimizing water resources, applying inputs judiciously, and minimizing soil disturbance, can lead to significant reductions in GHG emissions. Simultaneously, these changes promote C sequestration in both soil and vegetation, enhancing the process of capturing C from the atmosphere into the soil in a more sustainable manner. By implementing these strategies, agriculture can play a pivotal role in mitigating climate change while continuing to meet the critical demand for food security worldwide. Implementing "4 per mille" in global agriculture can sequester 2–3 Gt of C annually, offsetting 20–35% of anthropogenic GHG emissions, serving as a short-term climate change solution until advanced technologies emerge (Minasny et al. 2017).

Soil, often overlooked in comparison to air and water, is a vital resource with immense potential to mitigate agricultural emissions. It contains the highest concentration of soil organic C (SOC) on land, playing a pivotal role in the battle against climate change while offering a plethora of environmental and societal benefits. Storage of C within soil relies on intricate chemical and physical interactions with minerals, contributing significantly to its role as a major C reservoir, alongside oceans and forests. On a global scale, the soil C pool, containing approximately 2500 Pg C, holds three times more C than the atmosphere (760 Pg) and 4.5 times more than the biotic C (560 Pg) (Lal 2004). In fact, efficient utilization of soil has the potential to sequester a staggering 178 Pg C by the end of the twenty-first century (Lal et al. 2018). Peatlands, in particular, are exceptionally rich in C sinks—covering approximately 487 M ha and storing around 600 Pg of C (UNEP 2022). However, climate change is altering soil behavior, potentially triggering the release of C. Allowing this C to escape as CO_2 into the atmosphere risks initiating a self-perpetuating feedback loop, further escalating global temperatures and potentially exacerbating C emissions from soil. Yet, by safeguarding existing soil C and enhancing it through sustainable practices such as cover cropping, not only is climate change combatted but also soil quality is improved, and food production is sustained in the long term. Soil's role can transform from a potential source of GHG emissions, when mismanaged, to an effective sink when managed with wisdom and care.

Enhancing soil C is crucial for ecological balance, as agriculture-induced land use changes have led to a substantial loss of approximately 133 Pg of C from the top 2 m of global soils, with a sharp increase in loss rates over the past two centuries (Sanderman et al. 2017). Soil C exists in two primary forms: SOC and soil inorganic C (SIC), each bearing distinct characteristics. SOC, derived from decomposing plant and animal matter and inhabited by microorganisms, fungi, and roots, plays a dynamic role in soil processes, impacting water retention, nutrient availability, and structure. Conversely, SIC, mainly as stable carbonates formed through geological interactions, persists in soil for extended periods. In the top 1 m of soil, the SOC pool contains 1325 Pg C, while the SIC pool accounts for roughly 950 Pg C (Lal et al. 2015). Prioritizing SOC management, achieved through practices like incorporating organic amendments and minimizing soil disturbance, presents immediate opportunities to enhance soil health and re-carbonization of the terrestrial biosphere. While the direct manipulation of SIC is somewhat constrained, understanding how it interacts with SOC is crucial for managing C holistically. These two forms of C mutually influence each other, and their distribution varies along with the vertical profile of the soil (Zhao et al. 2019). Recognizing the presence and behavior of SIC in specific soil types provides valuable insights into their unique characteristics and functions (Tan et al. 2014). Recent research emphasizes that SIC role in C sequestration, particularly in drylands, and its significant influence on ecosystem services (ESs) such as nutrient cycling and soil buffering capacity (Dina Ebouel et al. 2024). Thus, an integrated approach that considers both SOC and SIC is essential for effective soil C management and sustainable land use strategies.

Recognizing how important soil C is for overall soil health, there's a big effort worldwide, from local to global levels to protect soil C stocks and reduce emissions. Sustainable agriculture (SA), conservation agriculture (CA), and regenerative agriculture (RA) all emphasize the critical importance of soil C preservation and emission mitigation in farming practices. These approaches prioritize minimizing soil disturbance, promoting soil cover, enhancing biodiversity, retaining plants in the soil, and integrating livestock into agricultural systems. By adopting these strategies, farmers uphold soil health, as illustrated by the shared principles of SA, RA, and CA in enhancing soil C and ecological resilience (Fig. 1.2). SA is defined by the American Society of Agronomy as "one that, over the long term, enhances the environmental quality and the resource base on which agriculture depends; provides for basic human food and fiber needs; is economically viable; and enhances the quality of life for farmers and society as a whole"(Francis and Youngberg 1990). This paradigm shift towards sustainable practices has bestowed significant benefits upon farmers, including enhanced soil fertility, increased crop yields, reduced soil erosion, elevated water retention, greater soil biodiversity, and improved water and air quality. However, there remain challenges in the widespread adoption of these sustainable methods due to various factors such as production, marketing, financial, human, and institutional risks (Kahan 2013; Dessalegn et al. 2018). Overcoming these hurdles requires providing farmers with comprehensive information about the potential effects of sustainable agricultural practices (SAPs) on crop yield, farm income, local resources, and the global environment. This knowledge empowers farmers to make informed decisions, considering both the risks and benefits, ultimately leading to the adoption of less C-intensive farming practices.

Several global initiatives have been established to promote SA to combat climate change. The COP28 declaration on SA, Resilient Food Systems, and Climate Action emphasizes the imperative to maximize environmental benefits in agriculture. It highlights key measures such as promoting conservation, protecting and restoring land and natural ecosystems, enhancing soil health, and transitioning to sustainable production and consumption approaches (Davey 2023). Additionally, the Global Soil Partnership (GSP), led by the FAO, addresses global soil issues, including C sequestration, while the Regenerative Organic Certification program prioritizes soil health and animal welfare. In the United States, the Soil Health Partnership focuses on implementing CA practices for enhanced soil health. Australia's C Farming Initiative encourages C sequestration practices. On a broader scale, global efforts such as the United Nations Framework Convention on Climate Change (UNFCCC), Global Alliance for Climate-Smart Agriculture (CSA), and SDGs, are actively working towards SA. The COP28 declaration aligns with these initiatives,

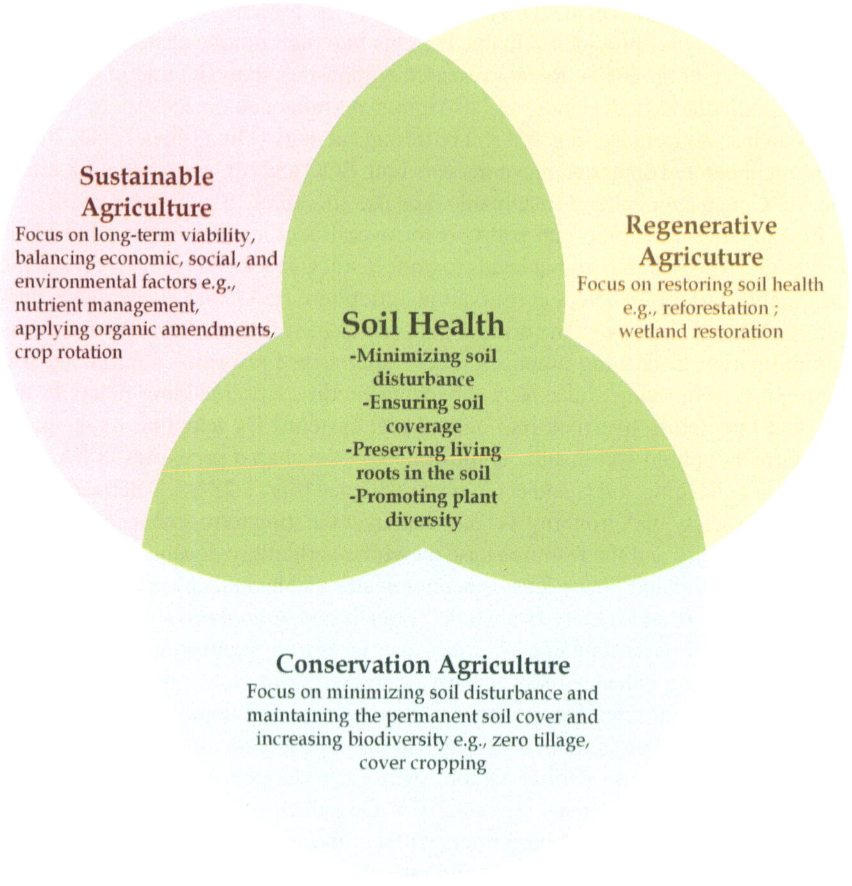

Fig. 1.2 Soil health management approaches in agriculture

emphasizing the importance of environmentally sustainable, economically viable, and socially inclusive agricultural practices. Collaborative initiatives, including the Agroecology Knowledge Hub, International Fund for Agricultural Development, Climate and Clean Air Coalition Agriculture Initiative, and Global Agriculture and Food Security Program, further contribute by facilitating knowledge exchange, reducing climate pollutants, and providing financial resources and technical assistance for SA projects in developing countries. The combined efforts of various stakeholders, including farmers, researchers, and agricultural organizations, are instrumental in advancing SAPs. This not only supports agricultural productivity but also contributes to the broader goal of mitigating climate change by sequestering C in soil.

SA is the central and widely adopted concept in every initiative aimed at addressing both food insecurity and concerns related to climate change. Widespread awareness exists about the significance of SA within farming systems. However, when the aim is to enhance C levels through SAPs, it prompts a series of vital considerations:

1. Site-specific assessment of farmlands: Tailoring practices to specific soil types and neighboring land management strategies ensure that SAPs are effectively implemented, maximizing their potential to sequester C.
2. Measuring C sequestration in transitional phases: Robust methods for quantifying C sequestration provide tangible evidence of the positive impact of SAPs on soil C levels, increasing confidence in their effectiveness.
3. Quantifying GHG emission reductions: Accurately measuring reductions in GHG emissions highlights the direct contribution of SAPs to mitigating environmental impact and enhancing soil C levels.
4. Evaluating farmers' comfort level and willingness to transition: Understanding farmers' comfort levels and addressing their concerns facilitates a smoother transition to SAPs, ensuring a higher likelihood of sustained implementation and C sequestration benefits (Buck and Palumbo-Compton 2022).
5. Short-term changes in soil behavior and patience levels: Demonstrating short-term changes in soil behavior highlights the initial benefits of SAPs, reinforcing farmers' confidence and commitment to the practices for long-term C sequestration.
6. Balancing benefits and risks: Analyzing the direct benefits, such as increased yield and soil health, provides clear incentives for farmers to adopt SAPs, leading to improved C in soils.
7. Scalability and broader impact: Evaluating the scalability of SAPs ensures that their positive impact on soil C levels can be extended across diverse agricultural organizations, contributing to a larger-scale reduction in GHG emissions.
8. Establishing collaborative mechanisms: Effective collaboration among stakeholders fosters knowledge exchange and facilitates the widespread adoption of SAPs, leading to a collective effort in enhancing soil C levels.

By considering these crucial aspects, a robust framework is established to support the sustainable increase of soil C levels. This, in turn, significantly contributes to environmental conservation and agricultural sustainability. In the pursuit of holistically addressing the concerns of the farming community, uncovering the intricacies of agriculture, and responding to the imperatives of the environment, all while emphasizing the significance of soil C, the journey leads through a thorough exploration of the transformative concept known as C farming. As a natural extension of SA, C farming actively removes CO_2 from the atmosphere and stores it securely within vegetation and soils, offering a powerful strategy to combat climate change while enhancing long-term resilience, productivity, and ecological health. C farming integrates science, policy, and practice to strengthen agricultural landscapes while contributing to climate mitigation, food security, and ecosystem restoration. It recognizes the intricate connections between soil C, agricultural productivity, environmental health, and farmer livelihoods. Recent global discussions have brought C farming to the center of climate change mitigation and food security dialogues (Sharma et al. 2021; Appunn 2022; Bumbiere et al. 2022; Green America

2023; CCI 2025). Rooted in principles of soil health, it not only reduces GHG emissions but also creates economic opportunities for farmers through C markets and credits.

The primary objective of this book is to provide a comprehensive, science-based understanding of C farming that connects field-level practices with policy frameworks, C markets, and digital monitoring systems. It aims to bridge local knowledge with scientific evidence, inform decision-making, and support the scaling of C farming practices through practical guidance, enabling technologies, and global experiences. This book is intended to inform, connect, and empower a wide audience, including farmers, researchers, practitioners, policymakers, students, and development professionals.

More than a set of techniques, C farming represents a comprehensive, systems-based solution that unites agriculture, the environment, and society. By integrating environmental stewardship, economic development, and social well-being, it embodies the essence of sustainability. The widespread adoption of C farming offers a practical and scalable pathway toward a more climate-resilient, productive, and sustainable future for agriculture and the planet.

1.2 Overview of the Book

Through the chapters that follow, the concept of C farming will be navigated, revealing its potential impact on soil health, its role in the C cycle, and its far-reaching implications for the global environment. Chapter 2 provides a comprehensive overview of C farming, an innovative approach to SA that enhances C sequestration through improved land management practices. It highlights the importance of key components such as vegetation, soil, soil microorganisms, and agricultural practices in capturing and storing C. The chapter outlines fundamental principles including promoting soil health, conserving ecosystems, monitoring soil C, ensuring policy alignment, and fostering sustainability and resilience. It also explores the wide-ranging benefits of C farming—ecological, environmental, social, and economic—such as climate change mitigation, biodiversity conservation, improved food security, and new income opportunities for farmers. Through these insights, the chapter underscores C farming's vital role in addressing climate challenges and promoting long-term agricultural sustainability.

Chapter 3 explores the intricate relationship between soil health and C sequestration, laying the foundation for a more holistic approach to C farming by emphasizing the crucial role of soil C in sustaining productive and resilient ecosystems. It begins by examining the concept of soil C and the various factors influencing its dynamics, including climate, vegetation, land management, and inherent soil properties. The chapter then delves into SOC stabilization, highlighting mechanisms such as spatial inaccessibility, chemical recalcitrance, and organo-mineral associations that protect C within the soil matrix. It further connects soil C to key soil health principles, including minimizing disturbance, maintaining cover, preserving living roots, and promoting plant diversity. Finally, it outlines the ecosystem benefits of soil C, such as enhanced nutrient cycling, microbial activity, water retention, pH buffering, and resistance to degradation, illustrating soil C's central role in sustaining healthy, climate-resilient agricultural landscapes.

Chapter 4 provides an in-depth exploration of diverse C farming techniques, including agroforestry, cover cropping, CA, crop rotation, and biochar. Each practice is examined with attention to its underlying principles, implementation methods, and potential for C sequestration. Additionally, the chapter presents compelling case studies showcasing successful implementations of these practices in real-world agricultural contexts. By combining scientific insight with practical examples, this comprehensive overview serves as a valuable resource for those seeking to adopt sustainable and climate-resilient farming practices.

Chapter 5 explores the versatile application of C farming across diverse agricultural systems, including row cropping, plantation farming, livestock, horticulture, specialty crops, and wetland agriculture. It highlights how C farming principles can be adapted to different contexts, offering flexible and scalable solutions for improving soil health and sequestering C. The chapter examines system-specific opportunities and challenges, showcases sustainable innovations like precision livestock farming, and emphasizes the role of specialty and wetland systems in C storage. It concludes by identifying key knowledge gaps and research priorities to guide future efforts in advancing C farming across diverse landscapes.

Chapter 6 investigates the critical aspect of measuring and monitoring C sequestration in the context of C farming. It examines a range of methodologies designed to accurately quantify SOC, beginning with traditional laboratory-based techniques such as wet oxidation and dry combustion, and extending to modern in-situ methods like spectroscopic analyses. The chapter also highlights the emerging role of remote sensing technologies in expanding spatial coverage and reducing monitoring costs. Further, it discusses SOC monitoring frameworks, including modeling approaches, and addresses key challenges such as spatial variability, data quality, and long-term consistency. By integrating conventional and emerging tools, this chapter equips readers with the knowledge required for reliable, science-based monitoring—empowering practitioners to optimize C farming strategies and enhance their environmental impact.

Chapter 7 focuses on the quantification of SOC stocks, a critical step in assessing the effectiveness of C farming. The chapter begins with the principles of SOC stock assessment, emphasizing the importance of accurately measuring key variables such as SOC concentration, bulk density, soil depth, and gravel content. It then outlines standardized sampling protocols and discusses appropriate sampling densities to ensure representativeness across heterogeneous landscapes. A key concept introduced is the equivalent soil mass approach, which improves comparability of SOC stock estimates by accounting for differences in soil mass rather than depth alone. The chapter concludes by addressing the challenges associated with SOC stock assessments, including variability in soil properties, methodological inconsistencies, and scaling from field to regional levels. By providing practical guidance and highlighting critical considerations, this chapter supports robust and reliable estimation of SOC stocks in diverse agroecosystems.

Chapter 8 provides a comprehensive overview of C trading as a mechanism to incentivize soil C sequestration within agricultural systems. It begins by discussing the societal value of C and the concept of C credits, detailing the key steps involved in C generation—project design, baseline establishment, measurement, reporting, verification, issuance, and retirement of credits. The chapter distinguishes between regulatory and voluntary C markets and explores how C pricing influences market dynamics. It critically examines the challenges specific to soil C credit programs, including low C prices, issues of additionality, permanence, leakage, and measurement reliability. The chapter concludes by outlining a path forward through payment for ecosystem services, removal of key barriers, enhanced policy and governance frameworks, and the need for continued research and innovation to ensure credible, scalable, and equitable C trading systems.

Chapter 9 highlights the transformative role of technology in enhancing the effectiveness and scalability of C farming. It examines a suite of advanced tools and innovations—including precision agriculture, smart sensors, remote sensing, artificial intelligence (AI), the Internet of Things (IoT), and blockchain technology—that collectively enable improved soil C sequestration, real-time monitoring, and data-driven decision-making. Blockchain is explored for its potential in ensuring transparency and traceability in C credit systems. The chapter also addresses key technological challenges, such as sensor accuracy, data integration, high initial costs, connectivity gaps, and regulatory and socio-economic barriers. Concluding with strategic directions for overcoming these challenges, the chapter equips readers with a forward-looking perspective on leveraging digital technologies to advance resilient, efficient, and climate-smart C farming systems.

Chapter 10 broadens the scope to present a global perspective on C farming, emphasizing its relevance across diverse agricultural and ecological regions. It explores how C farming practices are being adopted and adapted in different parts of the world, with regional insights from South Asia, Africa, Australia, Brazil, the United States, and Europe. The chapter highlights region-specific opportunities, challenges, and policy frameworks, offering a comparative understanding of implementation pathways. It also discusses the need for a unified global approach to C farming, grounded in shared principles but responsive to local realities. Emerging research priorities are outlined to guide future innovation, and the chapter concludes by emphasizing the importance of translating scientific knowledge into scalable, real-world action. By showcasing global experiences and pathways forward, this final chapter empowers stakeholders with the insights needed to promote C farming as a climate solution with global impact.

References

Appunn K (2022) Carbon farming explained: the pros, the cons and the EU's plans. EU

Buck HJ, Palumbo-Compton A (2022) Soil carbon sequestration as a climate strategy: what do farmers think? Biogeochemistry 161:59–70. https://doi.org/10.1007/s10533-022-00948-2

Bumbiere K, Diaz Sanchez FA, Pubule J, Blumberga D (2022) Development and assessment of carbon farming solutions. Environ Clim Technol 26:898–916. https://doi.org/10.2478/rtuect-2022-0068

Calvin K, Dasgupta D, Krinner G et al (2023) IPCC, 2023: climate change 2023: synthesis report. In: Core Writing Team, Lee H, Romero J (eds) Contribution of working groups I, II and III to the sixth assessment report of the intergovernmental panel on climate change. IPCC, Geneva

Carlisle L, Montenegro de Wit M, DeLonge MS et al (2019) Transitioning to sustainable agriculture requires growing and sustaining an ecologically skilled workforce. Front Sustain Food Syst 3:96. https://doi.org/10.3389/fsufs.2019.00096

CCI (2025) What is carbon farming? https://www.carboncycle.org/what-is-carbon-farming. Accessed 27 May 2025

Crippa M, Solazzo E, Guizzardi D et al (2021) Food systems are responsible for a third of global anthropogenic GHG emissions. Nat Food 2:198–209. https://doi.org/10.1038/s43016-021-00225-9

Davey EL (2023) 6 major food breakthroughs at COP28 – and what comes next. https://www.wri.org/insights/food-system-breakthroughs-cop28-whats-next. Accessed 17 Dec 2023

Dessalegn B, Kiktenko L, Zhumagazina B et al (2018) Explaining farmers' reluctance to adopt recommendations for sustainable ecosystem management. Ecol Process 7:24. https://doi.org/10.1186/s13717-018-0133-9

Dina Ebouel FJ, Betsi TB, Eze PN (2024) Soil inorganic carbon: a review of global research trends, analytical techniques, ecosystem functions and critical knowledge gaps. Catena (Amst) 242:108112. https://doi.org/10.1016/j.catena.2024.108112

Francis CA, Youngberg G (1990) Sustainable agriculture-an overview. In: Francis CA, Flora CB, King LD (eds) Sustainable Agriculture in Temperate Zones. Wiley, New York, pp 1–23

Green America (2023) What is carbon farming? https://www.greenamerica.org/food-climate/what-carbon-farming. Accessed 27 May 2025

Hayhoe K, Wuebbles DJ, Easterling DR et al (2018) Chapter 2: Our changing climate. In: Impacts, risks, and adaptation in the United States: the fourth National Climate Assessment, vol II. Washington, DC

IPCC (2019) Climate Change and Land: An IPCC Special Report on Climate Change, Desertification, Land Degradation, Sustainable Land Management, Food Security, and Greenhouse Gas Fluxes in Terrestrial Ecosystems. IPCC, Geneva, Switzerland

Kahan D (2013) Managing risk in farming. FAO, Rome

Lal R (2004) Soil carbon sequestration to mitigate climate change. Geoderma 123:1–22. https://doi.org/10.1016/j.geoderma.2004.01.032

Lal R, Negassa W, Lorenz K (2015) Carbon sequestration in soil. Curr Opin Environ Sustain 15:79–86. https://doi.org/10.1016/j.cosust.2015.09.002

Lal R, Smith P, Jungkunst HF et al (2018) The carbon sequestration potential of terrestrial ecosystems. J Soil Water Conserv 73:145A–152A. https://doi.org/10.2489/jswc.73.6.145A

Marvel K, Su W, Delgado R et al (2023) Climate trends. In: Crimmins AR, Avery CW, Easterling DR et al (eds) Fifth national climate assessment. U.S. Global Change Research Program, Washington, DC

Minasny B, Malone BP, McBratney AB et al (2017) Soil carbon 4 per mille. Geoderma 292:59–86. https://doi.org/10.1016/j.geoderma.2017.01.002

Pathak M, Slade R, Shukla PR et al (2023) Technical summary. In: Climate change 2022 – mitigation of climate change. Cambridge University Press, pp 51–148

Sanderman J, Hengl T, Fiske GJ (2017) Soil carbon debt of 12,000 years of human land use. Proc Natl Acad Sci 114:9575–9580. https://doi.org/10.1073/pnas.1706103114

Sharma M, Kaushal R, Kaushik P, Ramakrishna S (2021) Carbon farming: prospects and challenges. Sustainability 13:11122. https://doi.org/10.3390/su131911122

Tan W-F, Zhang R, Cao H et al (2014) Soil inorganic carbon stock under different soil types and land uses on the loess plateau region of China. Catena (Amst) 121:22–30. https://doi.org/10.1016/j.catena.2014.04.014

Tubiello FN, Karl K, Flammini A et al (2022) Pre- and post-production processes increasingly dominate greenhouse gas emissions from agri-food systems. Earth Syst Sci Data 14:1795–1809. https://doi.org/10.5194/essd-14-1795-2022

UNEP (2022) Global peatlands assessment: the state of the world's peatlands—evidence for action toward the conservation, restoration, and sustainable Management of Peatlands. United Nations Environment Programme, Nairobi

van Dijk M, Morley T, Rau ML, Saghai Y (2021) A meta-analysis of projected global food demand and population at risk of hunger for the period 2010–2050. Nat Food 2:494–501. https://doi.org/10.1038/s43016-021-00322-9

WMO (2024) WMO Greenhouse Gas Bulletin No. 20. Geneva, Switzerland. https://library.wmo.int/idurl/4/69057

Zhao W, Zhang R, Cao H, Tan W (2019) Factor contribution to soil organic and inorganic carbon accumulation in the Loess Plateau: structural equation modeling. Geoderma 352:116–125. https://doi.org/10.1016/j.geoderma.2019.06.005

Chapter 2
Carbon Farming

Abstract This chapter presents carbon (C) farming as a practical and powerful climate solution that helps transform agriculture into part of the answer to climate change. Using the four 'C' approach (Capture, Convert, Conserve, and Cultivate), C farming focuses on capturing C from the atmosphere and storing it in plants, soils, and ecosystems. The chapter explains how healthy soils, diverse vegetation, and improved land management can work together to store more C, reduce greenhouse gas emissions, and strengthen the land's natural resilience. Beyond helping the climate, C farming also brings important benefits such as protecting biodiversity, improving food security, and creating new income opportunities for farmers. This chapter shows how C farming can support both productive agriculture and a healthier planet.

2.1 Introduction

Considering the increasing awareness about climate change and its far-reaching consequences, nearly every nation is now deeply committed to minimizing their Carbon (C) emissions. This dedication is manifested in their Nationally Determined Contributions (NDCs) under the Paris Agreement, signaling a collective willingness to engage in climate change mitigation efforts and adhere to the imperative of limiting global warming to 1.5 °C above pre-industrial level (UNFCCC 2016). The COP28 Declaration calls for the integration of agriculture and food systems into NDCs, national adaptation plans, biodiversity plans, and long-term climate strategies (Davey 2023). Given the historical disparity in emissions, where developed nations have contributed significantly more compared to their developing countries counterparts, there exists a heightened responsibility for the former in climate change mitigation (Baumert et al. 2005). To fulfill these commitments and demonstrate accountability, the primary focus has shifted towards not only reducing greenhouse gas (GHG) emissions through improved technologies but also actively extracting excess C from the atmosphere and sequestering it in terrestrial resources. Given that CO_2 retains its position as the primary emitter with a global warming

N. Loria, R. Lal, *Carbon Farming*, https://doi.org/10.1007/978-3-032-00842-8_2

potential (GWP) of 1, and soil organic C (SOC) stands as the principal terrestrial C pool, a crucial approach is to augment C levels in land resources, which include both soil and vegetation, through sustainable land management practices. This concerted effort is encapsulated by the term "C farming".

C sequestration and C farming both are fundamentally oriented towards increasing C storage in natural systems but differ in their broader objectives and scope. C sequestration broadly refers to the process of capturing and storing CO_2 from the atmosphere, often involving methods like afforestation, reforestation, and C capture technologies. It's a critical component of climate change mitigation strategies, with the aim of directly reducing GHG concentrations in the atmosphere. On the other hand, C farming is a more comprehensive approach that not only encompasses C sequestration but also involves integrating sustainable land management practices within agricultural systems. It aims to prevent or minimize the release of GHGs while actively capturing C in vegetation and soils (Evans et al. 2015). This strategy seeks to optimize C sequestration within the context of agricultural operations and create a favorable environment for a diverse range of species and habitats promoting biodiversity. Beyond its climate and ecological benefits, C farming also enhances soil health and fertility, improves water retention, and can even lead to increased crop yields.

C farming is structured around the "Four C" approach—Capture, Convert, Conserve, and Cultivate—which outlines the sequential processes essential for maximizing C sequestration and ecosystem benefits (Fig. 2.1). It is a land management approach that captures atmospheric CO_2 through photosynthesis, converts it

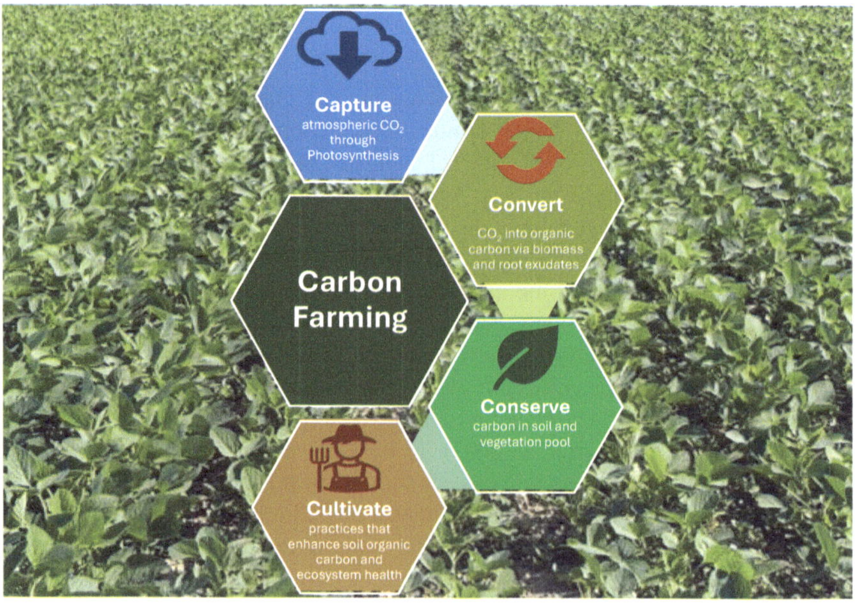

Fig. 2.1 Four 'C' approach to carbon farming

into organic C via plant biomass accumulation and root exudates, conserves it in stable soil and vegetation pools, and cultivates practices that enhance SOC levels and support overall ecosystem health.

C farming operates in tandem with conventional crop farming, which is primarily directed at meeting global food demands. However, unlike traditional farming, which yields essential food resources, C farming focuses on enriching the soil and above ground vegetation with C content. This establishes a foundation for healthier and more productive crops. This benefit not only contributes to long-term agricultural sustainability but can also be economically advantageous for farmers, as this innovative approach is recognized and valued at a financial level through the allocation of C offset credits. These credits serve a dual purpose: first, they acknowledge the significant environmental contributions made by C farming, and second, they provide a tangible incentive for farmers to adopt and invest in these practices. C farming entails incentivizing landholders, particularly through methods like assisted natural regeneration, to increase forest cover on private lands, while also delivering a range of socio-economic and biodiversity benefits and addressing deforestation challenges (Evans 2018).

C farming is a method focused on enhancing soil and vegetation C stocks through a series of agricultural practices and land-use strategies that align with soil health principles. The effectiveness of these agronomic practices hinges on a multitude of variables, encompassing climate, soil composition, and the unique features of the landscape. Moreover, the influence of local economic and cultural contexts is pivotal in ascertaining which practices are most suitable, underscoring the importance of a versatile technological approach. This diversity guarantees that C farming techniques can be precisely customized to distinct regions and situations, ultimately amplifying their positive influence on both the environment and economy. In essence, C farming emerges as a flexible and responsive strategy in efforts to combat climate change and advance SAPs. C farming serves as a crucial pivot in maintaining equilibrium between C-emitting practices, such as deforestation and intensive agriculture, and C-mitigating activities like afforestation, cover cropping, and agroforestry, underscoring its pivotal role in achieving environmental sustainability (Fig. 2.2).

While the concept of C farming has been around for a few decades, it gained significant attention and recognition in the early twenty-first century. Several countries have adopted diverse interpretations of C farming and have applied it in distinct ways within their respective climate change mitigation strategies. This concept has likely emerged within the broader discussion surrounding climate change and sustainable agriculture (SA). It's worth emphasizing that the concepts underpinning C farming have received longstanding support from a diverse community of environmentalists, scientists, researchers, and agricultural experts. The terminology likely evolved through collaboration within these communities.

C farming as defined by the European Commission is "a green business model that rewards land managers for taking up improved land management practices, resulting in the increase of C sequestration in living biomass, dead organic matter (OM) and soils by enhancing C capture and/or reducing the release of C into the

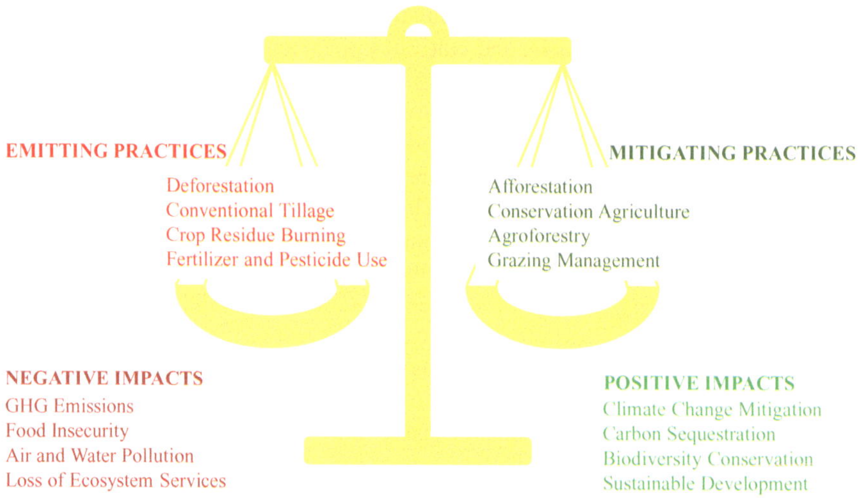

Fig. 2.2 Carbon balance: Assessing global impacts of emitting and mitigating practices

atmosphere, in respect of ecological principles favorable to biodiversity and the natural capital overall" (Appunn 2022). The introduction of C Farming Initiative (CFI) by the Australian government in 2011 marked a significant milestone, offering a comprehensive framework for offsetting C emissions. By aligning with ecological principles and international GHG accounting rules, the initiative not only helps meet national GHG reduction targets but also fosters environmental protection and climate resilience. The potential for generating certified offsets, both domestically and internationally, underscores its importance in the global C market (CCA 2014).

2.2 Importance of Carbon Farming

CO_2 constitutes a mere 0.04% of today's Earth's atmosphere, yet it holds a pivotal role in regulating the planet's temperature (Fleming 2018; Friedlingstein et al. 2023). An in-depth analysis of the C isotopes (13C) in specific fossils has revealed that during the glacial age approximately 200 million years ago, atmospheric CO_2 levels were significantly lower compared to the present concentrations, aligning with earlier scientific observations (Shackleton et al. 1983). The C-cycle provides valuable insights into how CO_2 levels fluctuated in ancient times. During the glacial age, atmospheric CO_2 levels were much lower than those found in plants, soil, and oceans, with the deep ocean serving as the primary C reservoir (Sigman and Boyle 2000). In interglacial periods (around 11,000 years ago), the Earth experiences a shift in temperature, ocean circulation patterns, as well as changes in the chemical and biological makeup of the oceans. These alterations collectively result in the release of CO_2 from the oceans into the atmosphere, contributing to an increase in CO_2 levels to around 280 parts per million (ppm) in the late 1700s (Latif 2017).

The concentration of CO_2 in the atmosphere is determined by a delicate balance between C sinks and sources. The surge in CO_2 emissions, predominantly attributed to global fossil fuel combustion and industrial processes that gained momentum with the onset of the industrial revolution around 1750 AD, drew attention to the imperative of enhancing C sinks as a pivotal strategy for mitigating climate change. To establish a benchmark for measuring emissions, the Intergovernmental Panel on Climate Change (IPCC) designates the period before 1750 AD as the pre-industrial era. The Kyoto Protocol, a significant milestone in international climate efforts, aimed for a 5% reduction in GHG emissions from 1990 levels in its first commitment period (2008–2012) and a more ambitious 18% reduction in the subsequent period (2013–2020). However, a transformative moment arrived in 2015 with the adoption of The Paris Agreement under the United Nations Framework Convention on Climate Change (UNFCCC). The Paris Agreement embodies a collective global commitment to combat climate change, reflecting a monumental step towards a sustainable, low-C future. It seeks to limit global warming to well below 2 °C above pre-industrial levels, while also aspiring to cap the temperature increase at a more stringent 1.5 °C (UNFCCC 2016). This can be achieved by either removing excess CO_2 from the atmosphere or storing it in natural reservoirs like oceans, vegetation, and soil. Presently, terrestrial C sequestration stands out as the most cost-effective and readily accessible approach for reducing GHG emissions in the short term. Lal (2004) indicated that over the next five decades, the implementation of best management practices (BMPs) could result in a moderate global SOC sequestration potential, estimated to be in the range of 30–60 Pg C, with an annual sequestration rate of about 0.9 ± 0.3 PgC/year. This represents roughly one-fourth to one-third of the annual increase in atmospheric CO_2 levels, which totals 3.3 Pg C/year during 2010s (Patle et al. 2013) and as much as 5.2 PgC/year during 2020s (Friedlingstein et al. 2022). C farming plays a significant role in this effort, actively sequestering C in terrestrial resources. However, the success of C farming hinges on the synergistic contributions of vegetation, soil, and soil-dwelling microorganisms, and thoughtfully implemented agricultural practices. These elements work in harmony to maximize C sequestration potential and minimize C emissions, making C farming a crucial component in the global effort to combat climate change.

2.2.1 Vegetation

Vegetation plays a crucial role in C sequestration, primarily through the process of photosynthesis and subsequent C allocation. Photosynthesis is the fundamental process through which plants, algae, and certain bacteria convert sunlight and inorganic substances into organic compounds. This process involves the absorption of CO_2 through specialized pores called stomata on plant leaves. Within the chloroplasts of these organisms, chlorophyll—a green pigment—plays a central role by capturing light energy from the sun. Through a series of chemical reactions, CO2 is transformed, with its C becoming an integral part of the plant's structure. Simultaneously, oxygen is released into the atmosphere (Hall and Rao 1999). The resulting glucose serves as a vital energy source for the plant's growth and development. Moreover, the C stored within plant tissues represents a form of long-term C sequestration, remaining in the plant until it undergoes decomposition or ceases to exist (Hartmann et al. 2020). This process not only sustains plant life but also plays a pivotal role in regulating the planet's climate.

2.2.2 Soil

Soil, a reservoir teeming with life, embodies a crucial facet of C farming. Comprising remnants of plants and animals at different stages of decomposition, roots, and a bustling community of microorganisms, it encapsulates sequestered C, positioning soil as an indispensable component in the realm of C farming. In this underground ecosystem, bacteria and fungi work together to break down complex organic substances, liberating essential nutrients and transforming matter into more stable forms. Amidst this transformative process, a portion of the organic material undergoes a remarkable metamorphosis into humus, a resilient and steadfast substance (Ponge 2013). Rich in C, humus is a long-standing repository for this essential element within the soil matrix. Beyond its role in nutrient cycling and soil health, humus significantly contributes to climate change mitigation by sequestering C (Berg and McClaugherty 2003). This dynamic relationship between soil, microbes, and humus holds significant implications for various aspects of the environment. In agriculture, it fosters healthier, more productive soils. For ecosystems, it supports biodiversity and overall vitality.

2.2.3 Soil Microorganisms

Microorganisms in soil play a pivotal role in the C- cycle of terrestrial ecosystems and can be in the form of bacteria, fungi, algae, nematodes etc. They are metabolic engines, driving processes like nutrient recycling and OM decomposition. While they release C into the atmosphere through their metabolic activities, they also convert external C into specific compounds, effectively storing it within the soil (Jiang et al. 2022). Additionally, mycorrhizal fungi, which form symbiotic partnerships with plant roots, enhance nutrient absorption and contribute to C- sequestration through their life cycle (Malyan et al. 2019). The presence and vitality of these fine, thread-like structures produced by fungi are critical factors in determining how effectively C can be stored in the soil (Solaiman and Mickan 2014). Overall, soil microorganisms are the linchpins of C sequestration, facilitating the conversion, storage, and cycling of C in soil ecosystems (Jiang et al. 2022). Their activities not only maintain soil health and fertility but also have far-reaching implications for mitigating climate change.

2.2.4 *Agricultural Practices*

Agricultural practices take on added significance as they directly influence the sequestration and retention of C in soil and vegetation. Practices like conservation tillage, cover cropping, crop rotation, adding biochar, and agroforestry are instrumental in this process (Kimble et al. 2016; Bai et al. 2019). Conservation tillage, for instance, minimizes soil disturbance during planting and cultivation, preserving soil structure and reducing C loss (Chan et al. 2003; Hussain et al. 2021). Cover cropping involves planting specific crops during non-growing seasons to protect the soil surface, enhance OM content, and promote microbial activity (Chan et al. 2003; Haruna et al. 2020). Crop rotation diversifies plant species, which can lead to increased biomass and OM, subsequently enriching the soil's C content (Holeplass et al. 2004; Poeplau and Don 2015). Additionally, the integration of OM, such as compost (Termorshuizen et al. 2004), manure (Mikha and Rice 2004), rice-wheat straw (Yu et al. 2020), or biochar (Ennis et al. 2012) enhances nutrient availability and encourages microbial populations. Agroforestry, with its integration of site-specific trees alongside crops or livestock, provides additional avenues for C sequestration (Lorenz and Lal 2014). Reviewing several research findings, Nair et al. (2010) highlighted that Agroforestry systems globally possess the capacity to sequester SOC. The C sequestration potential aboveground ranges from 0.29 to 12.04 Mg C per hectare annually, and belowground SOC storage ranges from 24 to 302 Mg C per hectare up to a depth of 100 cm (Nair et al. 2010). This variability is influenced by factors such as site conditions, species combination, land use practices, stand age, and soil types. Employing these practices collectively creates a favorable environment for C sequestration, optimizing the potential of C farming for long-term climate mitigation.

2.3 Carbon Farming Principles

C farming principles offer a holistic solution at the intersection of SA and climate action, with the added potential for financial incentives. These principles encompass a diverse range of strategies, including soil enrichment, ecosystem balance, robust monitoring, conservation and reforestation, sustainability practices, resilience and adaptation, soil security, and policy alignment (Fig. 2.3). Together, these principles strengthen ecosystems and contribute to a more sustainable future. In a world grappling with escalating GHG emissions, adopting C farming principles is crucial in collective endeavor to achieve C neutrality.

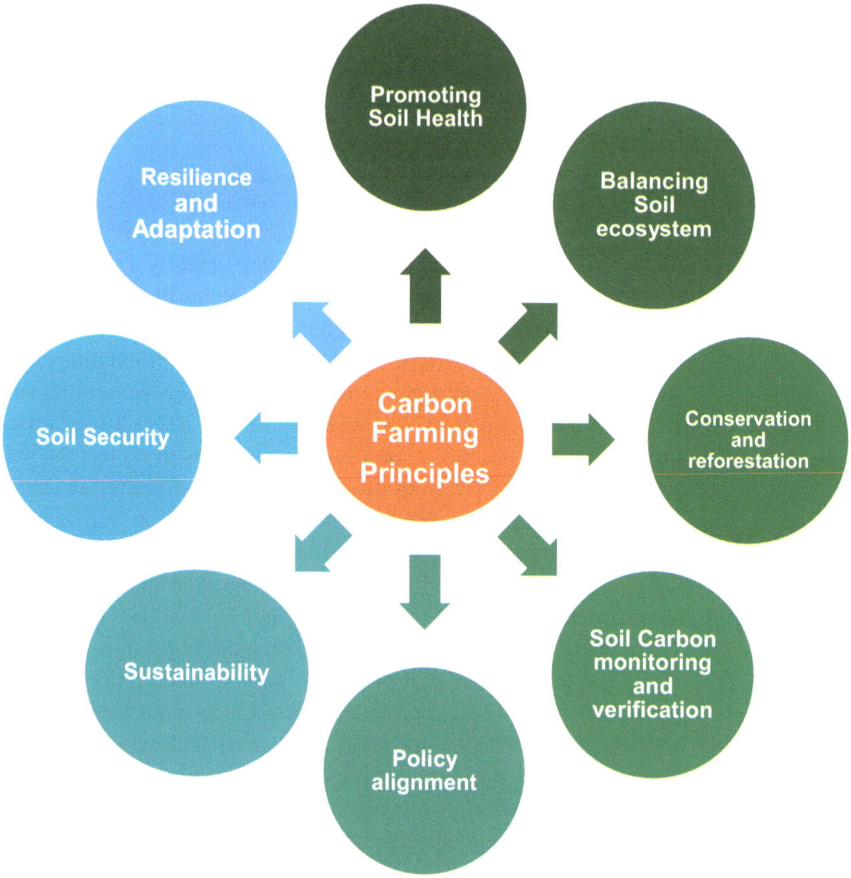

Fig. 2.3 Carbon farming principles

2.3.1 *Promoting Soil Health*

Enhancing soil health is a foundational step in successful C farming as this involves preparing the soil to effectively capture and securely store C by binding it with essential minerals over the long term. The principle emphasizes enriching the soil through a range of practices. This includes incorporating biomass like organic residues, such as biochar, compost, animal manure, and crop residues (Brichi et al. 2023). Additionally, it encourages the management of cover crops (Seepaul et al. 2023). Another vital aspect is minimizing tillage, as this helps maintain the natural structure of the soil (Pagliai et al. 2004). Furthermore, diversifying crop rotations is also a key component, as it encourages a diverse microbiome within the soil (Borase et al. 2020), which in turn aids in C sequestration efforts.

2.3.2 Balancing Soil Ecosystems

The principle of balancing soil ecosystems in C farming underscores the significance of maintaining a harmonious and diverse community of plants, microorganisms, fungi, and other soil organisms. This equilibrium is crucial for enhancing C sequestration. By fostering a healthy soil ecosystem, the various components work synergistically. For instance, beneficial microbes aid in decomposition, converting OM into stable forms of C. Additionally, plants and their associated mycorrhizal fungi form symbiotic relationships, facilitating nutrient uptake and further contributing to C storage. Implementing practices that support this balance, such as minimizing disturbance and avoiding excessive use of chemicals, fosters a thriving soil community that is more adept at sequestering C over the long term. Changes in plant community composition have a significant impact on soil community composition and functioning (Kardol et al. 2010). Different plant species have distinct root structures and patterns of OM deposition, influencing the type and activity of soil microbes (Hartmann et al. 2009). Additionally, specific plant-microbe interactions can stimulate the formation of stable soil aggregates, providing a protective environment for sequestered C (Hartmann and Six 2022). Therefore, balancing the soil ecosystem by considering plant community dynamics is essential for optimizing C sequestration and enhancing overall soil health.

2.3.3 Conservation and Reforestation

The principle of conservation and reforestation in C farming revolves around two key strategies for sequestering C: safeguarding existing vegetation and enhancing C levels through reforestation efforts. Preserving established vegetation ensures that these natural C sinks continue to thrive and effectively capture atmospheric CO_2 (Sha et al. 2022). Simultaneously, reforestation initiatives aim to replenish tree cover in areas where it has been depleted, significantly amplifying the capacity to sequester C (Nave et al. 2019). These efforts, seen in projects worldwide, have successfully restored C-rich environments and enhanced urban C sequestration. Agroforestry practices, like silvopasture, alley cropping, and windbreak plantings, integrate trees into farming systems, not only benefiting agriculture but also increasing C storage (Udawatta and Jose 2011). Moreover, the restoration of peatlands and wetlands, which are significant C reservoirs, through practices like halting drainage and wetland rehabilitation, helps conserve and even augment existing C stores (Harenda et al. 2018). Community-driven reforestation programs, involving local communities in tree planting and forest management, empower communities and lead to more sustainable ecosystems (Jong et al. 1996; Jindal et al. 2008). These collective efforts in conservation and reforestation embody the transformative potential of C farming, addressing climate change while promoting sustainable land management practices and biodiversity conservation.

2.3.4 Soil Carbon Monitoring and Verification

Soil C monitoring and verification is a critical component of C farming strategies. It involves systematic and periodic assessment of soil C levels within a given area. This assessment employs a combination of ex-situ techniques, which involve laboratory-based methods like wet and dry combustion, and in-situ approaches such as spectroscopic analysis and remote sensing technologies (Chatterjee et al. 2009). The collected data provides valuable insights into the effectiveness of C sequestration efforts. Regular monitoring allows for the tracking of changes in soil C content over time, helping farmers, and land managers understand the impact of their practices (Nevalainen et al. 2022). It also facilitates the identification of areas where additional efforts may be needed. Furthermore, independent verification by third-party experts adds a layer of credibility and transparency to the process. Accurate and reliable measurements are essential for ensuring that C farming initiatives are meeting their intended goals. By adhering to robust monitoring and verification protocols, stakeholders can confidently report their C sequestration achievements (Smith et al. 2020). This not only supports climate change mitigation efforts but also opens opportunities for C credit trading, providing a financial incentive for sustainable land management practices.

2.3.5 Resilience and Adaptation

This principle recognizes that climate change brings new and unpredictable challenges to agriculture. By prioritizing resilience and adaptation, C farming seeks to equip farmers with the tools and knowledge needed to thrive in a changing climate while contributing to C sequestration and sustainable land management (Young and Thomson 2020). Resilience in C farming involves implementing strategies that help farms endure and recover from various challenges. This can include measures like diversifying crops and livestock (Szymczak et al. 2020), employing agroforestry techniques (Sauer et al. 2021), and incorporating climate resilient crops (Rivero et al. 2022) that are better suited to changing climate conditions. By diversifying their operations, farmers can spread risks and reduce vulnerability to extreme weather events, pests, and diseases. Adaptation, on the other hand, involves the deliberate adjustment of farming practices in response to changing environmental conditions. This may include modifying planting schedules (Vashisht et al. 2019), adjusting irrigation techniques (Ojeda-Bustamante et al. 2017), or adopting conservation tillage methods (Hussain et al. 2021). Additionally, integrating climate-smart technologies and practices, such as precision agriculture and water-efficient irrigation systems, can further enhance the farm's adaptive capacity (Balasundram et al. 2023). Furthermore, this principle underscores the importance of ongoing monitoring, research, and knowledge exchange within the agricultural community. This helps identify emerging challenges and provides farmers with access to the latest scientific insights and innovative practices that can enhance their resilience.

2.3.6 Policy Alignment

Policy alignment in C farming refers to the synchronization of agricultural practices with established government policies and regulations aimed at mitigating climate change. It involves ensuring that the strategies employed on farms are consistent with broader climate objectives set by local, national, and international governing bodies. A prime illustration of this is the Conservation Compliance provisions embedded within the United States Farm Bill (Stubbs 2016). These provisions necessitate that farmers must adhere to specific requirements to be eligible for various benefits extended through programs administered by the US Department of Agriculture (USDA). This principle recognizes that effective C farming is not solely the responsibility of individual farmers but requires a supportive policy framework. It encourages policymakers to create incentives, subsidies, and regulations that promote and reward C sequestration practices (Richards 2004). This may include measures like C credits, tax breaks, or financial support for adopting SAPs as payment for ecosystem services. Furthermore, policy alignment fosters collaboration between agricultural stakeholders, environmental organizations, and governmental agencies. It facilitates knowledge sharing, research initiatives, and the dissemination of best practices, creating a conducive environment for the widespread adoption of C farming. By aligning policies with C farming goals, governments can harness the potential of agriculture as a powerful tool in the fight against climate change. In essence, policy alignment ensures that C farming becomes an integral part of broader climate strategies, amplifying its impact on reducing GHG emissions and enhancing soil health.

2.3.7 Sustainability

Sustainability in C farming represents a holistic approach to agricultural practices that not only sequester C but also prioritize the long-term health and resilience of agricultural systems. It acknowledges the intricate interplay between ecological, social, and economic factors in farming operations. At its core, sustainable C farming embraces regenerative agricultural practices that align with natural processes. Techniques like agroecology, which mimic natural ecosystems, are key components. By leveraging ecological interactions, farmers reduce their dependence on external inputs and enhance the overall resilience of their farming systems.

Furthermore, sustainability in C farming places a significant emphasis on preserving and enhancing ecosystem services. This entails conserving natural habitats like wetlands, riparian zones, and grasslands, which serve as vital C sinks and provide essential services such as water purification, pollination, and habitat for beneficial organisms (Power 2010). Safeguarding and restoring these ecosystems contribute to the overall sustainability of the agricultural landscape. Social sustainability is another crucial aspect, recognizing the importance of equitable resource access, fair labor practices, and community involvement. Collaboration among farmers, researchers, policymakers, and local communities is encouraged to collectively work towards sustainable land management practices. Economically, sustainability in C farming aims to enhance the profitability and resilience of farming operations. This can be achieved through diversification of income sources, value-added processing, and participation in C credit markets. Diversifying revenue streams helps farmers mitigate market volatility and increase financial stability (Antle & Stoorvogel 2008). Prioritizing sustainability in C farming is a powerful tool in addressing climate change while fostering healthy and resilient food systems.

2.3.8 Soil Security

Soil Security pertains to the preservation and enhancement of the global soil resource to facilitate the production of food, fiber, and freshwater (McBratney et al. 2014). Additionally, it contributes to sustainability in energy and climate contexts, while also safeguarding biodiversity and overall ecosystem health. Soil security is crucial for soil C sequestration as it is closely linked to soil quality, soil health, and soil protection. It recognizes the fundamental role that healthy, fertile soils play in the process of C sequestration. A healthy and secure soil possesses the capacity to effectively capture and store CO_2 from the atmosphere, aiding in the mitigation of climate change. By prioritizing soil security, C farming endeavors to enhance the overall health, structure, and productivity of soils. Furthermore, soil security is crucial for sustaining agricultural productivity and food security (Thangavel and Sridevi 2017). Healthy soils are better equipped to support robust plant growth, leading to higher crop yields. This not only ensures food availability but also contributes to the economic well-being of farming communities. Additionally, secure soils are less susceptible to erosion and degradation, safeguarding valuable topsoil and preserving the long-term viability of agricultural lands.

2.4 Carbon Farming Benefits

C farming bestows a diverse range of benefits, positively impacting the land (soil), air, water, and climate. These elements are intricately connected, working in harmony to support the principles of C Farming. The advantages of C farming are far-reaching and are most needed to address global issues like food security and the climate crisis. Starting from C storage in soil and vegetation, these benefits have the potential to transform agricultural practices with no penalty to plant yield and increase the resilience of soil to support food production and mitigate the effects of climate change. C farming touches all aspects of sustainability, including social, environmental, ecological, and economic dimensions (Fig. 2.4). This categorization

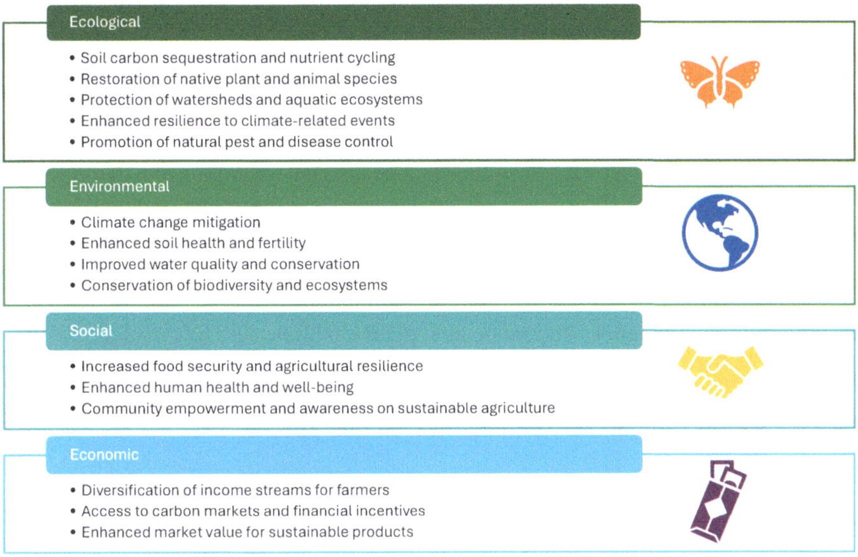

Fig. 2.4 Carbon farming benefits

helps highlight the multifaceted benefits that C farming brings to various aspects of society and the environment. It provides a comprehensive framework for understanding how C farming contributes to a more sustainable and resilient future.

2.4.1 Ecological Benefits

2.4.1.1 Soil Carbon Sequestration and Nutrient Cycling

C farming is primarily centered around soil C sequestration, with all other benefits stemming from the increased C storage in both soil and vegetation over time. This approach hinges on principles like minimizing soil disturbance, enhancing biodiversity, providing ground cover, and optimizing root presence to enhance C sequestration. By adopting practices such as no tillage or reduced tillage, incorporating cover crops, crop rotations, soil health is improved, further contributing to C storage. Notably, site-specific nutrient management combined with no-tillage practices significantly amplifies C sequestration efforts (Parihar et al. 2020).

Additionally, the inclusion of OM like crop residues and supplementary nutrients like nitrogen (N), phosphorus (P), and sulfur (S) has a substantial impact on enhancing the C sequestration potential of soil. This is achieved by increasing net humification efficiency, regardless of soil type and initial C input levels (Kirkby et al. 2013). Elevated C levels have a cascading effect on soil health. They foster improved soil structure, boosting water retention, aeration, and nutrient availability. This, in turn, leads to heightened crop productivity and greater resilience to environmental stresses. Moreover, C-rich soils exhibit enhanced stability, making them less susceptible to erosion. They also possess greater water-holding capacity, reducing runoff and aiding in the mitigation of extreme weather impacts.

Furthermore, C farming practices stimulate a dynamic nutrient cycling process by increasing OM content. This results in a greater availability of essential nutrients such as N, and potassium (K) for plants. Consequently, there is a reduced reliance on synthetic fertilizers, leading to lower input costs for farmers and a decreased environmental impact associated with fertilizer production and application. Additionally, well-managed nutrient cycling promotes a balanced nutrient uptake by plants, resulting in healthier and more resilient crops. This lowers the risk of nutrient-related imbalances and associated plant health issues (Reeve et al. 2016).

2.4.1.2 Restoration of Native Plant and Animal Species

Restoration of native plant and animal species is a significant benefit associated with C sequestration efforts. When C farming practices are implemented, they often involve the reintroduction or preservation of native vegetation and ecosystems. The practices, such as reforestation, afforestation, and the establishment of cover crops, create habitats that closely resemble natural ecosystems. This provides a conducive environment for native plant species to flourish. These plants, in turn, offer essential resources like food and shelter for native animal species. The presence of diverse flora and fauna contributes to a healthier and more balanced ecosystem. They are often better equipped to withstand local climate variations, pests, and diseases, making the ecosystem more stable and sustainable in the long run. Many native plant species have specialized relationships with native pollinators, like bees, butterflies, and birds (Stout and Tiedeken 2017). By restoring these native plants, C farming indirectly supports native pollinator populations. This is critical for maintaining pollination services that are essential for agricultural productivity. When native vegetation is restored, it can act as a natural barrier against the spread of invasive species. Native plants often compete more effectively with non-native species, helping to maintain the integrity of the ecosystem (Mappin et al. 2022).

2.4.1.3 Protection of Watersheds and Aquatic Ecosystems

C farming practices, such as cover cropping and minimal tillage, enhance soil structure and stability. This, in turn, reduces surface runoff and erosion (Dabney 1998). As a result, fewer sediments, nutrients, and pollutants are carried into nearby water bodies. This protects the watersheds from pollution, thereby preserving water quality. By reducing the introduction of pollutants and maintaining stable water flow, C farming indirectly supports the health and diversity of aquatic species. Effective nutrient cycling, a key aspect of C farming, helps regulate nutrient levels in the soil.

This may, in turn, reduces the risk of nutrient runoff into water bodies, which can lead to water pollution and eutrophication (Uribe et al. 2018; Speir et al. 2022; Waring et al. 2024). By maintaining balanced nutrient levels, C farming supports healthier aquatic ecosystems. C farming practices often emphasize the preservation or restoration of riparian zones—the transitional areas between land and water. These zones are critical for stabilizing streambanks, filtering pollutants, and providing habitat for aquatic organisms (Wantzen and Junk 2008). Wetlands are vital components of watersheds, serving as natural filters for pollutants and providing essential habitat for a diverse range of species. C farming initiatives often include measures to protect and restore wetland ecosystems, which in turn, benefits water quality and aquatic life (Mitsch et al. 2012). Protecting watersheds and aquatic ecosystems is crucial for mitigating the impacts of climate change as healthy ecosystems are more resilient to changes in temperature, precipitation patterns, and extreme weather events.

2.4.1.4 Enhanced Resilience to Climate-Related Events

Enhanced resilience to climate-related events refers to the increased capacity of agricultural systems to withstand and recover from extreme weather events and other climate-related challenges. Healthy soils enriched with C play a crucial role in this resilience. They act as a natural buffer against extreme weather conditions, such as heavy rainfall and floods. These soils have greater capacity to absorb excess water, minimizing the risk of flooding and its associated impacts on above and below ground ecosystem processes (Unger et al. 2009). Moreover, strategic interventions like windbreak and shelterbelt structures play a crucial role in shaping the local microclimate within agricultural landscapes. Especially in regions prone to severe weather events like storms or heatwaves, windbreaks offer a protective shield. They effectively moderate wind speeds and temperature fluctuations in sheltered areas, thereby mitigating the direct impact of these events on crops (Brandle et al. 2004). This increased resilience leads to greater stability and reliability in agricultural production. Farmers can better anticipate and manage potential disruptions, ultimately resulting in higher sustainability and profitability in their operations. Additionally, Conservation tillage systems have demonstrated their effectiveness in optimizing the infiltration and retention of rainfall within the soil (Hussain et al. 2021). This capability is particularly invaluable for crops that may not have access to additional irrigation during extended dry periods. By reducing soil disturbance, these systems maintain a protective cover on the soil surface, which prevents water runoff and enhances moisture retention. This allows plants to access stored water reserves, ensuring their survival through prolonged droughts (Rockström 2003).

2.4.1.5 Promotion of Natural Pest and Disease Control

Promotion of natural pest and disease control is a significant ecological benefit of C farming (Schmitz et al. 2017). This approach encourages the natural predators and beneficial organisms present in the ecosystem to regulate pest populations. By maintaining diverse and healthy ecosystems, C farming provides habitats and food sources for these beneficial organisms, such as ladybugs, predatory beetles, and birds. These natural predators feed on harmful crop pests, such as aphids, thrips, caterpillars, fungus gnats, and spider mites, effectively regulating pest populations and reducing the need for chemical pesticides (Raj 2023). Additionally, healthy soils and diverse plant communities foster a robust and balanced ecosystem, making it more resilient to pest outbreaks. This ecological approach not only minimizes the environmental impact of chemical pesticides but also supports the long-term sustainability of agricultural systems (Rusch et al. 2017). It promotes a harmonious coexistence between crops, beneficial organisms, and the surrounding environment, ultimately leading to more resilient and ecologically balanced agricultural landscapes.

2.4.2 Environmental Benefits

2.4.2.1 Climate Change Mitigation

C farming has emerged as a crucial climate change mitigation strategy, prioritizing soil health to enhance water and air quality while concurrently reducing GHG emissions through conservation practices. By discouraging forest clearing for agriculture (Lin et al. 2013), C farming crucially prevents the release of significant C stores, preserving vital C sinks. Minimal tillage practices drastically reduce fuel requirements for farm machinery (Sijtsma et al. 1998), resulting in a substantial reduction in emissions associated with fossil fuel combustion and contributing to a more sustainable agricultural model. Additionally, C farming minimizes reliance on synthetic fertilizers and pesticides, which release GHGs during production and application. Tree planting and agroforestry further sequester C in above-ground biomass (Nair et al. 2009), effectively removing CO_2 from the atmosphere and contributing to global C reduction efforts. In livestock management, C farming promotes practices like rotational grazing, instrumental in reducing CH_4 emissions from enteric fermentation (Zubieta et al. 2021). Optimizing nutrient levels and employing integrated nutrient management (INM) strategies effectively mitigates the release of N_2O (Ussiri and Lal 2013), a potent GHG produced from nitrogen-based fertilizers.

2.4.2.2 Enhanced Soil Health and Fertility

C farming is instrumental in enhancing soil health and fertility by improving soil properties and enabling vegetation to efficiently store C. Practices like cover cropping, composting, and minimal tillage lead to a higher level of OM in the soil, which is particularly rich in C compounds (Nelson and Sommers 1996). This, in turn, enhances soil aggregation, resulting in a crumbly and well-structured soil. Such improved structure facilitates better root penetration, water infiltration, and air exchange within the soil. Consequently, plants can access nutrients more effectively, ultimately leading to increased crop yield. Vital nutrients like NPK are stored efficiently within the OM, significantly reducing the risk of nutrient leaching (Rice 2002). This ensures that nutrients remain available to plants rather than being washed away by rain or irrigation. Additionally, the stability of pH levels in C-rich soils creates an environment where nutrients are readily accessible and easily absorbed by plants. Furthermore, the presence of OM acts as a food source for beneficial microorganisms, crucial for nutrient cycling. These microbes play a vital role in breaking down OM into forms readily usable by plants, contributing to overall soil health (Van Veen and Kuikman 1990). By augmenting the OM content, C farming establishes a robust foundation for long-term soil fertility, reducing the dependence on synthetic fertilizers. This comprehensive approach underscores the transformative potential of C farming in SA.

2.4.2.3 Improved Water Quality and Conservation

Soil health and water quality share a vital relationship, where improvements in one directly benefit the other. C farming practices play an important role in this synergy. By enhancing water-holding capacity through techniques like the addition of OM via cover cropping and agroforestry, the soil's ability to retain and utilize moisture is significantly enhanced. Furthermore, practices such as reduced tillage, crop rotation, and thoughtful grazing management contribute to improved soil structure. They achieve this by influencing pore size distribution, facilitating efficient water infiltration, and preventing runoff (Yang et al. 2015; Owuor et al. 2016). This natural process helps recharge groundwater aquifers and sustains the flow of streams and rivers during dry periods. A notable aspect of C farming is the shift away from chemical inputs, replaced with organic alternatives and natural predators to manage pests. This transition reduces the risk of water pollution, as fewer synthetic chemicals are introduced into the soil. As a result, there is a reduced likelihood of these substances leaching into water bodies. This dual approach, focusing on both soil health and water quality, underscores the holistic nature of C farming and its profound impact on agroecosystems. Innovative practices within C farming, such as the integration of biochar into sustainable drainage systems, further amplify these benefits. By enhancing soil structure and boosting pollutant removal capabilities, this integration significantly mitigates the impact of stormwater (He et al. 2022).

2.4.2.4 Conservation of Biodiversity and Ecosystems

Biodiversity conservation stands as a pivotal benefit of C farming, intimately tied to the enhancement of both above-ground vegetation and below-ground soil health. A key strategy in achieving this is diversifying plant species through practices like cover cropping and agroforestry. This not only amplifies the variety of microorganisms in the soil but also fosters a rich diversity of fauna above ground, including various animals and birds (Araujo et al. 2012; Dollinger and Jose 2018). It's important to highlight that the extent of this biodiversity enhancement can be influenced by species selection and residue management strategies (Muhammad et al. 2021). The improvement of SOC leads to a proliferation of beneficial microorganisms such as bacteria and fungi. These invaluable organisms, in turn, forge a symbiotic relationship with plants, bolstering their health and resilience (Lehmann et al. 2020). This enhanced vitality allows plants to thrive even in challenging environmental conditions, consequently attracting a myriad of bird and animal species in search of habitat and sustenance. Moreover, the augmentation of soil quality, achieved through C sequestration, leads to the restoration of native plant and animal species. These indigenous species often possess unique genetic traits finely tuned to specific local conditions. Preserving and revitalizing these species safeguards a reservoir of genetic diversity, poised to play a crucial role in adapting to future shifts in environmental dynamics (Hajjar et al. 2008). This intricate interplay between C farming practices and biodiversity conservation underscores the profound positive impact this approach has on the intricate web of life within agricultural landscapes.

2.4.3 Social Benefits

2.4.3.1 Increased Food Security and Agricultural Resilience

The relationship between soil health, food production, and climate resilience is a symbiotic one. Investing in C farming practices ensures that we continue to produce an abundance of nutritious food and safeguards agricultural systems against the uncertainties of a changing climate (Umesha et al. 2018). These practices, ranging from cover cropping and composting to minimal tillage, work in harmony to enrich the soil with vital OM. This, in turn, enhances the soil's ability to support robust plant growth and facilitate the uptake of essential nutrients (Reeve et al. 2016). This leads to higher crop yields, ensuring a steadier food supply for communities (Ma et al. 2023). Additionally, the emphasis on crop diversification acts as a safeguard against the vulnerabilities associated with mono-cropping, providing a vital defense against pests, diseases, and unpredictable weather patterns (Ijaz et al. 2019). By promoting local, small-scale agriculture, C farming nurtures self-reliant food systems, reducing reliance on precarious global supply chains (Altieri et al. 2012; Masuku et al. 2017). The conservation of agrobiodiversity through practices like agroforestry further enhances adaptability in agriculture (Thrupp 2000; Eddy and Yang 2022). The integration of perennial crops and grasses further establishes a sustainable basis for food production, with their extensive root systems enhancing C storage and improving soil structure (Toensmeier 2016; Johansson et al. 2022). This results in a consistent, year-round supply of nutritious food, reducing the need for frequent replanting and providing a safety net against unpredictable weather patterns. Through its comprehensive approach, it pioneers a path toward a more robust, sustainable, and food-secure future, where communities are better equipped to withstand the uncertainties posed by a rapidly changing world.

2.4.3.2 Enhanced Human Health and Well-Being

C farming extends its positive impact beyond the fields, enhancing human health, livelihood, and overall well-being (Welch and Graham 1999). Through SAPs, it ensures communities have access to a diverse array of fresh and nutritious produce. By reducing the reliance on synthetic fertilizers and pesticides, C farming minimizes the exposure to harmful chemicals, benefiting both farmers and consumers (Mustafa et al. 2019). Engaging with nature in biodiverse, sustainable landscapes, as promoted by C farming, has been linked to improved mental well-being, fostering a sense of purpose and connection to the land. Additionally, the physical labor required by many C farming practices, such as manual weeding and agroforestry, encourages regular exercise, contributing to better physical health for those involved. The inclusion of afforestation and agroforestry practices in C farming leads to improved air quality, creating a healthier environment for local communities (Peckham 2015). Moreover, the collaborative nature of C farming fosters community building and social cohesion, further supporting overall well-being. As articulated by Lal (2024), the guiding principle for sustaining human health is: "Healthy Soil = Healthy Diet = Healthy People = Healthy Ecosystems = Healthy Planetary Processes = World Peace". Ultimately, by mitigating climate change through C sequestration, C farming indirectly reduces health risks associated with climate-related events, contributing to a healthier and more resilient future for all. Furthermore, the integration of digitalization in agriculture, aligning with the principles of C farming, holds immense potential to strike a harmonious balance between food security and SAPs (Stoian et al. 2022; Balasundram et al. 2023). This shift substantially minimizes adverse effects on both the environment and human health (Lal 2025).

2.4.3.3 Community Empowerment and Awareness on Sustainable Agriculture

C farming empowers communities and builds their capacity for sustainable development by encouraging local participation and knowledge-sharing. This equips individuals with valuable skills and expertise in SA. As communities engage in C farming practices, they gain a deeper understanding of ecosystem dynamics and the importance of soil health, biodiversity, and C sequestration. This knowledge empowers them to take active roles in environmental conservation, enhancing their sense of ownership and stewardship over local natural resources (Kostyuchenko et al. 2015). Furthermore, C farming often involves collaborative efforts, bringing community members together towards a common goal of environmental sustainability, leading to more resilient and self-reliant communities (Johansson et al. 2022). Workshops, training sessions, and outreach programs associated with C farming further disseminate knowledge on SAPs. Moreover, C farming showcases the economic viability of SA, dispelling the notion that profitability is at odds with environmental responsibility (Sen et al. 2017). This shift in perception encourages a broader adoption of sustainable practices across the agricultural sector. As awareness spreads, communities become more informed advocates for SA, advocating for policies and practices that prioritize environmental conservation. In essence, C farming becomes a catalyst for a paradigm shift towards more sustainable and ecologically conscious approaches to agriculture, with education and awareness serving as its cornerstone.

2.4.4 Economic Benefits

2.4.4.1 Diversification of Income Streams for Farmers

C farming extends a transformative hand to farmers and rural communities, redefining their livelihoods in profound ways. Through a diverse range of BMPs, C farming introduces new streams of income. This includes opportunities like selling forest products, participating in C credit programs, and engaging in eco-tourism ventures linked to SA (Bishop and Landell-Mills 2012). Beyond financial gains, C farming trims operational costs by reducing dependency on synthetic inputs, like fertilizers and pesticides, making agriculture more financially accessible, particularly for smallholders and resource-constrained farmers (Han et al. 2007). The resilience of agricultural systems is amplified through practices like intercropping and agroforestry, leading to more stable yields and reducing vulnerability to crop failures and market fluctuations (Lithourgidis et al. 2011). By enriching soil health, C farming leads to higher quality and more productive soils, translating to increased yields and better-quality crops, providing farmers with a stronger market position. Moreover, the adoption of C farming practices often necessitates additional labor, resulting in the creation of employment opportunities for local communities, stimulating economic growth in rural areas (Carlisle et al. 2019). Implementing energy-saving measures in C farming, such as employing energy-efficient irrigation technologies like drip irrigation or micro-sprinklers, empowers farmers to trim their energy expenses, contributing to overall cost savings (Chel and Kaushik 2011). This holistic approach exemplifies C farming as a forward-thinking solution, fostering economic and environmental sustainability in agriculture.

2.4.4.2 Access to Carbon Markets and Financial Incentives

Access to C markets and financial incentives is a crucial economic benefit of C farming (Badgery et al. 2021). This involves participating in programs and markets where C credits are traded (Oldfield et al. 2021). A single C credit permits the release of one metric ton of CO_2 or an equivalent quantity of other GHGs into the atmosphere (Gupta 2011). Through C farming practices, farmers can sequester C in their soils and vegetation, which can be quantified and verified (Paustian et al. 2019; Guan et al. 2023). This sequestered C can then be converted into C credits and sold on the C market (Keenor et al. 2021). This financial support can help offset the costs associated with implementing and maintaining SAPs. It incentivizes farmers to continue and expand their C farming efforts, contributing to long-term environmental benefits. This can be particularly valuable for smallholder farmers and those in economically disadvantaged regions (Thangata and Hildebrand 2012; Havemann 2013; Lee 2017). The revenue generated from C credits can be reinvested into the farm, supporting further sustainable initiatives, or facilitating the adoption of new technologies. This can lead to increased agricultural productivity and overall farm profitability. Furthermore, there are often financial incentives and subsidies provided by governments or environmental organizations to encourage the adoption of C farming practices (Lin and Li 2011). These incentives may come in the form of grants, tax credits, or subsidies for implementing specific C-sequestering practices (Hickey et al. 2023).

2.4.4.3 Enhanced Market Value for Sustainable Products

As individuals grasp the socio-economic advantages of adopting C farming practices, their awareness of SOC and its significance escalates. This newfound understanding prompts people to consider how their purchasing decisions impact land resources, and subsequently, influence SOC dynamics. This heightened awareness leads to a substantial surge in the demand for environmentally friendly and sustainable products (Kumar et al. 2017). Consumers increasingly recognize that their choices hold the power to support practices that not only benefit the environment but also contribute to the long-term health and productivity of agricultural landscapes. C farming, being a sustainable agricultural approach, aligns with this consumer demand. Products that are grown or produced through C farming methods, such as organic food (Akter et al. 2023; Parashar et al. 2023), agroforestry products (Leakey 1999; Sileshi et al. 2023), soil conditioners (Babla et al. 2022), biopesticides (Ortiz and Sansinenea 2023), and biofertilizers (Borriss 2015; Backer et al. 2018), carry a premium in the market. This increased market value can translate into higher prices and better profitability for farmers engaged in C farming. Moreover, it can open opportunities for farmers to access niche markets and specialty retailers that prioritize sustainably produced goods. The incorporation of the Life Cycle Assessment (LCA) approach in measuring the C footprint of a product is paramount in establishing its sustainability (Roy et al. 2009; Pandey and Agrawal 2014). Additionally, many companies and retailers are setting sustainability goals and prefer to source products from suppliers engaged in climate-resilient farming practices. This can lead to long-term contracts and partnerships for farmers practicing C farming, providing them with a stable and reliable market for their products (Gahukar 2007).

2.5 Conclusion

C farming, structured around the Four 'C' approach (Capture, Convert, Conserve, and Cultivate), operates through a set of core principles that collectively drive its success: promoting soil health, balancing soil ecosystems, conserving and restoring natural resources, implementing robust monitoring and verification systems, building resilience and adaptive capacity, ensuring policy alignment, advancing sustainability, and securing soil resources. These principles enable the synergistic functioning of vegetation, soil, microorganisms, and best management practices to maximize C sequestration across diverse agricultural systems.

The implementation of these principles generates multiple co-benefits that extend beyond C storage. Ecologically, C farming enhances nutrient cycling, restores native species, protects watersheds, and builds resilience against climate-related disturbances. Environmentally, it mitigates GHG emissions, improves water quality, and safeguards ecosystems. Socially, it strengthens food security, human health, community empowerment, and knowledge-sharing around SA. Economically, it diversifies farmer incomes, provides access to C markets and financial incentives, and enhances market value for sustainably produced goods.

Collectively, these interconnected principles and benefits position C farming as a transformative, scalable, and sustainable solution for addressing global climate challenges, restoring degraded ecosystems, and building resilient, productive, and economically viable agricultural systems for future generations.

References

Akter S, Ali S, Fekete-Farkas M et al (2023) Why organic food? Factors influence the organic food purchase intension in an emerging country (Study from northern part of Bangladesh). Resources 12:5. https://doi.org/10.3390/resources12010005

Altieri MA, Funes-Monzote FR, Petersen P (2012) Agroecologically efficient agricultural systems for smallholder farmers: contributions to food sovereignty. Agron Sustain Dev 32:1–13. https://doi.org/10.1007/s13593-011-0065-6

Antle JM, Stoorvogel JJ (2008) Agricultural carbon sequestration, poverty, and sustainability. Environ Dev Econ 13:327–352. https://doi.org/10.1017/S1355770X08004324

Appunn K (2022) Carbon farming explained: the pros, the cons and the EU's plans. https://www.cleanenergywire.org/factsheets/carbon-farming-explained-pros-cons-and-eus-plans. Accessed 12 Sept 2023

Araujo ASF, Leite LFC, Iwata B de F, et al (2012) Microbiological process in agroforestry systems. A review. Agron Sustain Dev 32:215–226. https://doi.org/10.1007/s13593-011-0026-0

Babla M, Katwal U, Yong M-T et al (2022) Value-added products as soil conditioners for sustainable agriculture. Resour Conserv Recycl 178:106079. https://doi.org/10.1016/j.resconrec.2021.106079

Backer R, Rokem JS, Ilangumaran G et al (2018) Plant growth-promoting Rhizobacteria: context, mechanisms of action, and roadmap to commercialization of biostimulants for sustainable agriculture. Front Plant Sci 9. https://doi.org/10.3389/fpls.2018.01473

Badgery W, Murphy B, Cowie A et al (2021) Soil carbon market-based instrument pilot—the sequestration of soil organic carbon for the purpose of obtaining carbon credits. Soil Res 59:12. https://doi.org/10.1071/SR19331

Bai X, Huang Y, Ren W et al (2019) Responses of soil carbon sequestration to climate-smart agriculture practices: a meta-analysis. Glob Chang Biol 25:2591–2606. https://doi.org/10.1111/gcb.14658

Balasundram SK, Shamshiri RR, Sridhara S, Rizan N (2023) The role of digital agriculture in mitigating climate change and ensuring food security: an overview. Sustainability 15:5325. https://doi.org/10.3390/su15065325

Baumert KA, Herzog T, Pershing J (2005) Navigating the numbers: greenhouse gas data and international climate policy. Washington, DC

Berg B, McClaugherty C (2003) Plant litter: decomposition, humus formation, carbon sequestration. Springer Berlin Heidelberg, Berlin, Heidelberg

Bishop J, Landell-Mills N (2012) Forest environmental services: an overview. In: Bishop J, Pagiola S (eds) Selling forest environmental services: market-based mechanisms for conservation and development. Taylor & Francis

Borase DN, Nath CP, Hazra KK et al (2020) Long-term impact of diversified crop rotations and nutrient management practices on soil microbial functions and soil enzymes activity. Ecol Indic 114:106322. https://doi.org/10.1016/j.ecolind.2020.106322

Borriss R (2015) Towards a new generation of commercial microbial disease control and plant growth promotion products. In: Lugtenberg B (ed) Principles of plant-microbe interactions. Springer International Publishing, Cham, pp 329–337

Brandle JR, Hodges L, Zhou XH (2004) Windbreaks in North American agricultural systems. In: Nair PKR, Rao MR, Buck LE (eds) Advances in agroforestry. Springer, Dordrecht, pp 65–78

Brichi L, Fernandes JVM, Silva BM et al (2023) Organic residues and their impact on soil health, crop production and sustainable agriculture: a review including bibliographic analysis. Soil Use Manag 39:686–706. https://doi.org/10.1111/sum.12892

Carlisle L, Montenegro de Wit M, DeLonge MS et al (2019) Transitioning to sustainable agriculture requires growing and sustaining an ecologically skilled workforce. Front Sustain Food Syst 3:96. https://doi.org/10.3389/fsufs.2019.00096

CCA (2014) Carbon Farming Initiative Review

Chan KY, Heenan DP, So HB (2003) Sequestration of carbon and changes in soil quality under conservation tillage on light-textured soils in Australia: a review. Aust J Exp Agric 43:325. https://doi.org/10.1071/EA02077

Chatterjee A, Lal R, Wielopolski L et al (2009) Evaluation of different soil carbon determination methods. CRC Crit Rev Plant Sci 28:164–178. https://doi.org/10.1080/07352680902776556

Chel A, Kaushik G (2011) Renewable energy for sustainable agriculture. Agron Sustain Dev 31:91–118. https://doi.org/10.1051/agro/2010029

Dabney SM (1998) Cover crop impacts on watershed hydrology. J Soil Water Conserv 53:207–213

Davey EL (2023) 6 Major Food Breakthroughs at COP28 — and What Comes Next. https://www.wri.org/insights/food-system-breakthroughs-cop28-whats-next. Accessed 17 Dec 2023

de Jong BHJ, Montoya-Gomez G, Nelson K et al (1996) Community forest management and carbon sequestration: a feasibility study from Chiapas, Mexico. Interciencia 20:409–416

Dollinger J, Jose S (2018) Agroforestry for soil health. Agrofor Syst 92:213–219. https://doi.org/10.1007/s10457-018-0223-9

Eddy WC, Yang WH (2022) Improvements in soil health and soil carbon sequestration by an agroforestry for food production system. Agric Ecosyst Environ 333:107945. https://doi.org/10.1016/j.agee.2022.107945

Ennis CJ, Evans AG, Islam M et al (2012) Biochar: carbon sequestration, land remediation, and impacts on soil microbiology. Crit Rev Environ Sci Technol 42:2311–2364. https://doi.org/10.1080/10643389.2011.574115

Evans MC (2018) Effective incentives for reforestation: lessons from Australia's carbon farming policies. Curr Opin Environ Sustain 32:38–45. https://doi.org/10.1016/j.cosust.2018.04.002
Evans MC, Carwardine J, Fensham RJ et al (2015) Carbon farming via assisted natural regeneration as a cost-effective mechanism for restoring biodiversity in agricultural landscapes. Environ Sci Pol 50:114–129. https://doi.org/10.1016/j.envsci.2015.02.003
Fleming RJ (2018) An updated review about carbon dioxide and climate change. Environ Earth Sci 77:262. https://doi.org/10.1007/s12665-018-7438-y
Friedlingstein P, O'Sullivan M, Jones MW et al (2022) Global carbon budget 2022. Earth Syst Sci Data 14:4811–4900. https://doi.org/10.5194/essd-14-4811-2022
Friedlingstein P, O'Sullivan M, Jones MW et al (2023) Global Carbon Budget 2023. Earth Syst Sci Data 15:5301–5369. https://doi.org/10.5194/essd-15-5301-2023
Gahukar R (2007) Contract farming for organic crop production in India. Curr Sci 93:1661–1663
Guan K, Jin Z, Peng B et al (2023) A scalable framework for quantifying field-level agricultural carbon outcomes. Earth Sci Rev 243:104462. https://doi.org/10.1016/j.earscirev.2023.104462
Gupta Y (2011) Carbon credit: a step towards green environment. Global Journal of Management and Business Research 11:16–19
Hajjar R, Jarvis DI, Gemmill-Herren B (2008) The utility of crop genetic diversity in maintaining ecosystem services. Agric Ecosyst Environ 123:261–270. https://doi.org/10.1016/j.agee.2007.08.003
Hall DO, Rao KK (1999) Photosynthesis, Sixth. Cambridge University Press
Han FX, Lindner JS, Wang C (2007) Making carbon sequestration a paying proposition. Naturwissenschaften 94:170–182. https://doi.org/10.1007/s00114-006-0170-6
Harenda KM, Lamentowicz M, Samson M, Chojnicki BH (2018) The role of peatlands and their carbon storage function in the context of climate change. In: Zielinski T, Sagan I, Surosz W (eds) Interdisciplinary approaches for sustainable development goals. geoPlanet: earth and planetary sciences. Springer, Cham, pp 169–187
Hartmann M, Six J (2022) Soil structure and microbiome functions in agroecosystems. Nat Rev Earth Environ 4:4–18. https://doi.org/10.1038/s43017-022-00366-w
Hartmann A, Schmid M, van Tuinen D, Berg G (2009) Plant-driven selection of microbes. Plant Soil 321:235–257. https://doi.org/10.1007/s11104-008-9814-y
Hartmann H, Bahn M, Carbone M, Richardson AD (2020) Plant carbon allocation in a changing world—challenges and progress: introduction to a virtual issue on carbon allocation. New Phytol 227:981–988. https://doi.org/10.1111/nph.16757
Haruna SI, Anderson SH, Udawatta RP et al (2020) Improving soil physical properties through the use of cover crops: a review. Agrosyst Geosci Environ 3. https://doi.org/10.1002/agg2.20105
Havemann T (2013) Financing mitigation in smallholder agricultural systems: issues and opportunities. In: Climate change mitigation and agriculture. pp 131–143
He M, Xu Z, Hou D et al (2022) Waste-derived biochar for water pollution control and sustainable development. Nat Rev Earth Environ 3:444–460. https://doi.org/10.1038/s43017-022-00306-8
Hickey C, Fankhauser S, Smith SM, Allen M (2023) A review of commercialisation mechanisms for carbon dioxide removal. Front Clim 4. https://doi.org/10.3389/fclim.2022.1101525
Holeplass H, Singh BR, Lal R (2004) Carbon sequestration in soil aggregates under different crop rotations and nitrogen fertilization in an inceptisol in southeastern Norway. Nutr Cycl Agroecosyst 70:167–177. https://doi.org/10.1023/B:FRES.0000048483.94397.b6
Hussain S, Hussain S, Guo R et al (2021) Carbon sequestration to avoid soil degradation: a review on the role of conservation tillage. Plants 10:2001. https://doi.org/10.3390/plants10102001
Ijaz M, Nawaz A, Ul-Allah S et al (2019) Crop diversification and food security. In: Agronomic crops. Springer Singapore, Singapore, pp 607–621
Jiang P, Xiao LQ, Wan X et al (2022) Research Progress on microbial carbon sequestration in soil: a review. Eurasian Soil Sci 55:1395–1404. https://doi.org/10.1134/S1064229322100064
Jindal R, Swallow B, Kerr J (2008) Forestry-based carbon sequestration projects in Africa: potential benefits and challenges. Nat Res Forum 32:116–130. https://doi.org/10.1111/j.1477-8947.2008.00176.x

Johansson EL, Brogaard S, Brodin L (2022) Envisioning sustainable carbon sequestration in Swedish farmland. Environ Sci Pol 135:16–25. https://doi.org/10.1016/j.envsci.2022.04.005
Kardol P, Cregger MA, Campany CE, Classen AT (2010) Soil ecosystem functioning under climate change: plant species and community effects. Ecology 91:767–781. https://doi.org/10.1890/09-0135.1
Keenor SG, Rodrigues AF, Mao L et al (2021) Capturing a soil carbon economy. R Soc Open Sci 8:rsos.202305. https://doi.org/10.1098/rsos.202305
Kimble JM, Lal R, Follett RF (eds) (2016) Agricultural practices and policies for carbon sequestration in soil. CRC Press, Boca Raton
Kirkby CA, Richardson AE, Wade LJ et al (2013) Carbon-nutrient stoichiometry to increase soil carbon sequestration. Soil Biol Biochem 60:77–86. https://doi.org/10.1016/j.soilbio.2013.01.011
Kostyuchenko N, Petrushenko Y, Smolennikov D, Danko Y (2015) Community-based approach to local development as a basis for sustainable agriculture: experience from Ukraine. Int J Agric Resour Gov Ecol 11:178. https://doi.org/10.1504/IJARGE.2015.072901
Kumar B, Manrai AK, Manrai LA (2017) Purchasing behaviour for environmentally sustainable products: a conceptual framework and empirical study. J Retail Consum Serv 34:1–9. https://doi.org/10.1016/j.jretconser.2016.09.004
Lal R (2004) Soil carbon sequestration to mitigate climate change. Geoderma 123:1–22. https://doi.org/10.1016/j.geoderma.2004.01.032
Lal R (2024) Soil Degradation Effects on Human Malnutrition and Under-Nutrition. Med Res Arch 12. https://doi.org/10.18103/mra.v12i10.5753
Lal R (2025) Soil Degradation and Pollution as the Global Public Health Emergency. Med Res Arch 13. https://doi.org/10.18103/mra.v13i5.6474
Latif M (2017) Climate change: the point of no return. In: A sustainable future. Haus Publishing, pp 53–79
Leakey RRB (1999) Potential for novel food products from agroforestry trees: a review. Food Chem 66:1–14. https://doi.org/10.1016/S0308-8146(98)00072-7
Lee J (2017) Farmer participation in a climate-smart future: evidence from the Kenya Agricultural Carbon Project. Land Use Policy 68:72–79. https://doi.org/10.1016/j.landusepol.2017.07.020
Lehmann J, Bossio DA, Kögel-Knabner I, Rillig MC (2020) The concept and future prospects of soil health. Nat Rev Earth Environ 1:544–553. https://doi.org/10.1038/s43017-020-0080-8
Lin B, Li X (2011) The effect of carbon tax on per capita CO2 emissions. Energy Policy 39:5137–5146. https://doi.org/10.1016/j.enpol.2011.05.050
Lin BB, Macfadyen S, Renwick AR et al (2013) Maximizing the environmental benefits of carbon farming through ecosystem service delivery. Bioscience 63:793–803. https://doi.org/10.1525/bio.2013.63.10.6
Lithourgidis AS, Dordas CA, Damalas CA, Vlachostergios D (2011) Annual intercrops: an alternative pathway for sustainable agriculture. Aust J Crop Sci 5:396–410
Lorenz K, Lal R (2014) Soil organic carbon sequestration in agroforestry systems. A review. Agron Sustain Dev 34:443–454. https://doi.org/10.1007/s13593-014-0212-y
Ma Y, Woolf D, Fan M et al (2023) Global crop production increase by soil organic carbon. Nat Geosci 16:1159–1165. https://doi.org/10.1038/s41561-023-01302-3
Malyan SK, Kumar A, Baram S et al (2019) Role of fungi in climate change abatement through carbon sequestration, pp 283–295
Mappin B, Ward A, Hughes L et al (2022) The costs and benefits of restoring a continent's terrestrial ecosystems. J Appl Ecol 59:408–419. https://doi.org/10.1111/1365-2664.14008
Masuku M, Selepe M, Ngcobo N (2017) Small scale agriculture in enhancing household food security in rural areas. J Hum Ecol 58:153–161. https://doi.org/10.1080/09709274.2017.1317504
McBratney A, Field DJ, Koch A (2014) The dimensions of soil security. Geoderma 213:203–213. https://doi.org/10.1016/j.geoderma.2013.08.013
Mikha MM, Rice CW (2004) Tillage and manure effects on soil and aggregate-associated carbon and nitrogen. Soil Sci Soc Am J 68:809–816. https://doi.org/10.2136/sssaj2004.8090
Mitsch WJ, Zhang L, Stefanik KC et al (2012) Creating wetlands: primary succession, water quality changes, and self-design over 15 years. Bioscience 62:237–250. https://doi.org/10.1525/bio.2012.62.3.5

Muhammad I, Wang J, Sainju UM et al (2021) Cover cropping enhances soil microbial biomass and affects microbial community structure: a meta-analysis. Geoderma 381:114696. https://doi.org/10.1016/j.geoderma.2020.114696

Mustafa MA, Mateva KI, Massawe F (2019) Sustainable crop production for environmental and human health—the future of agriculture. In: Annual Plant Reviews online. Wiley, pp 1117–1140

Nave LE, Walters BF, Hofmeister KL et al (2019) The role of reforestation in carbon sequestration. New For (Dordr) 50:115–137. https://doi.org/10.1007/s11056-018-9655-3

Nelson DW, Sommers LE (1996) Total carbon, organic carbon, and organic matter. In: Sparks DL, Page AL, Helmke PA et al (eds) Methods of soil analysis: part 3 chemical methods. SSSA, pp 961–1010

Nevalainen O, Niemitalo O, Fer I et al (2022) Towards agricultural soil carbon monitoring, reporting, and verification through the Field Observatory Network (FiON). Geosci Instrum Methods Data Syst 11:93–109. https://doi.org/10.5194/gi-11-93-2022

Ojeda-Bustamante W, Ontiveros-Capurata RE, Flores-Velázquez J, Iñiguez-Covarrubias M (2017) Changes in water demands under adaptation actions to climate change in an irrigation district. J Water Clim Change 8:191–202. https://doi.org/10.2166/wcc.2016.028

Oldfield EE, Eagle AJ, Rubin RL et al (2021) Agricultural soil carbon credits: making sense of protocols for carbon sequestration and net greenhouse gas removals, New York

Ortiz A, Sansinenea E (2023) Genetically modified plants based on bacillus genes and commercial bacillus-based biopesticides for sustainable agriculture. Horticulturae 9:963. https://doi.org/10.3390/horticulturae9090963

Owuor SO, Butterbach-Bahl K, Guzha AC et al (2016) Groundwater recharge rates and surface runoff response to land use and land cover changes in semi-arid environments. Ecol Process 5:16. https://doi.org/10.1186/s13717-016-0060-6

Pagliai M, Vignozzi N, Pellegrini S (2004) Soil structure and the effect of management practices. Soil Tillage Res 79:131–143. https://doi.org/10.1016/j.still.2004.07.002

Pandey D, Agrawal M (2014) Carbon footprint estimation in the agriculture sector. In: Muthu S (ed) Assessment of carbon footprint in different industrial sectors. Springer, Singapore, pp 25–47

Parashar S, Singh S, Sood G (2023) Examining the role of health consciousness, environmental awareness and intention on purchase of organic food: a moderated model of attitude. J Clean Prod 386:135553. https://doi.org/10.1016/j.jclepro.2022.135553

Parihar CM, Singh AK, Jat SL et al (2020) Soil quality and carbon sequestration under conservation agriculture with balanced nutrition in intensive cereal-based system. Soil Tillage Res 202:104653. https://doi.org/10.1016/j.still.2020.104653

Patle GT, Bandyopadhyay KK, Singh DK (2013) Impact of conservation agriculture and resource conservation technologies on carbon sequestration-a review. Indian J Agric Sci 83:3–13

Paustian K, Collier S, Baldock J et al (2019) Quantifying carbon for agricultural soil management: from the current status toward a global soil information system. Carbon Manag 10:567–587. https://doi.org/10.1080/17583004.2019.1633231

Peckham R (2015) Hygienic nature: afforestation and the greening of colonial Hong Kong. Mod Asian Stud 49:1177–1209. https://doi.org/10.1017/S0026749X13000620

Poeplau C, Don A (2015) Carbon sequestration in agricultural soils via cultivation of cover crops—a meta-analysis. Agric Ecosyst Environ 200:33–41. https://doi.org/10.1016/j.agee.2014.10.024

Ponge J-F (2013) Plant–soil feedbacks mediated by humus forms: a review. Soil Biol Biochem 57:1048–1060. https://doi.org/10.1016/j.soilbio.2012.07.019

Power AG (2010) Ecosystem services and agriculture: tradeoffs and synergies. Philos Trans R Soc B 365:2959–2971. https://doi.org/10.1098/rstb.2010.0143

Raj NG (2023) Insect ecological interactions: from trophic cascades to ecosystem services. In: ARM N, Venukumar S, Kumar SV et al (eds) Entomology redefined: current trends and future directions, pp 304–317

Ramachandran Nair PK, Mohan Kumar B, Nair VD (2009) Agroforestry as a strategy for carbon sequestration. J Plant Nutr Soil Sci 172:10–23. https://doi.org/10.1002/jpln.200800030

Ramachandran Nair PK, Nair VD, Mohan Kumar B, Showalter JM (2010) Carbon sequestration in agroforestry systems. In: Advances in agronomy, pp 237–307

Reeve JR, Hoagland LA, Villalba JJ et al (2016) Organic farming, soil health, and food quality: considering possible links. In: Sparks DL (ed) Advances in agronomy, pp 319–367
Rice CW (2002) Organic matter and nutrient dynamics. In: Encyclopedia of soil science. pp 1180–1183
Richards KR (2004) A brief overview of carbon sequestration economics and policy. Environ Manag 33:545–558. https://doi.org/10.1007/s00267-003-9111-8
Rivero RM, Mittler R, Blumwald E, Zandalinas SI (2022) Developing climate-resilient crops: improving plant tolerance to stress combination. Plant J 109:373–389. https://doi.org/10.1111/tpj.15483
Rockström J (2003) Resilience building and water demand management for drought mitigation. Phys Chem Earth, Parts A/B/C 28:869–877. https://doi.org/10.1016/j.pce.2003.08.009
Roy P, Nei D, Orikasa T et al (2009) A review of life cycle assessment (LCA) on some food products. J Food Eng 90:1–10. https://doi.org/10.1016/j.jfoodeng.2008.06.016
Rusch A, Bommarco R, Ekbom B (2017) Conservation biological control in agricultural landscapes. In: Advances in botanical research. pp 333–360
Sauer TJ, Dold C, Ashworth AJ et al (2021) Agroforestry practices for soil conservation and resilient agriculture. In: Agroforestry and ecosystem services. Springer International Publishing, Cham, pp 19–48
Schmitz OJ, Buchkowski RW, Smith JR et al (2017) Predator community composition is linked to soil carbon retention across a human land use gradient. Ecology 98:1256–1265. https://doi.org/10.1002/ecy.1794
Seepaul R, George S, Love J et al (2023) Managing cover crops for improved soil health. EDIS 2023. https://doi.org/10.32473/edis-ag277-2023
Sen B, Venkatesh P, Jha GK et al (2017) Agricultural diversification and its impact on farm income: a case study of Bihar. Agric Econ Res Rev 30:77. https://doi.org/10.5958/0974-0279.2017.00023.4
Sha Z, Bai Y, Li R et al (2022) The global carbon sink potential of terrestrial vegetation can be increased substantially by optimal land management. Commun Earth Environ 3:8. https://doi.org/10.1038/s43247-021-00333-1
Shackleton NJ, Hall MA, Line J, Shuxi C (1983) Carbon isotope data in core V19-30 confirm reduced carbon dioxide concentration in the ice age atmosphere. Nature 306:319–322. https://doi.org/10.1038/306319a0
Sigman DM, Boyle EA (2000) Glacial/interglacial variations in atmospheric carbon dioxide. Nature 407:859–869. https://doi.org/10.1038/35038000
Sijtsma CH, Campbell AJ, McLaughlin NB, Carter MR (1998) Comparative tillage costs for crop rotations utilizing minimum tillage on a farm scale. Soil Tillage Res 49:223–231. https://doi.org/10.1016/S0167-1987(98)00175-5
Sileshi GW, Dagar JC, Akinnifesi FK, Mng'omba SA (2023) Potentials of indigenous fruit trees in enhancing nutrition, income and biodiversity conservation in African agroforestry. In: Dagar JC, Gupta SR, Sileshi GW (eds) Agroforestry for sustainable intensification of agriculture in Asia and Africa. Springer, Singapore, pp 321–361
Smith P, Soussana J, Angers D et al (2020) How to measure, report and verify soil carbon change to realize the potential of soil carbon sequestration for atmospheric greenhouse gas removal. Glob Chang Biol 26:219–241. https://doi.org/10.1111/gcb.14815
Solaiman ZM, Mickan B (2014) Use of mycorrhiza in sustainable agriculture and land restoration. In: Solaiman Z, Abbott LK, Varma A (eds) Soil biology. Springer, pp 1–15
Speir SL, Tank JL, Trentman MT et al (2022) Cover crops control nitrogen and phosphorus transport from two agricultural watersheds at multiple measurement scales. Agric Ecosyst Environ 326:107765. https://doi.org/10.1016/j.agee.2021.107765
Stoian M, Dobre I, Popescu C et al (2022) Increasing sustainability of food production and ensuring human health through agriculture digitalization. Ekonomika poljoprivrede 69:1209–1223. https://doi.org/10.5937/ekoPolj2204209S
Stout JC, Tiedeken EJ (2017) Direct interactions between invasive plants and native pollinators: evidence, impacts and approaches. Funct Ecol 31:38–46. https://doi.org/10.1111/1365-2435.12751
Stubbs M (2016) Conservation compliance and U.S. farm policy

Szymczak LS, de Carvalho PCF, Lurette A et al (2020) System diversification and grazing management as resilience-enhancing agricultural practices: the case of crop-livestock integration. Agric Syst 184:102904. https://doi.org/10.1016/j.agsy.2020.102904
Termorshuizen AJ, Moolenaar SW, Veeken AHM, Blok WJ (2004) The value of compost. Rev Environ Sci Biotechnol 3:343–347. https://doi.org/10.1007/s11157-004-2333-2
Thangata PH, Hildebrand PE (2012) Carbon stock and sequestration potential of agroforestry systems in smallholder agroecosystems of sub-Saharan Africa: mechanisms for 'reducing emissions from deforestation and forest degradation' (REDD+). Agric Ecosyst Environ 158:172–183. https://doi.org/10.1016/j.agee.2012.06.007
Thangavel P, Sridevi G (2017) Soil security: a key role for sustainable food productivity. In: Dhanarajan A (ed) Sustainable agriculture towards food security. Springer Singapore, Singapore, pp 309–325
Thrupp LA (2000) Linking agricultural biodiversity and food security: the valuable role of agrobiodiversity for sustainable agriculture. Int Aff 76:265–281. https://doi.org/10.1111/1468-2346.00133
Toensmeier E (2016) The carbon farming solution: a global toolkit of perennial crops and regenerative agriculture practices for climate change mitigation and food security. Chelsea Green Publishing, Vermont
Udawatta RP, Jose S (2011) Carbon sequestration potential of agroforestry practices in temperate North America. In: Kumar B, Nair P (eds) Advances in agroforestry. Springer, Dordrecht, pp 17–42
Umesha S, Manukumar HMG, Chandrasekhar B (2018) Sustainable agriculture and food security. In: Biotechnology for sustainable agriculture. Elsevier, pp 67–92
UNFCCC (2016) The Paris agreement. United Nations Framework Convention on Climate Change (UNFCCC)
Unger IM, Kennedy AC, Muzika R-M (2009) Flooding effects on soil microbial communities. Appl Soil Ecol 42:1–8. https://doi.org/10.1016/j.apsoil.2009.01.007
Uribe N, Corzo G, Quintero M et al (2018) Impact of conservation tillage on nitrogen and phosphorus runoff losses in a potato crop system in Fuquene watershed, Colombia. Agric Water Manag 209:62–72. https://doi.org/10.1016/j.agwat.2018.07.006
Ussiri D, Lal R (2013) The role of fertilizer Management in Mitigating Nitrous Oxide Emissions. In: Soil emission of nitrous oxide and its mitigation. Springer Netherland, Dordrecht, pp 315–346
Van Veen JA, Kuikman PJ (1990) Soil structural aspects of decomposition of organic matter by micro-organisms. Biogeochemistry 11:213–233. https://doi.org/10.1007/BF00004497
Vashisht BB, Maharjan B, Jalota SK (2019) Management practice to optimize wheat yield and water use in changing climate. Arch Agron Soil Sci 65:1802–1819. https://doi.org/10.1080/03650340.2019.1578957
Wantzen K, Junk MW (2008) Riparian wetlands. In: Ecosystem ecology. p 342
Waring ER, Pederson C, Lagzdins A et al (2024) Water and soil quality respond to no-tillage and cover crops differently through 10 years of implementation. Agric Ecosyst Environ 360:108791. https://doi.org/10.1016/j.agee.2023.108791
Welch RM, Graham RD (1999) A new paradigm for world agriculture: meeting human needs. Field Crop Res 60:1–10. https://doi.org/10.1016/S0378-4290(98)00129-4
Yang X, Chen Y, Pacenka S et al (2015) Effect of diversified crop rotations on groundwater levels and crop water productivity in the North China Plain. J Hydrol (Amst) 522:428–438. https://doi.org/10.1016/j.jhydrol.2015.01.010
Young KM, Thomson A (2020) Growing carbon for sustainable and resilient soils. Crops Soils 53:37–42. https://doi.org/10.1002/crso.20032
Yu C, Li Y, Mo R et al (2020) Effects of long-term straw retention on soil microorganisms under a rice–wheat cropping system. Arch Microbiol 202:1915–1927. https://doi.org/10.1007/s00203-020-01899-8
Zubieta ÁS, Savian JV, de Souza Filho W et al (2021) Does grazing management provide opportunities to mitigate methane emissions by ruminants in pastoral ecosystems? Sci Total Environ 754:142029. https://doi.org/10.1016/j.scitotenv.2020.142029

Chapter 3
Soil Health and Carbon Sequestration

Abstract Soil health and carbon (C) sequestration are intricately linked, with soil C playing a key role in maintaining productive and resilient soils. This chapter explains the factors affecting soil C dynamics, the processes that stabilize organic C, and how soil C supports essential soil health functions like nutrient cycling, microbial activity, soil structure, water retention, and resistance to degradation, highlighting its importance for sustainable ecosystems.

3.1 Soil Carbon

Soil carbon (C) is a fundamental component of the Earth's C cycle and plays an important role in the storage and cycling of C between the atmosphere, plants, and soil (Fig. 3.1). To comprehend its role in C dynamics, understanding the fundamentals of soil C storage and dynamics become essential. It occurs in two primary forms: soil organic C (SOC) and soil inorganic C (SIC), collectively contributing to the total C (TC) content, and further classified based on their decomposition rates and nature of compounds present (Fig. 3.2). SOC originates from the decomposition of once-living organisms, including plant and animal residues. It includes plant material like roots, leaves, and organic residues from microbial activity. Plants, through photosynthesis, fix atmospheric CO_2 into organic compounds, which eventually become part of the soil organic matter (SOM). This form of C is present in complex organic molecules such as carbohydrates, proteins, lipids, and lignin (Lorenz et al. 2007). SOC functions in nutrient supply by serving as a vital energy source for soil microorganisms. As microbes break down SOM, they release nutrients that become available for plant uptake. SOC improves soil structure, creating pores that enhance water retention (Rawls et al. 2003). It acts like a sponge, holding water and preventing excessive runoff. Microorganisms, crucial for nutrient cycling and soil health, thrive on SOC (Wang et al. 2021). The microbial decomposition of SOM contributes to the formation of stable, long-lasting humus (Li et al. 2023c; Wang et al. 2022b). In contrast, SIC is derived from minerals and carbonate (CO_3^{2-}) compounds present in the soil. It can enter the soil through weathering of rocks

N. Loria, R. Lal, *Carbon Farming*, https://doi.org/10.1007/978-3-032-00842-8_3

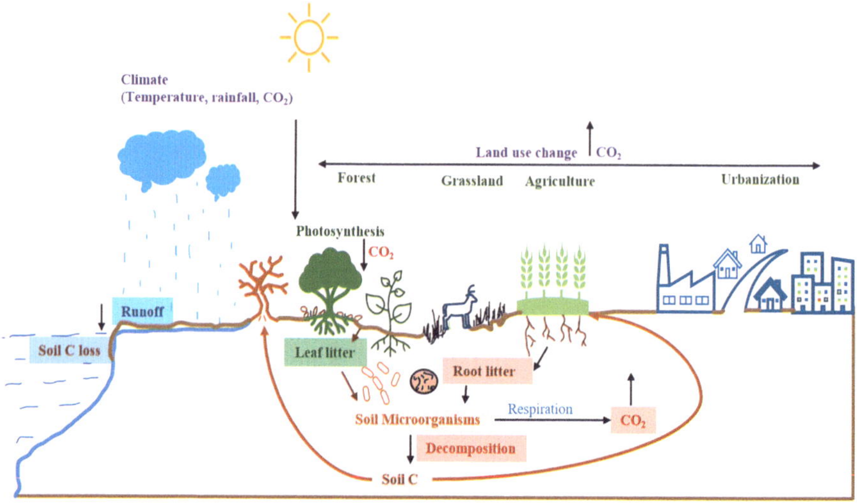

Fig. 3.1 Soil carbon cycling: Atmosphere, Plants, Soil. The cycle of soil carbon (C) involves dynamic interactions between the atmosphere, plants, and the soil. Atmospheric CO_2 is absorbed by plants through photosynthesis, leading to the creation of organic matter, including leaf and root litter. This C-rich organic matter is then introduced into the soil through plant roots and undergoes subsequent decomposition. Microorganisms play a crucial role in breaking down this SOM, releasing C back into the soil through decomposition and, to some extent, into the atmosphere as CO_2 through respiration. Climatic factors such as intensive rainfall can lead to the loss of some soil C into water bodies through runoff. Additionally, changes in land use can result in the release of soil C in the form of CO_2 into the atmosphere

containing CO_3^{2-} minerals, or from the dissolution of CO_2 in water (Bartmiński et al. 2022). SIC exists in several forms, with CO_3^{2-} and bicarbonate (HCO_3^-) being the most common (Ferdush and Paul 2021). These forms are often associated with calcium, magnesium, or other metal cations. Zamanian et al. (2016) classified SIC into three main groups: geogenic carbonates (GC) from soil parent materials, biogenic carbonates (BC) formed within living organisms, and pedogenic carbonates (PC) created and redistributed within soils. SIC serves as a crucial buffer in soil, maintaining a stable pH by neutralizing acidic or alkaline conditions, thereby influencing soil chemistry, and providing essential ecosystem services (Groshans et al. 2018). Carbonic acid, formed when CO_2 dissolves in water, participates in the weathering of minerals, releasing essential nutrients for plant uptake. Certain forms of SIC, notably CO_3^{2-} minerals, contribute to long-term C storage in soils, especially in arid and semiarid climates (Lal 2002; Lal 2021a). However, the increasing prevalence of atmospheric acid deposition and the intensive use of agricultural fertilizers and pesticides have led to enhanced CO_3^{2-} dissolution, consequently releasing CO_2 into the atmosphere (Song et al. 2022).

SOC and SIC interact dynamically in the soil. Soil microorganisms decompose SOM, releasing CO_2, which can react with water to form HCO_3^-, influencing soil pH. Both forms of C play a role in C sequestration (CS), with SOC being more responsive to management practices, while SIC, particularly carbonate minerals, offers stable, long-term storage. A comprehensive review by Naorem et al. (2022)

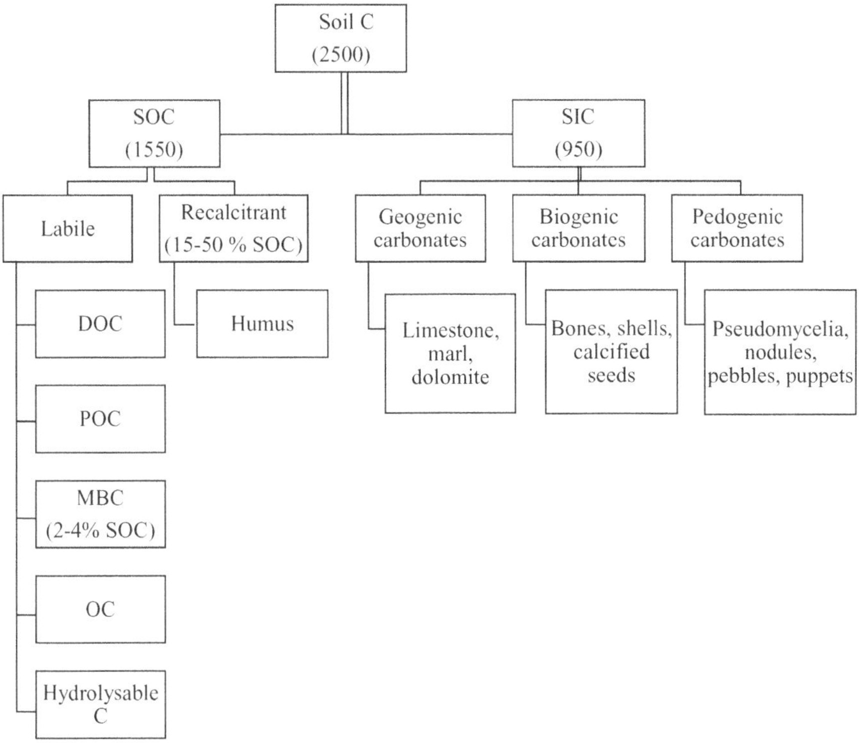

Fig. 3.2 Classification of soil carbon. C- carbon; SOC- Soil organic C; SIC- Soil inorganic C; DOC- Dissolved Organic C; POC- Particulate Organic C; MBC- Microbial biomass C; OC- oxidative C. Numbers in parenthesis represent Peta-grams of C present in top 1 m soil unless otherwise stated (Lorenz et al. 2007; Zamanian et al. 2016; Lorenz and Lal 2018; Zhang and Zhou 2018; Ferdush and Paul 2021; Bartmiński et al. 2022)

focused on the potential for long-term SIC storage in arid/dryland regions, identifying processes like silicate mineral weathering, carbonate dissolution, and pedogenic inorganic carbonate formation as key contributors to CO_2 exchange between soil and atmosphere in dryland soils. Factors influencing SIC dynamics include climate, microbial activity, edaphic conditions (soil depth, salinity, and acidification), land use types, and management practices. Effective soil C management requires a comprehensive understanding of the dynamics of both SOC and SIC within the context of sustainable agricultural practices (SAPs). While the impacts of SAPs, such as reduced tillage, crop rotation strategies, and the incorporation of SOM, are visibly evident in SOC, the significance of SIC is often overlooked due to its gradual and less conspicuous nature. SIC process occurs over extended periods, and changes in SIC are not as readily apparent through short-term observations. Balancing and optimizing the levels of SOC and SIC contribute to improved soil fertility, structure, and resilience, thereby enhancing overall ecosystem health. Therefore, alongside efforts to enhance SOC through SAPs, recognizing and safeguarding SIC is imperative.

3.2 Factors Influencing Soil Carbon Dynamics

3.2.1 *Climate*

Climate plays a pivotal role in influencing soil C dynamics through its impact on various biogeochemical processes. Temperature, precipitation, and other climatic factors significantly affect microbial activity, SOM decomposition, and vegetation growth, thereby influencing the accumulation and loss of soil C. Warmer temperatures generally accelerate microbial activity, leading to increased decomposition rates of SOM (Crohn and Valenzuela-Solano 2003; Rubenstein et al. 2017). This enhanced microbial activity results in a higher turnover of soil C, potentially leading to a release of CO_2 into the atmosphere (Dilekoğlu and Sakin 2017). Conversely, colder temperatures can slow down microbial processes, reducing decomposition rates and promoting CS (Kirschbaum 2010). Higher temperatures can also contribute to the formation of more recalcitrant C products (Berg et al. 1993; Dalias et al. 2001; Li et al. 2022a). Precipitation patterns, providing adequate moisture, facilitate microbial decomposition, supporting SOM breakdown and C release (Brye and Gbur 2010; Tomar and Baishya 2020; Qi et al. 2022). However, excessive rainfall may lead to soil erosion, transporting C-rich particles away from their original locations (Polyakov and Lal 2004; Shi and Schulin 2018). Climate-induced shifts in vegetation, including changes in species composition and growth patterns, further influence C inputs into the soil (Bardgett et al. 2013; Moravcová et al. 2023). Changes in precipitation regimes and temperature can affect plant productivity and, consequently, the amount of SOM returned to the soil (Chen et al. 2018; Luo et al. 2017).

3.2.2 *Vegetation*

Vegetation type is a crucial determinant of soil C dynamics, exerting significant influence on the quantity, quality, and cycling of SOM within terrestrial ecosystems. Different plant species exhibit distinct growth patterns, litter amounts and root physiology, and mycorrhizal associations. Influencing the rate of SOM decomposition and subsequent CS (Chen et al. 2018; Wang et al. 2020). For instance, the seasonal leaf shedding of deciduous trees contributes nutrient-rich litter, fostering rapid microbial activity and C release (Anderson et al. 2019; Nickels and Prescott 2021; Wu et al. 2023). In contrast, evergreen trees with year-round leaf retention produce more recalcitrant litter, leading to slower breakdown rates and a more stable C pool (Sohrabi et al. 2016; Lin et al. 2020; Bohara et al. 2020; Lu et al. 2021). Rooting depth plays a crucial role, as deep-rooted plants such as many grasses and certain tree species transport SOM into the soil profile, enhancing long-term CS (Schrumpf et al. 2013; Hicks Pries et al. 2018; Zhou et al. 2019). Shallow-rooted plants, common in herbaceous ecosystems, may contribute to faster turnover rates of SOC, as their fine root-derived SOM is more accessible to microbial decomposers near the soil surface (Solly et al. 2014; Lai et al. 2017; Dietzel et al. 2017). Root litter quality further shapes SOC storage by influencing microbial diversity, where high-quality litter decomposition declines with increasing bacterial diversity, while low-quality litter contributes to soil C with rising fungal diversity (Heitkötter et al. 2017; Xu et al. 2023). Additionally, rhizosphere processes, especially the symbiotic relationships between plants and mycorrhizal fungi, significantly impact soil C dynamics (Finzi et al. 2015). Mycorrhizal fungi form mutually beneficial associations with plant roots, facilitating nutrient uptake for the plants while receiving C compounds in return (Huang et al. 2021; Orwin et al. 2011; Talbot et al. 2008; Treseder 2016; Zak et al. 2019). This belowground exchange influences the allocation of C, affecting microbial communities and C storage in the soil. The diversity of mycorrhizal associations across vegetation types contributes to variations in soil CS and nutrient cycling (Beugnon et al. 2023; Liu et al. 2022; Luo et al. 2023; Soudzilovskaia et al. 2019). Sustainable land management practices that consider the intricate relationships between plant communities and soil C dynamics are crucial for mitigating climate change, enhancing ecosystem resilience, and promoting soil health.

3.2.3 Land Management Practices

Land use exerts a significant influence on soil C dynamics, shaping the quantity, quality, and stability of SOC. Various land management practices, such as agriculture, forestry, urbanization, and restoration efforts, significantly impact the cycling of C within terrestrial ecosystems. Agricultural activities, including tillage, crop rotations, and fertilizer applications, directly influence SOC dynamics (Batlle-Bayer et al. 2010; Tu et al. 2018). Intensive tillage disrupts soil structure, accelerates microbial decomposition, and enhances C mineralization (Álvaro-Fuentes et al. 2007; Yadav et al. 2019). Additionally, continuous cropping without sufficient cover crops can lead to decreased C inputs into the soil, further affecting SOC levels (Blanco-Canqui and Lal 2008). Conversely, conservation practices like reduced tillage, cover cropping, and residue management promote soil health and contribute to CS by minimizing soil disturbance and enhancing organic inputs (Frasier et al. 2016; Hussain et al. 2021; Li et al. 2013; Veenstra et al. 2007). Forestry practices, such as logging and afforestation, play a crucial role. Deforestation leads to a rapid release of C stored in trees and soil, contributing to elevated atmospheric CO_2 (Huang and Asner 2010; Arias-Ortiz et al. 2021; Damian et al. 2021). Afforestation efforts, on the other hand, can promote CS by establishing new C sinks in the form of growing forests (Epron et al. 2009; Hong et al. 2020; Li et al. 2023b). However, the type of tree species, management practices, and disturbances like wildfires can influence the effectiveness of afforestation in enhancing SOC (Huang et al. 2023; Nolan et al. 2022; Siahaan et al. 2020). Urbanization poses challenges with soil compaction and C stability leading to a decline in SOC (Pouyat et al. 2002; Devigne et al. 2016). Impervious surfaces, common in urban environments, reduce water infiltration and exacerbate soil erosion, further affecting C storage (Miyajima et al.

2015; Stewart et al. 2019; Yu and Song 2023). However, sustainable urban planning practices, such as green infrastructure and urban forestry, can help mitigate these impacts by promoting vegetation cover and minimizing soil disturbance (Lorenz and Lal 2015; Tresch et al. 2019; Rivers et al. 2021). Restoration efforts, including reforestation and ecosystem rehabilitation, increase C inputs through plant litter and root biomass, fostering accelerated microbial turnover rates (Shao et al. 2019; Quartucci et al. 2023). These activities focus on restoring degraded ecosystems, promoting vegetation growth, and improving soil structure (Bin and Xin-Hua 2006; Wei et al. 2013; Matzek et al. 2020; Mutio et al. 2023; Hinshaw and Wohl 2023).

3.2.4 Soil Factors

Inherent soil properties, such as soil type and texture, play an important role in soil C dynamics. Soil type, for instance, the Inceptisol exhibited increased SOC accumulation on the soil surface due to its heavy texture, while the new alluvial Entisol showed lower CS attributed to partial biomass decomposition and macroaggregate formation with sand (Rakesh et al. 2023). The depth of the soil profile varies among different soil types (Kempen et al. 2011). Deeper soils, with more space for root growth and SOM accumulation, significantly influence the overall C storage capacity of a soil profile (Harper and Tibbett 2013; Callesen et al. 2016; Lal 2021b).

Soil texture, determined by the relative proportions of sand, silt, and clay particles, is a stable factor influencing soil biophysical properties and long-term soil fertility (Bauer and Black 1992; Lehmann et al. 2003; Bacq-Labreuil et al. 2018; Hao et al. 2020; Siebielec et al. 2020; Li et al. 2021b; Zhu et al. 2021; Pauwels et al. 2023). Sandy soils, like Alfisols, exhibiting lower SOM content and faster decomposition rates, often result in lower SOC levels, whereas clayey soils, such as Oxisols, with their ability to physically protect SOM within aggregates, contribute to higher C retention (Chivenge et al. 2011; Bonetti et al. 2017; Liao et al. 2021; Possignolo-Vitti et al. 2017; Tumer et al. 2013).

Furthermore, soil mineralogy, determined by the types and amounts of minerals present, have impact on C dynamics. Clay minerals, such as montmorillonite and vermiculite, exhibit high surface area, influencing water and nutrient interactions, thereby regulating soil structure, microbial activity, and nutrient availability, ultimately impacting C cycling (Das et al. 2023; Singh et al. 2019; Yu et al. 2022). CEC, influenced by soil texture and SOM, determines the soil's ability to hold and exchange nutrients, allowing soils with higher CEC to retain more SOM and nutrients, thus contributing to increased soil C storage (Gao and Chang 1996; Gruba and Mulder 2015). Multiple factor interactions, including the interaction of iron oxides with clay minerals forming macroaggregates, stabilize SOM and contribute to C accumulation (Wiseman and Püttmann 2006). Significantly, research led by de Andrade Bonetti et al. (2017) highlights increased soil resilience in subtropical regions, emphasizing the impact of factors such as SOM content, texture, and mineralogy, potentially allowing for long-term C storage. Moreover, the intricate interplay between plant litter chemistry and soil texture in influencing soil C dynamics is exemplified in a study by Huys et al. (2022), revealing that in coarse-textured soils, the stable mineral-associated SOM pool exhibits resilience despite substantial C fluxes.

3.3 Soil Organic Carbon Stabilization

3.3.1 *The Role of Soil Organic Matter*

Achieving stable SOC necessitates a comprehensive understanding of SOM formation processes, functions, biochemical compositions, and the underlying mechanisms of stabilization. SOM is closely linked to SOC, originating from the microbial decomposition of plant and animal residues, whether living or dead. Various studies highlight that SOM constitutes 58% of SOC (Bernoux and Cerri 2005; Bianchi et al. 2008; Pribyl 2010; Manns et al. 2016). The decomposition of SOM is influenced by climate (Puissant et al. 2015; Osanai et al. 2015; Holden et al. 2015), soil texture (Hassink et al. 1997; Angst et al. 2021), soil mineral composition (Vogel et al. 2015; Zacháry et al. 2023), fertilization (Li et al. 2023a; Li et al. 2020; Shaghaleh et al. 2023), and land management practices (Gunapala et al. 1998; Liu et al. 2006; Masoudi et al. 2023). Among the biochemical components of SOM are lipids, mono- and polycyclic aromatics, phenolics, polysaccharides, lignins, and N-bearing compounds. (Chen et al. 2023; Yan et al. 2022). SOM is often used interchangeably with humus (Stevenson 1995; Paul 2016), the dark colored stable fraction of SOM that is formed from decomposed plant and animal residues (Ponge 2003). Different ecosystems develop specific humus forms, including mor, mull, and moder based on their unique resource-use strategies and ecological traits (Ponge 2003, 2013). These three forms offer insights into the complex dynamics of SOM interaction with mineral components along soil profiles, driven by biological, fungal, and mechanical factors in soil ecosystems (Ponge 2013).

SOM primarily consists of two humus forms: humic and non-humic substances (Bernoux and Cerri 2005; Ghosh 2017). Humic substances, comprising ~60–80% of SOM in most soils (Stevenson 1994; Sanderman and Amundson 2014), are heterogenous, high molecular weight organic compounds formed through chemical transformations (Tiwari et al. 2023; Zavarzina et al. 2021). They are categorized into humic and fulvic acids, and humin (Table 3.1). In contrast, non-humic substances, being more labile and readily decomposed, belong to a simpler class of organic compounds such as naturally occurring polymers like proteins, lipids, and carbohydrates. The exact structure of humic substances remains unknown, and there is not a clear distinction between humic and non-humic substances (Simpson et al. 2002; Sutton and Sposito 2005). Evidence suggests that humic substances may include non-humic components such as esters, alcohols, lignin-derived products, and polypeptides (Theng 2012). Functioning as a stable component in soil, humus offers physical benefits, enhancing aggregate stability and water-holding capacity

Table 3.1 Humic substances fractions

	Humic	Fulvic	Humin
Source	Derived from breakdown of plant and microbial residues during the humidification	Formed through further decomposition of humic acids and other organic materials by microbial activity	Represents the most stable and recalcitrant fraction of SOM, consisting of partially decomposed plant and microbial residues.
Color	Dark brown to black	Yellowish	Non extractable
Solubility	Soluble in alkaline solutions; precipitated at pH < 2	Soluble in both acidic and alkaline solutions	Insoluble in both acidic and alkaline
Characteristics	Contribute to soil structure, CEC, and nutrient retention	Nutrient transport, chelation of metals, and increasing the bioavailability of nutrients to plants	Contributes to SOM stability, forms complexes with clay minerals, and helps in soil aggregation.
Examples	Quinone, phenol, catechol	–	–

(Havlin 2005; Pintaldi et al. 2018; Yu et al. 2020), thereby facilitating the tillage of clay soils. It also contributes to the soil's chemical properties, increasing cation exchange capacity (CEC) (Shoba and Chudnenko 2014; Ramos et al. 2018), pH buffering capacity (Stevenson 1995; Zhang et al. 2016), and nutrient availability (Horwath 2007). Furthermore, humus provides biological advantages by acting as a nutrient source for soil microorganisms and fostering microbial diversity (Russell 1977; Ponge 2013).

In soil ecosystems, SOM is categorized into distinct pools based on its stability, composition, and turnover rates, providing insights into the dynamic processes of decomposition and nutrient cycling (von Lützow et al. 2008; Dungait et al. 2012; Chertov and Komarov 2013). The active pool encompasses recently added plant and animal residues, along with microbial biomass, characterized by high turnover rates (MRT ~ 1–10 years) as these materials actively decompose to release nutrients. The intermediate pool consists of partially decomposed organic materials, such as humus and decomposed plant residues, with an intermediate turnover rate (MRT ~ 10–100 years), contributing to nutrient storage and microbial activity. In contrast, the passive pool comprises highly stabilized or recalcitrant SOM, including well-decomposed materials like humic substances and resistant plant compounds. This pool experiences a low turnover rate (MRT > 100 years), acting as a long-term reservoir of SOC that contributes to soil structure, stability, and nutrient retention. SOM can be classified into physical and chemical fractions through various fractionalization methods that target different aspects of its composition (Hassink et al. 1997; von Lützow et al. 2007; Helfrich et al. 2007) (Fig. 3.3). Physical fractionalization involves separating SOM based on particle size or density, while chemical fractionalization focuses on specific chemical properties. Examples of physical methods include sieving, density fractionation, and particle size separation, while chemical methods include acid hydrolysis, oxidation, and extraction techniques.

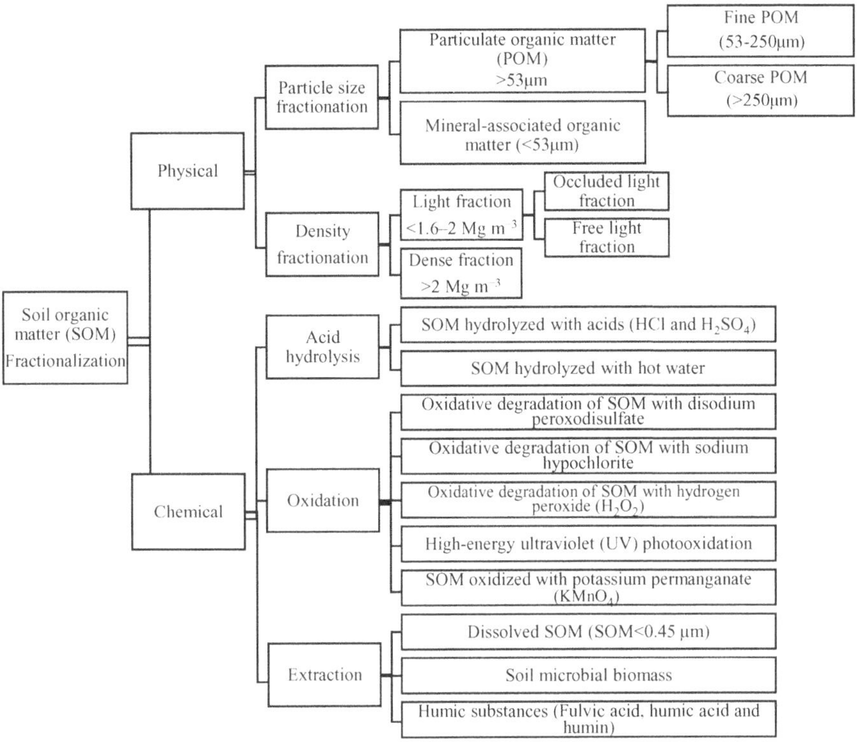

Fig. 3.3 Soil organic matter fractionation methods

3.3.2 Soil Organic Matter Accessibility to Microbes

SOM serves as a vital reservoir for soil microbial communities, influencing microbial access to nutrients and energy. The accessibility of SOM to microbes is a complex process influenced by the soil minerology, nutrient availability as well as environmental conditions (Kästner and Miltner 2018; Inagaki et al. 2020; Cotrufo and Lavallee 2022; Kim et al. 2022). The quality, composition, and structure of SOM dictate the degree to which microbes can efficiently decompose and utilize organic substrates (Six and Jastrow 2002; Kleber and Johnson 2010). Spatial constraints, such as organo-mineral associations and protection within microaggregates, impact microbial mineralization (Lehmann et al. 2007). Particulate organic matter (POM) and Mineral-associated organic matter (MAOM) exhibit distinct characteristics in composition and decomposition resistance (Lavallee et al. 2020). POM quality and distribution are influenced by plant inputs, while MAOM accumulation is regulated by microbial by-products, irrespective of the ecosystem (Giannetta et al. 2018). MAOM significantly influence soil microbial communities, SOM composition, and C stabilization, with factors such as soil type and depth contributing to the variability observed across diverse soils (Kögel-Knabner et al. 2008). While

POM exhibits higher biochemical recalcitrance than MAOM (Filley et al. 2008; Angst et al. 2017; Giannetta et al. 2019), potentially leading to decreased microbial activity (Savy et al. 2020), the prolonged persistence of MAOM compared to POM suggests that increased recalcitrance may not be the primary factor influencing SOM stability (Dungait et al. 2012; Poeplau et al. 2017; Lavallee et al. 2020). Instead, the physical protection of SOM within microaggregates, limiting microbial access, becomes a crucial factor (Balesdent et al. 2000; Kögel-Knabner et al. 2008; Wiesmeier et al. 2012). This protection requires enzymatic activities and microbial processes to break down complex organic molecules within the confined microenvironments (Lagomarsino et al. 2012). Spatial distribution is a key factor, with subsoils favoring greater SOM persistence in distributed forms, while topsoils, rich in POM, show reduced sensitivity to spatial dynamics (Inagaki et al. 2023). Soil management practices further contribute to soil-microorganism interaction, as evident in a Mediterranean horticultural system where reduced tillage and higher compost input over a 9-year period enhance C accessibility to microorganisms (Laudicina et al. 2011). Environmental factors such as moisture (Bachmann et al. 2008; Védère et al. 2020) and temperature (Kleber et al. 2011; Frey et al. 2013; Pold et al. 2017) exert additional influence on microbial activity, impacting their ability to access and decompose SOM. A comprehensive understanding of these intricate SOM-microbe interactions is indispensable for stabilizing SOC and optimizing sustainable soil management practices.

3.3.3 *Soil Organic Carbon Stabilization Mechanisms*

C stabilization in soil depends on its biotic and abiotic environment. The stability of organic compounds is notably influenced by soil mineralogy, which governs the mechanisms of decomposition and preservation (Thaymuang et al. 2013; Zhao et al. 2022). In addition, soil texture plays a pivotal role in stabilizing SOC by regulating the association of SOM with soil minerals (Hassink et al. 1997; Feng et al. 2013). The soil CS largely depends on SOC stability and persistence of SOC in turns depends on decomposition and mineralization processes (Schrumpf et al. 2013). There are three main mechanisms for C stabilization in soil: spatial inaccessibility, chemical recalcitrance, and organo-mineral association (Lal 2018; Christensen 2001; Six et al. 2002; Pisani et al. 2014).

3.3.3.1 Spatial Inaccessibility

Spatial inaccessibility refers to the physical unavailability of SOC to decomposer organisms due to specific location or association with the soil matrix. The SOC may be physically protected within aggregates or mineral surfaces, making it spatially inaccessible to microbes. The slow decomposition observed within soil aggregates can be attributed to a complex interplay of several key mechanisms (Balesdent et al.

2000). Firstly, SOM is often subject to sorption onto solid surfaces within aggregates, forming associations that hinder microbial access and enzymatic breakdown. Additionally, sequestration of SOM in small pores within aggregates creates a protective environment, limiting the exposure of organic materials to microbial decomposers. The intricate control exerted by microfauna (nematodes and protozoa) on microbial turnover further contributes to the preservation of SOM within aggregates. Predatory organisms regulate the abundance and activity of microbial communities, influencing the rate at which organic materials are processed (Saleem et al. 2013). Furthermore, oxygen limitation within the dense microenvironments of soil aggregates creates anaerobic conditions, slowing down decomposition processes that are typically aerobic in nature (Smith 1980). Collectively, these mechanisms act synergistically to impede the decomposition of SOM within soil aggregates, highlighting the importance of microscale processes in shaping overall soil C dynamics. Notably, short-term SOM stabilization may not be solely controlled by physical protection in microaggregates (Fernández et al. 2014). Enhanced aggregate stability through measures like straw mulch (Rovira et al. 2010; Xue et al. 2020) or controlled grazing (Wiesmeier et al. 2012) contributes to long-term CS. Research emphasizes the importance of physical protection over recalcitrance for ensuring C stability and persistence (Mikutta et al. 2006; Marschner et al. 2008; Dungait et al. 2012; Lavallee et al. 2019).

3.3.3.2 Chemical Recalcitrance

Chemical recalcitrance refers to the inherent resistance of certain organic compounds, like lignin, lipids, and polyphenols, to decomposition. With complex molecular structures, such as phenolics with hydroxyl functional groups and lipids with alkyl C chains, these compounds resist microbial breakdown (Lorenz et al. 2007). The hydrophobic nature of organic compounds like lipids slows decomposition in polar solvents (water, methanol) due to clustering and restricted enzyme access (Kleber and Johnson 2010). The recalcitrance of SOM, especially lignocellulosic biomass, is controlled by a complex interaction of structural and chemical factors. These include cellulose-specific surface area, crystallinity, polymerization degree, pore characteristics (size and volume), as well as the chemical composition, presence of lignin, hemicelluloses, and acetyl groups (Zhao et al. 2012; Zoghlami and Paës 2019). The stability of organic molecules in soil over time is significantly influenced by microbe-iron-lignin interactions, as indicated in studies by Huang et al. (2019), Liao et al. (2022), and Liu et al. (2023). Chemical processes contribute to biochemical stabilization of SOC. For instance, abiotic condensation reactions facilitate the recombination of sugars (glucose) and amino compounds (glycine), resulting in the production of substances like melanoidins, which exhibit recalcitrance (Kleber and Johnson 2010; Essington 2015). Changes in the distribution of labile and recalcitrant SOC pools, especially in deeper soil layers where C is stored for extended periods, play a crucial role in determining the extent of biochemical recalcitrance (Rovira and Vallejo 2002; Liu et al. 2020). Moreover, the proportion of SOC as labile or recalcitrant undergoes alterations during the mineralization process, influenced by variations in mean residence time (Mills et al. 2014).

3.3.3.3 Organo-Mineral Associations

Organo-mineral associations involve the binding or association of organic molecules with mineral surfaces in the soil. Organic compounds, particularly those with functional groups like carboxyls (–COOH) and hydroxyls (–OH), establish bonds with minerals such as clay and silt, influenced by differences in mineral particle size (Ortner et al. 2022). Notably, sand and silt interact less actively than clay minerals (von Lützow et al. 2007). Various mechanisms drive the adsorption of organic molecules onto mineral surfaces, including electrostatic interactions (Tombácz et al. 1998), hydrogen bonding (Ndzana et al. 2022), and chemical processes like ligand exchange and redox reactions (Chen et al. 2014; Adhikari and Yang 2015; Hong et al. 2019; Das et al. 2023), creating stable associations. Soil fauna, especially earthworms, are soil engineers, enhancing organo-mineral associations through the production of mucus (Guhra et al. 2020). This mucus, consisting of proteins, lipids, and polysaccharides, effectively binds with soil minerals like illite, kaolinite, and goethite (Guhra et al. 2020). Factors influencing organo-mineral associations include types of organic compounds, clay mineral properties (surface area, crystallinity), soil conditions (pH, ionic strength), and microbial content (Angst et al. 2017; Ndzana et al. 2022; Chi et al. 2022). External factors affecting these associations include climate (Poeplau et al. 2017; Inagaki et al. 2020; Monhonval et al. 2023), agriculture management practices (Das et al. 2019; Veloso et al. 2022), and soil erosion processes (Ayoubi et al. 2012; Lorenz and Lal 2018; Wang et al. 2019). During mineral formation or precipitation events, SOM can co-precipitate with minerals, leading to their association in the soil matrix (Kleber et al. 2015; Lenhardt et al. 2022). The positive correlations between specific surface area, CEC, and soil C dynamics emphasize the critical relationship between mineralogical composition and C stabilization mechanisms (Kaiser and Guggenberger 2003; Ndzana et al. 2022; Das et al. 2023).

3.4 Soil Carbon and Soil Health

Soil C plays a fundamental role in soil health, and comprehending the concept of soil health is essential to fully grasp the significance of soil C within this context. While "soil health" and "soil quality" are frequently used interchangeably in scientific discussions, it's important to note that they represent distinct concepts (Lal 2016). Soil quality is tied to the functions performed by the soil, highlighting its roles and capabilities. In contrast, soil health portrays the soil as a dynamic and finite living resource, directly influencing plant health, and consequently, human well-being. The term 'soil quality' typically describes the characteristics and dynamics of soil properties (Wander et al. 2019) and the term 'soil health' entails soil physical, chemical, and biological characteristics that specify the soils' ability to function as a living, dynamic system to sustain plant and animal productivity and maintain environmental quality (Doran and Zeiss 2000).

Soil health is a cornerstone of sustainable agriculture, shaped by land management practices and climatic factors. Climate variations, such as irregular rainfall and wind erosion, can strip away the fertile topsoil layer, compromising the soil's ability to sustain plant life. Temperature fluctuations impact crucial underground microbial activity vital for nutrient cycling and fertility. Challenging the notion that land can endlessly yield abundant crops without consequence, it's crucial to recognize soil's finite capacity to produce quality food. Intensive agricultural practices can disrupt the intricate soil ecosystem, leading to degradation that not only affects soil vitality but also impacts crops and human well-being (Kopittke et al. 2019). Soil degradation encompasses both quantity and quality loss. Instances like SOM loss, greenhouse gas (GHG) emissions, erosion, nutrient imbalances, acidification, salinization, and declining genetic diversity jeopardize soil's sustained ability to support human needs, particularly future food production (Lal 2015). Over the past century, agricultural soil has lost 60% of its original C content (Lal 2004). The economic consequences of soil degradation are evident, as it results in the loss of essential soil ecosystem services. The annual valuation of these services, encompassing provisioning, regulating, supporting, and cultural functions, is estimated at an impressive $11.4 trillion (McBratney et al. 2017). This figure reflects the cumulative value of the diverse benefits provided by soils, including their role in supplying resources, regulating environmental processes, offering support to biodiversity, and contributing to cultural and aesthetic aspects. The Nature Conservancy (TNC 2016) estimates annual costs of $85.1 billion due to unintended consequences of losing healthy soils on agricultural productivity, human health, biodiversity, and resilience, highlighting the urgency for sustainable solutions. Recognizing the imperative role of healthy soils in sustaining ecosystem services, SAPs are essential (Table 3.2). However, it is crucial to acknowledge that maintaining soil health is not a one-size-fits-all endeavor. SAPs must be thoughtfully tailored to specific regions, acknowledging the inherent variations in soil characteristics, climate conditions, and topography (Miner et al. 2020).

Table 3.2 Healthy soils contribution to ecosystem services

Provisioning	Regulating	Supporting	Cultural
Crop productivity	Nutrient cycling Water retention Disease and pest suppression GHG regulation Water filtrations Soil structure improvement Clean air and water	Biodiversity support Habitat for organisms	Aesthetic value Recreational value Scientific value Spiritual value Educational value

3.4.1 Soil Health Principles

Advancing soil health is a gradual process that necessitates sustained commitment to established management practices rooted in four key principles. Conservation agriculture, regenerative agriculture, sustainable agriculture, and related terms all revolve around enhancing soil health by adhering to these principles (Magdoff 2007; Khangura et al. 2023; Francaviglia et al. 2023; Mulimbi et al. 2023). These include minimizing soil disturbance, ensuring soil coverage, preserving living roots in the soil, and promoting plant diversity.

3.4.1.1 Minimizing Soil Disturbance

Achieved by minimizing manual activities and restricting the use of heavy equipment, minimum soil disturbance involves techniques such as direct placing of seed or planting material and emphasizes practices like no-till or controlled traffic farming, and direct drilling (FAO 2025). This approach aims to reduce the penetration of equipment into the soil, keeping the seed slot and associated soil movement to an absolute minimum. Mechanized tillage with a versatile multi-crop planter is considered within the realm of minimized soil disturbance (Haque et al. 2017). Transparent plastic minirhizotron tubes proposed for studying root dynamics in sandy rocky soils also align with the principle of minimizing soil disturbance (Phillips et al. 2000). In China's slope farmland, Jia et al. (2019) explore practices that minimize soil disturbance to reduce sediment yield and control soil erosion, including no-tillage, no-tillage with mulch, collecting soil to form ridges with no tillage, contour tillage, and contour tillage with hedge row micro basins. Improper mechanical soil disturbance, such as heavy machinery use (Schäffer et al. 2007; Wenting Han 2011; Picchio et al. 2012; McGeary et al. 2022), harvesting operations (Bianchini et al. 2022), and intensive grazing (de Andrade Bonetti et al. 2019; Houlbrooke et al. 2011; Shawver et al. 2021; Twerdoff et al. 1999) may lead to the compression of soil particles, reducing pore space and increasing soil density. Soil compaction can negatively impact soil structure, impede water infiltration, and hinder root penetration, ultimately affecting soil health and fertility (Batey 2009). Effective soil disturbance management involves minimizing issues like clods, soil support loss, and poaching, while strategically creating fissures for water drainage in scenarios like severe wheel rutting (Spoor 2006). Balancing disturbance based on cropping needs is crucial for maintaining soil health and preventing undesired consequences.

3.4.1.2 Ensuring Soil Coverage

Strategic soil coverage, achieved through the application of mulch or organic materials such as rice and wheat straw, coffee husk, saw dust, and wood chips, offers a myriad of benefits crucial for soil health and agricultural sustainability (Zhao et al. 2017). This practice plays a pivotal role in moderating soil temperature (Kader et al.

2017), conserving water (Ramadan et al. 2021), and ensuring a continuous supply of nutrients for soil biota (Dong et al. 2023). Additionally, it acts as a protective shield, preventing soil erosion (Zhang et al. 2022) and stabilizing GHG emissions (da Silva et al. 2018). Plant residues contribute to improved soil compaction by maintaining soil structure and minimizing soil disturbance caused by agricultural machinery (Shahgholi and Moinfar 2021). In regions where above-ground biomass production is limited, such as highly eroded or arid areas, the incorporation of cover crops becomes essential (Johnson et al. 2021). Cover crops not only safeguard the soil during fallow periods but also aid in retaining water and nutrients. Legume cover crops like crimson clover, winter peas when combined with the main crop, play a crucial role in recycling nutrients and mobilizing them within the soil profile (Hoyt and Hargrove 1986). This dynamic interaction enhances soil fertility, structure, and overall resilience, underscoring the significance of soil coverage in SAPs.

3.4.1.3 Preserving Living Roots in Soil

Preserving living roots in the soil is a fundamental strategy for promoting soil health and enhancing overall ecosystem resilience. Agricultural practices such as cover cropping, crop rotation, the integration of perennial crops, or forage plants, and agroforestry play pivotal roles in maintaining a continuous presence of living roots throughout the year. Roots contribute to soil dynamics by creating biopores through decomposition, facilitated by earthworms and plant roots' biological activity. These biopores enhance nutrient and water absorption from the subsoil, benefiting both the soil and plants (Xiong et al. 2022). Studies, such as Tiunov and Scheu (1999), emphasize the importance of biopores by demonstrating increased SOC, total nitrogen, moisture, and pH in burrow walls compared to control soil. Additionally, roots influence rhizosphere hydraulic properties by exuding mucilage, impacting water absorption and surface tension (Marin et al. 2022). Planted soils consistently exhibit superior properties, including aggregate stability, porosity, infiltrability, and SOM presence (Lucas 2022), with significant implications for restoring disturbed heavy-textured soils (Hao et al. 2020). The length density of roots is a critical factor shaping water potential and exerting a unique influence on plant behavior (Shein and Pachepsky 1995). Differences in root density, where cereals typically exhibit higher densities compared to legumes, contribute significantly to enhanced drought tolerance (Osborne et al. 2020). This variation underscores the adaptability of different plant types to water stress, underscoring the need to comprehend root system characteristics for effective soil health management and sustainable crop production.

3.4.1.4 Promoting Plant Diversity

Plant diversification, encompassing practices such as intercropping, crop rotation, and agroforestry, involves cultivating a variety of crops on the same land to foster biodiversity and support soil microbial populations. Beillouin et al. (2021) define crop diversification as any type of agricultural diversification practice that involves increasing the diversity of cultivated annual or perennial crops (including non-harvested crops like cover crops or catch crops) at genetic or species level. This strategic approach brings multiple advantages to agricultural systems. By diversifying the sources of income for farmers and mitigating the risk of crop failure, it contributes to economic resilience (Helmers et al. 2001; Selvarani and Alagumani 2018; Li et al. 2021a; Akber et al. 2022). Simultaneously, plant diversification plays a pivotal role in enhancing soil health by suppressing diseases, improving structure, and promoting microbial activity (Baldwin-Kordick et al. 2022; Kumar et al. 2021; Wang et al. 2022a). To ensure comprehensive soil benefits, it is essential to incorporate a variety of crops, including C4 (e.g., maize, sugarcane), C3 (e.g., wheat, fodder trees), leguminous (e.g., soybean), and non-leguminous (e.g., cereal rye) species (Vukicevich et al. 2016; Choudhary et al. 2023). This diverse combination ensures that different crops with varying needs are planted, preventing undue pressure on soil resources both temporally and spatially throughout the year.

Diversified planting not only provides economic stability against extreme weather conditions but also reduces the dependence on agrochemicals, fostering environmental sustainability (Bonnet et al. 2021). Diversifying crops, especially with drought and salt-tolerant varieties, holds the key to sustainable agricultural growth in regions with salt-affected soils and supports climate-smart practices (Hussain et al. 2020). Several studies have indicated the potential of increased plant diversity to enhance crop productivity, attributed to the presence of natural enemies that effectively control pests and contribute to improved soil health (Brooker et al. 2023; Poveda et al. 2008; Redlich et al. 2021; Staton et al. 2019). Beyond agriculture, promoting plant diversity extends its benefits to essential ecosystem services. This includes enhancements in water and soil quality, increased pollination services, the maintenance of continuous biomass cover, CS, prevention of soil erosion, and the provision of recreational and cultural benefits (Brockerhoff et al. 2017; Hajjar et al. 2008; Osseni et al. 2020; Quijas et al. 2010; Windsor et al. 2021)

3.4.2 Role of Soil Carbon in Soil Health and Ecosystem Dynamics

Soil C, a fundamental component of SOM, plays a central role in shaping the overall health of the soil. Interactions with microorganisms, nutrients, water, and soil structure position soil C as a key influencer, as illustrated in Fig. 3.4. Serving as the foremost indicator, soil C, when coupled with various biological indicators, provides valuable insights into the intricate dynamics of C and microbial activity within the soil (Liptzin et al. 2022). The relationship between soil health and soil C is multifaceted and influences various aspects of the soil ecosystem including nutrient cycling, microbial activity, soil structure, water retention, pH buffering, and resistance to soil degradation.

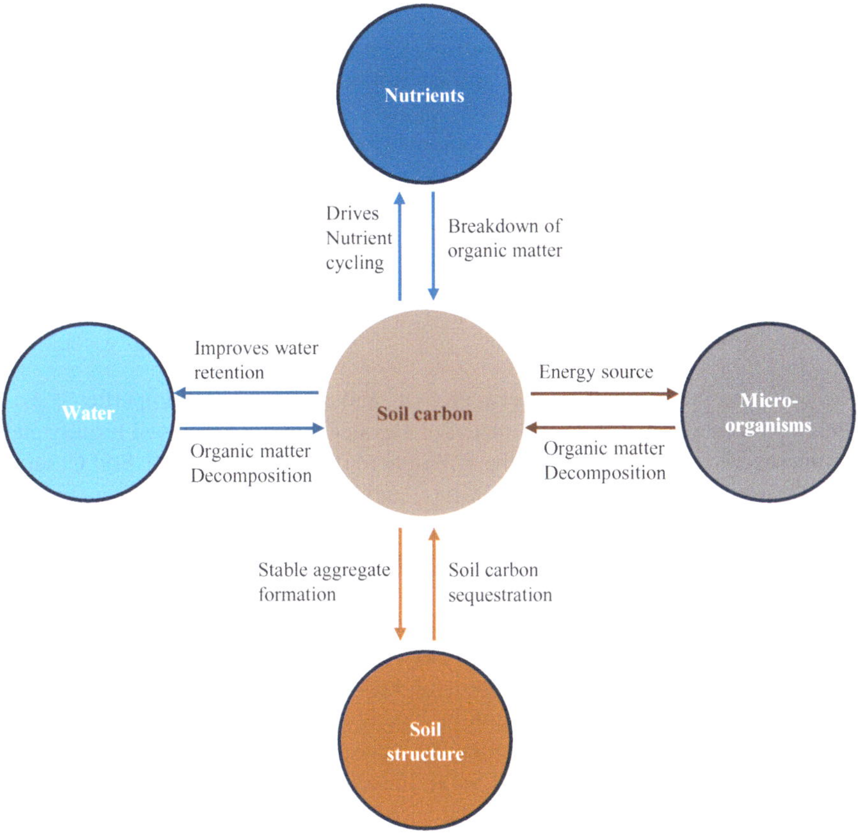

Fig. 3.4 Soil carbon's integral role in soil health

3.4.2.1 Nutrient Cycling

Approximately 71% of the total annual value of ecosystem services offered by soil is attributable to nutrient cycling (McBratney et al. 2017), and soil C stands as a central component in the complex network of nutrient cycling within ecosystems. As a primary component of SOM, soil C (58% of SOM) serves as a pivotal reservoir for vital nutrients, encompassing nitrogen (N), phosphorus (P), sulfur (S), and micronutrients (Chen et al. 2003; Kirkby et al. 2013; Powlson et al. 2015). The microbial decomposition of SOM acts as a gradual nutrient release mechanism, ensuring a continuous supply available for plant uptake. Changes in soil C content have cascading effects on the physical, chemical, and biological properties of soil, including aggregate stability, water-holding capacity, soil erodibility, color, CEC, acidity, salinity, and microbial activity (Murphy 2015). Wang et al. (2021) have underscored the significant impact of C inputs, particularly SOC, on microbial-mediated nutrient cycling in paddy soils. Their findings illuminate how shifts in microbial communities and phosphorus dynamics are indicative of the critical role played by C in influencing nutrient transformations. Moreover, soil C, by enhancing CEC, facilitates the adsorption of positively charged nutrients such as iron, calcium, and magnesium, thus improves the nutrient-holding capacity of the soil, promoting a more fertile environment for plant development. Notably, the content of C emerges as the primary factor explaining variations in CEC, as demonstrated in a tropical soil context (Oorts et al. 2003). This complex interaction between soil C and nutrient cycling highlights the fundamental importance of C dynamics in sustaining soil health and ecosystem functionality.

3.4.2.2 Microbial Activity

The profound interdependence between SOC and soil microbial properties is evident in studies that highlight the impact of elevated CO_2 levels on soil fertility and the microecological cycle within the rhizosphere (Huang et al. 2020). Soil C, serving as a food source for microorganisms, undergoes increased turnover through moderate soil treatments like fructose, oxalic acid, and catechol, known as priming. This process enhances the growth of a diverse suite of microorganisms, leading to increased microbial enzyme production (Hamer and Marschner 2005). Notably, this heightened enzyme production is more dominant in the rhizosphere than in the bulk soil (Singh and Sharma 2020). The increased microbial activity facilitated by sufficient soil C plays a crucial role in improving soil structure, enhancing nutrient availability, and contributing to the overall biological activity in the soil (da Silva et al. 2021). Concurrently, microorganisms play a pivotal role in supplying nutrients to plants through the decomposition of SOM. Easily degradable C emerges as a versatile and reliable indicator of soil degradation, offering broad applicability for assessing soil health and guiding effective management strategies (Wolińska et al. 2018). Furthermore, soil CS is intricately tied to the microbial community's role in C cycling (Bhattacharyya et al. 2022). However, an excess of C may intensify microbial competition with plants for nutrients, underscoring the importance of a balanced nutrient management approach (Bertini and Azevedo 2022).

3.4.2.3 Soil Structure and Water Retention

The assessment of soil health relies heavily on physical properties such as soil structure and water retention. A critical factor influencing soil structure is aggregate stability, measuring how well soil aggregates withstand external stresses such as rainfall, irrigation, or tillage. The SOM, particularly SOC, significantly contributes to enhancing soil structure. Soil aggregates, comprising a mix of organic and mineral materials, form the foundational structure of the soil (Malamoud et al. 2009). SOC acts as a crucial binding agent, leading to the formation of large and stable aggregates (Xue et al. 2019), preventing soil particles from easily breaking apart during rainfall. This is achieved by slowing down water infiltration, thereby mitigating soil erosion (Phil-Eze 2010; Yang et al. 2013) and reducing the risk of drought stress as increased SOC improves water retention (Iizumi and Wagai 2019). A strong relationship has been observed between the rise in SOC content and the stability of macro- and microaggregate under continuous wetting process (Al-Kaisi et al. 2014). Furthermore, SOC plays a vital role in fortifying aggregate stability against other external forces like wind and cultivation activities, resulting in the accumulation of particulate organic C (POC) and dissolved organic C (DOC) (Chaplot and Cooper 2015). This mechanism involves improved water infiltration and reduced runoff, underscoring the integral role of soil C in strengthening soil structure and influencing water dynamics in the soil (Chaplot and Cooper 2015).

3.4.2.4 pH Buffering

Soil C contributes significantly to the soil's buffering capacity, acting as a stabilizing force against pH fluctuations (Ritchie and Dolling 1985; Pocknee and Sumner 1997; Wang et al. 2013). Soil C can impact soil pH through its interactions with other soil components, including buffering capacity, microbial activity, CEC, and nutrient availability. Microbial activity involved in the decomposition of SOM releases organic acids, subtly influencing a gradual decrease in soil pH over time (Momma et al. 2006; Rukshana et al. 2014). The CEC impacts the soil's ability to hold and exchange positively charged ions, such as hydrogen (H+) and aluminum (Al3+), influencing the availability of H+ ions and thereby affecting soil pH (Lemire et al. 2006). Moreover, the decomposition of SOM, including C, releases nutrients that can alter the balance of basic and acidic ions in the soil, further shaping soil pH (Hartemink and Barrow 2023). While soil C is a key player in these processes, it's essential to recognize the collaborative influence of factors like soil minerals (Li et al. 2022b), parent material (Zhang et al. 2019), climate (Ma et al. 2022), vegetation (Zhao et al. 2011), and human interventions, such as organic amendments or fertilizers (Jayalath et al. 2016; Gillespie et al. 2021), all contributing to the intricate dynamics of soil pH regulation and improving soil health.

3.4.2.5 Resistance to Soil Degradation

Soil degradation, a complex challenge, manifests through erosive processes, nutrient depletion, waterlogging, acidification, salinization, pollution, biodiversity loss, and compaction (Karlen and Rice 2015). SOM is a key player in mitigating these forms of degradation by exerting positive influences on critical soil properties like structure, water retention, microbial activity, and nutrient availability (Noordwijk et al. 2015; O'Rourke et al. 2015). These combined effects enhance the overall resilience of the soil, making it more resistant to diverse forms of degradation. Recognizing the importance of SOC as a reliable indicator of soil degradation, numerous studies have utilized SOC levels to assess the health of soils (Obalum et al. 2017; Wolińska et al. 2018; Lorenz et al. 2019). Increased soil C stocks contribute to enhanced soil aggregate stability, leading to improved soil quality and serving as a preventive measure against degradation in dry, semi-arid soils (Ameer et al. 2023). Addressing the global issue of soil degradation has become a priority for nations worldwide, and many are implementing strategies to increase SOC stocks in their soils as a means of restoring and maintaining soil health (Lal 2015; Hussain et al. 2021). By focusing on the enhancement of SOC, nations aim to fortify their soils against the myriad challenges posed by degradation, fostering sustainability in agriculture and environmental conservation.

3.5 Conclusion

Soil C is a foundational component of terrestrial ecosystems, playing a critical role in regulating nutrient cycling, sustaining microbial communities, improving soil structure and water retention, buffering pH fluctuations, and enhancing resistance to degradation. Its central function in maintaining soil health, agricultural productivity, and ecosystem resilience underscores its broader significance in global environmental sustainability. Existing in both organic (SOC) and inorganic (SIC) forms, soil C contributes to long-term CS, with SOC supporting microbial activity, nutrient dynamics, and structural stability, while SIC, formed through mineral processes, provides stable C storage, particularly in arid and semi-arid regions. The dynamics of both forms are governed by complex interactions among climatic conditions, vegetation, land management practices, and intrinsic soil properties such as texture, mineralogy, and CEC.

SOC stabilization is controlled by biological, chemical, and physical processes operating across spatial and temporal scales. The persistence of soil C is regulated by three principal mechanisms: spatial inaccessibility, chemical recalcitrance, and organo-mineral associations. These stabilization pathways are further modulated by the quality of SOM, microbial accessibility, environmental variability, and soil mineralogy.

Maintaining and enhancing soil C stocks necessitates the implementation of site-specific SAPs that integrate climatic, edaphic, and management considerations. Key principles, including minimizing soil disturbance, maintaining continuous soil cover, preserving living roots, and promoting plant diversity, are essential for improving soil C storage and delivering multiple ecosystem services. These strategies not only enhance soil resilience and productivity but also contribute to addressing global challenges such as climate change, land degradation, and food insecurity.

As scientific understanding of soil C dynamics continues to advance, targeted soil management strategies can be better optimized to maximize CS, support ecosystem function, and contribute significantly to climate mitigation, biodiversity conservation, and the achievement of global sustainability goals.

References

Adhikari D, Yang Y (2015) Selective stabilization of aliphatic organic carbon by iron oxide. Sci Rep 5:11214. https://doi.org/10.1038/srep11214

Akber MA, Islam MA, Rahman MM, Rahman MR (2022) Crop diversification in southwest coastal Bangladesh: insights into farming adaptation. Agroecol Sustain Food Syst 46:316–324. https://doi.org/10.1080/21683565.2021.1989105

Al-Kaisi MM, Douelle A, Kwaw-Mensah D (2014) Soil microaggregate and macroaggregate decay over time and soil carbon change as influenced by different tillage systems. J Soil Water Conserv 69:574–580. https://doi.org/10.2489/jswc.69.6.574

Álvaro-Fuentes J, Arrúe JL, Gracia R, López MV (2007) Soil management effects on aggregate dynamics in semiarid Aragon (NE Spain). Sci Total Environ 378:179–182. https://doi.org/10.1016/j.scitotenv.2007.01.046

Ameer I, Kubar KA, Ali Q et al (2023) Land degradation resistance potential of a dry, semiarid region in relation to soil organic carbon stocks, carbon management index, and soil aggregate stability. Land Degrad Dev 34:624–636. https://doi.org/10.1002/ldr.4480

Anderson J, Prescott CE, Grayston SJ (2019) Organic matter accumulation in reclaimed soils under spruce, poplar and grass in the Alberta Oil Sands. New For (Dordr) 50:307–322. https://doi.org/10.1007/s11056-018-9646-4

Angst G, Mueller KE, Kögel-Knabner I et al (2017) Aggregation controls the stability of lignin and lipids in clay-sized particulate and mineral associated organic matter. Biogeochemistry 132:307–324. https://doi.org/10.1007/s10533-017-0304-2

Angst G, Pokorný J, Mueller CW et al (2021) Soil texture affects the coupling of litter decomposition and soil organic matter formation. Soil Biol Biochem 159:108302. https://doi.org/10.1016/j.soilbio.2021.108302

Arias-Ortiz A, Masqué P, Glass L et al (2021) Losses of soil organic carbon with deforestation in mangroves of Madagascar. Ecosystems 24:1–19. https://doi.org/10.1007/s10021-020-00500-z

Ayoubi S, Mokhtari Karchegani P, Mosaddeghi MR, Honarjoo N (2012) Soil aggregation and organic carbon as affected by topography and land use change in western Iran. Soil Tillage Res 121:18–26. https://doi.org/10.1016/j.still.2012.01.011

Bachmann J, Guggenberger G, Baumgartl T et al (2008) Physical carbon-sequestration mechanisms under special consideration of soil wettability. J Plant Nutr Soil Sci 171:14–26. https://doi.org/10.1002/jpln.200700054

Bacq-Labreuil A, Crawford J, Mooney SJ et al (2018) Effects of cropping systems upon the three-dimensional architecture of soil systems are modulated by texture. Geoderma 332:73–83. https://doi.org/10.1016/j.geoderma.2018.07.002

Baldwin-Kordick R, De M, Lopez MD et al (2022) Comprehensive impacts of diversified cropping on soil health and sustainability. Agroecol Sustain Food Syst 46:331–363. https://doi.org/10.1080/21683565.2021.2019167

Balesdent J, Chenu C, Balabane M (2000) Relationship of soil organic matter dynamics to physical protection and tillage. Soil Tillage Res 53:215–230. https://doi.org/10.1016/S0167-1987(99)00107-5

Bardgett RD, Manning P, Morriën E, De Vries FT (2013) Hierarchical responses of plant–soil interactions to climate change: consequences for the global carbon cycle. J Ecol 101:334–343. https://doi.org/10.1111/1365-2745.12043

Bartmiński P, Świtoniak M, Drewnik M et al (2022) Methodological problems with the classification and measurement of soils containing carbonates. Soil Sci Annu 73:1–10. https://doi.org/10.37501/soilsa/149235

Batey T (2009) Soil compaction and soil management – a review. Soil Use Manag 25:335–345. https://doi.org/10.1111/j.1475-2743.2009.00236.x

Batlle-Bayer L, Batjes NH, Bindraban PS (2010) Changes in organic carbon stocks upon land use conversion in the Brazilian Cerrado: a review. Agric Ecosyst Environ 137:47–58. https://doi.org/10.1016/j.agee.2010.02.003

Bauer A, Black AL (1992) Organic carbon effects on available water capacity of three soil textural groups. Soil Sci Soc Am J 56:248–254. https://doi.org/10.2136/sssaj1992.03615995005600010038x
Beillouin D, Ben-Ari T, Malézieux E et al (2021) Positive but variable effects of crop diversification on biodiversity and ecosystem services. Glob Chang Biol 27:4697–4710. https://doi.org/10.1111/gcb.15747
Berg B, McClaugherty C, Johansson M-B (1993) Litter mass-loss rates in late stages of decomposition at some climatically and nutritionally different pine sites. Long-term decomposition in a Scots pine forest. VIII. Canad J Bot 71:680–692. https://doi.org/10.1139/b93-078
Bernoux M, Cerri CEP (2005) GEOCHEMISTRY | soil, organic components. In: Encyclopedia of analytical science. Elsevier, pp 203–208
Bertini SCB, Azevedo LCB (2022) Soil microbe contributions in the regulation of the global carbon cycle. In: Microbiome under changing climate. Elsevier, pp 69–84
Beugnon R, Bu W, Bruelheide H et al (2023) Abiotic and biotic drivers of tree trait effects on soil microbial biomass and soil carbon concentration. Ecol Monogr 93. https://doi.org/10.1002/ecm.1563
Bhattacharyya SS, Ros GH, Furtak K et al (2022) Soil carbon sequestration – an interplay between soil microbial community and soil organic matter dynamics. Sci Total Environ 815:152928. https://doi.org/10.1016/j.scitotenv.2022.152928
Bianchi SR, Miyazawa M, de Oliveira EL, Pavan MA (2008) Relationship between the mass of organic matter and carbon in soil. Braz Arch Biol Technol 51:263–269. https://doi.org/10.1590/S1516-89132008000200005
Bianchini L, Alemanno R, Di Stefano V et al (2022) Soil compaction in harvesting operations of Phalaris arundinacea L. Land (Basel) 11:1031. https://doi.org/10.3390/land11071031
Bin Z, Xin-Hua P (2006) Organic matter enrichment and aggregate stabilization in a severely degraded ultisol after reforestation. Pedosphere 16:699–706. https://doi.org/10.1016/S1002-0160(06)60105-7
Blanco-Canqui H, Lal R (2008) No-tillage and soil-profile carbon sequestration: an on-farm assessment. Soil Sci Soc Am J 72:693–701. https://doi.org/10.2136/sssaj2007.0233
Bohara M, Acharya K, Perveen S et al (2020) In situ litter decomposition and nutrient release from forest trees along an elevation gradient in Central Himalaya. Catena (Amst) 194:104698. https://doi.org/10.1016/j.catena.2020.104698
Bonnet C, Gaudio N, Alletto L et al (2021) Design and multicriteria assessment of low-input cropping systems based on plant diversification in southwestern France. Agron Sustain Dev 41:65. https://doi.org/10.1007/s13593-021-00719-7
Brockerhoff EG, Barbaro L, Castagneyrol B et al (2017) Forest biodiversity, ecosystem functioning and the provision of ecosystem services. Biodivers Conserv 26:3005–3035. https://doi.org/10.1007/s10531-017-1453-2
Brooker RW, Hawes C, Iannetta PPM et al (2023) Plant diversity and ecological intensification in crop production systems. J Plant Ecol 16. https://doi.org/10.1093/jpe/rtad015
Brye KR, Gbur EE (2010) Regional differences in soil carbon and nitrogen storage as affected by land use and soil moisture regime. Soil Sci 175:339–348. https://doi.org/10.1097/SS.0b013e3181e83db2
Callesen I, Harrison R, Stupak I et al (2016) Carbon storage and nutrient mobilization from soil minerals by deep roots and rhizospheres. For Ecol Manag 359:322–331. https://doi.org/10.1016/j.foreco.2015.08.019
Chaplot V, Cooper M (2015) Soil aggregate stability to predict organic carbon outputs from soils. Geoderma 243–244:205–213. https://doi.org/10.1016/j.geoderma.2014.12.013
Chen C, Dynes JJ, Wang J, Sparks DL (2014) Properties of Fe-organic matter associations via coprecipitation versus adsorption. Environ Sci Technol 48:13751–13759. https://doi.org/10.1021/es503669u
Chen G, Zhu H, Zhang Y (2003) Soil microbial activities and carbon and nitrogen fixation. Res Microbiol 154:393–398. https://doi.org/10.1016/S0923-2508(03)00082-2

Chen S, Wang W, Xu W et al (2018) Plant diversity enhances productivity and soil carbon storage. Proc Natl Acad Sci 115:4027–4032. https://doi.org/10.1073/pnas.1700298114
Chen Y, Liu K, Hu N et al (2023) Biochemical composition of soil organic matter physical fractions under 32-year fertilization in Ferralic Cambisol. Carbon Res 2:1. https://doi.org/10.1007/s44246-022-00034-0
Chertov OG, Komarov AS (2013) Theoretical approaches to modelling the dynamics of soil organic matter. Eurasian Soil Sci 46:845–853. https://doi.org/10.1134/S1064229313080012
Chi J, Fan Y, Wang L et al (2022) Retention of soil organic matter by occlusion within soil minerals. Rev Environ Sci Biotechnol 21:727–746. https://doi.org/10.1007/s11157-022-09628-x
Chivenge P, Vanlauwe B, Gentile R, Six J (2011) Comparison of organic versus mineral resource effects on short-term aggregate carbon and nitrogen dynamics in a sandy soil versus a fine textured soil. Agric Ecosyst Environ 140:361–371. https://doi.org/10.1016/j.agee.2010.12.004
Choudhary D, Datt N, Sharma SK et al (2023) Impact of crop diversification on soil physical properties and maize grain equivalent yield in a Typic Hapludalf of Himachal Pradesh. Biol Forum 15:1162–1167
Christensen BT (2001) Physical fractionation of soil and structural and functional complexity in organic matter turnover. Eur J Soil Sci 52:345–353. https://doi.org/10.1046/j.1365-2389.2001.00417.x
Cotrufo MF, Lavallee JM (2022) Soil organic matter formation, persistence, and functioning: a synthesis of current understanding to inform its conservation and regeneration. Adv Agron, In, pp 1–66
Crohn DM, Valenzuela-Solano C (2003) Modeling temperature effects on decomposition. J Environ Eng 129:1149–1156. https://doi.org/10.1061/(ASCE)0733-9372(2003)129:12(1149)
da Silva DMN, Venturim CHP, Valory Capucho MEO et al (2018) Impact of soil cover systems on soil quality and organic production of yacon. Sci Hortic 235:407–412. https://doi.org/10.1016/j.scienta.2018.03.024
da Silva GJ, Rivadavea WR, de Lima JD et al (2021) Microbial enzymes and soil health. In: Plant growth-promoting microbes for sustainable biotic and abiotic stress management. Springer International Publishing, Cham, pp 133–155
Dalias P, Anderson JM, Bottner P, Coûteaux M-M (2001) Long-term effects of temperature on carbon mineralisation processes. Soil Biol Biochem 33:1049–1057. https://doi.org/10.1016/S0038-0717(01)00009-8
Damian JM, Durigan MR, Cherubin MR et al (2021) Deforestation and land use change mediate soil carbon changes in the eastern Brazilian Amazon. Reg Environ Chang 21:64. https://doi.org/10.1007/s10113-021-01796-w
Das A, Purakayastha TJ, Ahmed N et al (2023) Influence of clay mineralogy on soil organic carbon stabilization under tropical climate, India. J Soil Sci Plant Nutr 23:1003–1018. https://doi.org/10.1007/s42729-022-01099-x
Das R, Purakayastha TJ, Das D et al (2019) Long-term fertilization and manuring with different organics alter stability of carbon in colloidal organo-mineral fraction in soils of varying clay mineralogy. Sci Total Environ 684:682–693. https://doi.org/10.1016/j.scitotenv.2019.05.327
de Andrade Bonetti J, Anghinoni I, de Moraes MT, Fink JR (2017) Resilience of soils with different texture, mineralogy and organic matter under long-term conservation systems. Soil Tillage Res 174:104–112. https://doi.org/10.1016/j.still.2017.06.008
de Andrade Bonetti J, Anghinoni I, Ivonir Gubiani P et al (2019) Impact of a long-term crop-livestock system on the physical and hydraulic properties of an Oxisol. Soil Tillage Res 186:280–291. https://doi.org/10.1016/j.still.2018.11.003
Devigne C, Mouchon P, Vanhee B (2016) Impact of soil compaction on soil biodiversity – does it matter in urban context? Urban Ecosyst 19:1163–1178. https://doi.org/10.1007/s11252-016-0547-z
Dietzel R, Liebman M, Archontoulis S (2017) A deeper look at the relationship between root carbon pools and the vertical distribution of the soil carbon pool. Soil 3:139–152. https://doi.org/10.5194/soil-3-139-2017

Dilekoğlu MF, Sakin E (2017) Effect of temperature and humidity in soil carbon dioxide emission. J Anim Plant Sci 27:1596–1603

Dong R, Wang X, Wang Y et al (2023) Differences in soil microbial communities with successional stage depend on vegetation coverage and soil substrates in alpine desert shrublands. Plant Soil 485:549–568. https://doi.org/10.1007/s11104-022-05849-9

Doran JW, Zeiss MR (2000) Soil health and sustainability: managing the biotic component of soil quality. Appl Soil Ecol 15:3–11. https://doi.org/10.1016/S0929-1393(00)00067-6

Dungait JAJ, Hopkins DW, Gregory AS, Whitmore AP (2012) Soil organic matter turnover is governed by accessibility not recalcitrance. Glob Chang Biol 18:1781–1796. https://doi.org/10.1111/j.1365-2486.2012.02665.x

Epron D, Marsden C, Thongo M'Bou A et al (2009) Soil carbon dynamics following afforestation of a tropical savannah with eucalyptus in Congo. Plant Soil 323:309–322. https://doi.org/10.1007/s11104-009-9939-7

Essington ME (2015) Soil and water chemistry: an integrative approach, Second. CRC Press

FAO (2025) Conservation agriculture. In: Food and Agriculture Organization of the United Nations. https://www.fao.org/conservation-agriculture/overview/what-is-conservation-agriculture/en/. Accessed 31 May 2025

Feng W, Plante AF, Six J (2013) Improving estimates of maximal organic carbon stabilization by fine soil particles. Biogeochemistry 112:81–93. https://doi.org/10.1007/s10533-011-9679-7

Ferdush J, Paul V (2021) A review on the possible factors influencing soil inorganic carbon under elevated CO2. Catena (Amst) 204:105434. https://doi.org/10.1016/j.catena.2021.105434

Fernández JM, López-de-Sá EG, Polo A, Plaza C (2014) Short- <scp>T</scp> erm stabilization of organic matter in physically, chemically, and biochemically protected pools in soils amended with municipal wastes. Clean (Weinh) 42:487–493. https://doi.org/10.1002/clen.201300127

Filley TR, Boutton TW, Liao JD et al (2008) Chemical changes to nonaggregated particulate soil organic matter following grassland-to-woodland transition in a subtropical savanna. J Geophys Res Biogeosci 113. https://doi.org/10.1029/2007JG000564

Finzi AC, Abramoff RZ, Spiller KS et al (2015) Rhizosphere processes are quantitatively important components of terrestrial carbon and nutrient cycles. Glob Chang Biol 21:2082–2094. https://doi.org/10.1111/gcb.12816

Francaviglia R, Almagro M, Vicente-Vicente JL (2023) Conservation agriculture and soil organic carbon: principles, processes, practices and policy options. Soil Syst 7:17. https://doi.org/10.3390/soilsystems7010017

Frasier I, Noellemeyer E, Figuerola E et al (2016) High quality residues from cover crops favor changes in microbial community and enhance C and N sequestration. Glob Ecol Conserv 6:242–256. https://doi.org/10.1016/j.gecco.2016.03.009

Frey SD, Lee J, Melillo JM, Six J (2013) The temperature response of soil microbial efficiency and its feedback to climate. Nat Clim Chang 3:395–398. https://doi.org/10.1038/nclimate1796

Gao G, Chang C (1996) Changes in CEC and particle size distribution of soils associated with long-term annual applications of cattle feedlot manure. Soil Sci 161:115–120. https://doi.org/10.1097/00010694-199602000-00006

Ghosh K (2017) Effect of sunlight on decomposition of soil humic and non-humic fractions. Proc Indian Natl Sci Acad 94. https://doi.org/10.16943/ptinsa/2017/49139

Giannetta B, Plaza C, Vischetti C et al (2018) Distribution and thermal stability of physically and chemically protected organic matter fractions in soils across different ecosystems. Biol Fertil Soils 54:671–681. https://doi.org/10.1007/s00374-018-1290-9

Giannetta B, Plaza C, Zaccone C et al (2019) Ecosystem type effects on the stabilization of organic matter in soils: combining size fractionation with sequential chemical extractions. Geoderma 353:423–434. https://doi.org/10.1016/j.geoderma.2019.07.009

Gillespie CJ, Antonangelo JA, Zhang H (2021) The response of soil pH and exchangeable Al to alum and lime amendments. Agriculture 11:547. https://doi.org/10.3390/agriculture11060547

Groshans GR, Mikhailova EA, Post CJ, Schlautman MA (2018) Accounting for soil inorganic carbon in the ecosystem services framework for United Nations sustainable development goals. Geoderma 324:37–46. https://doi.org/10.1016/j.geoderma.2018.02.009

Gruba P, Mulder J (2015) Tree species affect cation exchange capacity (CEC) and cation binding properties of organic matter in acid forest soils. Sci Total Environ 511:655–662. https://doi.org/10.1016/j.scitotenv.2015.01.013

Guhra T, Stolze K, Schweizer S, Totsche KU (2020) Earthworm mucus contributes to the formation of organo-mineral associations in soil. Soil Biol Biochem 145:107785. https://doi.org/10.1016/j.soilbio.2020.107785

Gunapala N, Venette RC, Ferris H, Scow KM (1998) Effects of soil management history on the rate of organic matter decomposition. Soil Biol Biochem 30:1917–1927. https://doi.org/10.1016/S0038-0717(98)00062-5

Hajjar R, Jarvis DI, Gemmill-Herren B (2008) The utility of crop genetic diversity in maintaining ecosystem services. Agric Ecosyst Environ 123:261–270. https://doi.org/10.1016/j.agee.2007.08.003

Hamer U, Marschner B (2005) Priming effects in different soil types induced by fructose, alanine, oxalic acid and catechol additions. Soil Biol Biochem 37:445–454. https://doi.org/10.1016/j.soilbio.2004.07.037

Hao H, Wei Y, Cao D et al (2020) Vegetation restoration and fine roots promote soil infiltrability in heavy-textured soils. Soil Tillage Res 198:104542. https://doi.org/10.1016/j.still.2019.104542

Haque ME, Bell RW, Islam AKMS et al (2017) An innovative versatile multi-crop planter for crop establishment using two-wheel tractors. Agric Mech Asia Afr Latin Am 48:33–37

Harper RJ, Tibbett M (2013) The hidden organic carbon in deep mineral soils. Plant Soil 368:641–648. https://doi.org/10.1007/s11104-013-1600-9

Hartemink AE, Barrow NJ (2023) Soil pH – nutrient relationships: the diagram. Plant Soil 486:209–215. https://doi.org/10.1007/s11104-022-05861-z

Hassink J, Whitmore AP, Kubát J (1997) Size and density fractionation of soil organic matter and the physical capacity of soils to protect organic matter. Eur J Agron 7:189–199. https://doi.org/10.1016/S1161-0301(97)00045-2

Havlin JL (2005) Fertility. Encyclopedia of Soils in the Environment. Elsevier, In, pp 10–19

Heitkötter J, Heinze S, Marschner B (2017) Relevance of substrate quality and nutrients for microbial C-turnover in top- and subsoil of a Dystric Cambisol. Geoderma 302:89–99. https://doi.org/10.1016/j.geoderma.2017.04.029

Helfrich M, Flessa H, Mikutta R et al (2007) Comparison of chemical fractionation methods for isolating stable soil organic carbon pools. Eur J Soil Sci 58:1316–1329. https://doi.org/10.1111/j.1365-2389.2007.00926.x

Helmers GA, Yamoah CF, Varvel GE (2001) Separating the impacts of crop diversification and rotations on risk. Agron J 93:1337–1340. https://doi.org/10.2134/agronj2001.1337

Hicks Pries CE, Sulman BN, West C et al (2018) Root litter decomposition slows with soil depth. Soil Biol Biochem 125:103–114. https://doi.org/10.1016/j.soilbio.2018.07.002

Hinshaw S, Wohl E (2023) Carbon sequestration potential of process-based river restoration. River Res Appl 39:1812–1827. https://doi.org/10.1002/rra.4183

Holden SR, Berhe AA, Treseder KK (2015) Decreases in soil moisture and organic matter quality suppress microbial decomposition following a boreal forest fire. Soil Biol Biochem 87:1–9. https://doi.org/10.1016/j.soilbio.2015.04.005

Hong H, Chen S, Fang Q et al (2019) Adsorption of organic matter on clay minerals in the Dajiuhu peat soil chronosequence, South China. Appl Clay Sci 178:105125. https://doi.org/10.1016/j.clay.2019.105125

Hong S, Yin G, Piao S et al (2020) Divergent responses of soil organic carbon to afforestation. Nat Sustain 3:694–700. https://doi.org/10.1038/s41893-020-0557-y

Horwath W (2007) Carbon cycling and formation of soil organic matter. In: Soil microbiology, ecology and biochemistry. Elsevier, pp 303–339

Houlbrooke DJ, Paton RJ, Littlejohn RP, Morton JD (2011) Land-use intensification in New Zealand: effects on soil properties and pasture production. J Agric Sci 149:337–349. https://doi.org/10.1017/S0021859610000821

Hoyt GD, Hargrove WL (1986) Legume cover crops for improving crop and soil management in the Southern United States. HortScience 21:397–402. https://doi.org/10.21273/HORTSCI.21.3.397

Huang C, Li S, He HS et al (2023) Effects of forest management practices on carbon dynamics of China's boreal forests under changing climates. J Environ Manag 335:117497. https://doi.org/10.1016/j.jenvman.2023.117497

Huang J, Liu W, Yang S et al (2021) Plant carbon inputs through shoot, root, and mycorrhizal pathways affect soil organic carbon turnover differently. Soil Biol Biochem 160:108322. https://doi.org/10.1016/j.soilbio.2021.108322

Huang M, Asner GP (2010) Long-term carbon loss and recovery following selective logging in Amazon forests. Global Biogeochem Cycles 24. https://doi.org/10.1029/2009GB003727

Huang S, Huang X, Fang B (2020) Elevated CO2 affects the soil organic carbon fractions and their relation to soil microbial properties in the rhizosphere of Robinia pseudoacacia L. seedlings in cd-contaminated soils. J Soil Sci Plant Nutr 20:1203–1214. https://doi.org/10.1007/s42729-020-00205-1

Huang W, Hammel KE, Hao J et al (2019) Enrichment of lignin-derived carbon in mineral-associated soil organic matter. Environ Sci Technol 53:7522–7531. https://doi.org/10.1021/acs.est.9b01834

Hussain MI, Farooq M, Muscolo A, Rehman A (2020) Crop diversification and saline water irrigation as potential strategies to save freshwater resources and reclamation of marginal soils—a review. Environ Sci Pollut Res 27:28695–28729. https://doi.org/10.1007/s11356-020-09111-6

Hussain S, Hussain S, Guo R et al (2021) Carbon sequestration to avoid soil degradation: a review on the role of conservation tillage. Plants 10:2001. https://doi.org/10.3390/plants10102001

Huys R, Poirier V, Bourget MY et al (2022) Plant litter chemistry controls coarse-textured soil carbon dynamics. J Ecol 110:2911–2928. https://doi.org/10.1111/1365-2745.13997

Iizumi T, Wagai R (2019) Leveraging drought risk reduction for sustainable food, soil and climate via soil organic carbon sequestration. Sci Rep 9:19744. https://doi.org/10.1038/s41598-019-55835-y

Inagaki TM, Possinger AR, Grant KE et al (2020) Subsoil organo-mineral associations under contrasting climate conditions. Geochim Cosmochim Acta 270:244–263. https://doi.org/10.1016/j.gca.2019.11.030

Inagaki TM, Possinger AR, Schweizer SA et al (2023) Microscale spatial distribution and soil organic matter persistence in top and subsoil. Soil Biol Biochem 178:108921. https://doi.org/10.1016/j.soilbio.2022.108921

Jayalath N, Fitzpatrick RW, Mosley L, Marschner P (2016) Type of organic carbon amendment influences pH changes in acid sulfate soils in flooded and dry conditions. J Soils Sediments 16:518–526. https://doi.org/10.1007/s11368-015-1240-1

Jia L, Zhao W, Fu B et al (2019) Effects of minimum soil disturbance practices on controlling water erosion in China's slope farmland: a meta-analysis. Land Degrad Dev 30:706–716. https://doi.org/10.1002/ldr.3258

Johnson AM, Gamble AV, Balkcom KS, Hull NR (2021) Influence of cover crop mixtures on soil health in southeastern crop production systems. Agrosyst Geosci Environ 4:e20202. https://doi.org/10.1002/agg2.20202

Kader MA, Senge M, Mojid MA, Ito K (2017) Recent advances in mulching materials and methods for modifying soil environment. Soil Tillage Res 168:155–166. https://doi.org/10.1016/j.still.2017.01.001

Kaiser K, Guggenberger G (2003) Mineral surfaces and soil organic matter. Eur J Soil Sci 54:219–236. https://doi.org/10.1046/j.1365-2389.2003.00544.x

Karlen D, Rice C (2015) Soil degradation: will humankind ever learn? Sustainability 7:12490–12501. https://doi.org/10.3390/su70912490

Kästner M, Miltner A (2018) SOM and microbes—what is left from microbial life. In: The future of soil carbon. Elsevier, pp 125–163

Kempen B, Brus DJ, Stoorvogel JJ (2011) Three-dimensional mapping of soil organic matter content using soil type–specific depth functions. Geoderma 162:107–123. https://doi.org/10.1016/j.geoderma.2011.01.010

Khangura R, Ferris D, Wagg C, Bowyer J (2023) Regenerative agriculture—a literature review on the practices and mechanisms used to improve soil health. Sustainability 15:2338. https://doi.org/10.3390/su15032338

Kim K, Daly EJ, Gorzelak M, Hernandez-Ramirez G (2022) Soil organic matter pools response to perennial grain cropping and nitrogen fertilizer. Soil Tillage Res 220:105376. https://doi.org/10.1016/j.still.2022.105376

Kirkby CA, Richardson AE, Wade LJ et al (2013) Carbon-nutrient stoichiometry to increase soil carbon sequestration. Soil Biol Biochem 60:77–86. https://doi.org/10.1016/j.soilbio.2013.01.011

Kirschbaum MUF (2010) The temperature dependence of organic matter decomposition: seasonal temperature variations turn a sharp short-term temperature response into a more moderate annually averaged response. Glob Chang Biol 16:2117–2129. https://doi.org/10.1111/j.1365-2486.2009.02093.x

Kleber M, Eusterhues K, Keiluweit M et al (2015) Mineral–organic associations: formation, properties, and relevance in soil environments. Adv Agron 130:1–140

Kleber M, Johnson MG (2010) Advances in understanding the molecular structure of soil organic matter. Adv Agron 106:77–142

Kleber M, Nico PS, Plante A et al (2011) Old and stable soil organic matter is not necessarily chemically recalcitrant: implications for modeling concepts and temperature sensitivity. Glob Chang Biol 17:1097–1107. https://doi.org/10.1111/j.1365-2486.2010.02278.x

Kögel-Knabner I, Guggenberger G, Kleber M et al (2008) Organo-mineral associations in temperate soils: integrating biology, mineralogy, and organic matter chemistry. J Plant Nutr Soil Sci 171:61–82. https://doi.org/10.1002/jpln.200700048

Kopittke PM, Menzies NW, Wang P et al (2019) Soil and the intensification of agriculture for global food security. Environ Int 132:105078. https://doi.org/10.1016/j.envint.2019.105078

Kumar M, Mitra S, Mazumdar SP et al (2021) Improvement of soil health and system productivity through crop diversification and residue incorporation under jute-based different cropping systems. Agronomy 11:1622. https://doi.org/10.3390/agronomy11081622

Lagomarsino A, Grego S, Kandeler E (2012) Soil organic carbon distribution drives microbial activity and functional diversity in particle and aggregate-size fractions. Pedobiologia (Jena) 55:101–110. https://doi.org/10.1016/j.pedobi.2011.12.002

Lai Z, Liu J, Zhang Y et al (2017) Introducing a shrub species in a degraded steppe shifts fine root dynamics and soil organic carbon accumulations, in Northwest China. Ecol Eng 100:277–285. https://doi.org/10.1016/j.ecoleng.2017.01.001

Lal R (2002) Soil carbon dynamics in cropland and rangeland. Environmental Pollution 116:353–362. https://doi.org/10.1016/S0269-7491(01)00211-1

Lal R (2004) Soil carbon sequestration impacts on global climate change and food security. Science 1979(304):1623–1627

Lal R (2015) Restoring soil quality to mitigate soil degradation. Sustainability 7:5875–5895. https://doi.org/10.3390/su7055875

Lal R (2016) Soil health and carbon management. Food Energy Secur 5:212–222. https://doi.org/10.1002/fes3.96

Lal R (2018) Digging deeper: A holistic perspective of factors affecting soil organic carbon sequestration in agroecosystems. Glob Chang Biol 24:3285–3301. https://doi.org/10.1111/gcb.14054

Lal R (2021a) Soil strength and carbon sequestration. In: Hunt A, Egli M, Faybishenko B (eds) Hydrogeology, Chemical Weathering, and Soil Formation. American Geophysical Union and John Wiley and Sons, pp 201–204.

Lal R (2021b) Soil management for carbon sequestration. S Afr J Plant Soil 38:231–237. https://doi.org/10.1080/02571862.2021.1891474

Laudicina VA, Badalucco L, Palazzolo E (2011) Effects of compost input and tillage intensity on soil microbial biomass and activity under Mediterranean conditions. Biol Fertil Soils 47:63–70. https://doi.org/10.1007/s00374-010-0502-8

Lavallee JM, Conant RT, Haddix ML et al (2019) Selective preservation of pyrogenic carbon across soil organic matter fractions and its influence on calculations of carbon mean residence times. Geoderma 354:113866. https://doi.org/10.1016/j.geoderma.2019.07.024

Lavallee JM, Soong JL, Cotrufo MF (2020) Conceptualizing soil organic matter into particulate and mineral-associated forms to address global change in the 21st century. Glob Chang Biol 26:261–273. https://doi.org/10.1111/gcb.14859
Lehmann J, Kern D, German L et al (2003) Soil fertility and production potential. In: Amazonian dark earths. Kluwer Academic Publishers, Dordrecht, pp 105–124
Lehmann J, Kinyangi J, Solomon D (2007) Organic matter stabilization in soil microaggregates: implications from spatial heterogeneity of organic carbon contents and carbon forms. Biogeochemistry 85:45–57. https://doi.org/10.1007/s10533-007-9105-3
Lemire E, Taillon KM, Hendershot WH (2006) Using pH-dependent CEC to determine lime requirement. Can J Soil Sci 86:133–139. https://doi.org/10.4141/S05-002
Lenhardt KR, Breitzke H, Buntkowsky G et al (2022) Interactions of dissolved organic matter with short-range ordered aluminosilicates by adsorption and co-precipitation. Geoderma 423:115960. https://doi.org/10.1016/j.geoderma.2022.115960
Li C, Chen X, Jiang A et al (2021a) Socioeconomic determinants of crop diversity and its effect on farmer income in Guangxi, Southern China. Agriculture 11:336. https://doi.org/10.3390/agriculture11040336
Li H, Liu G, Luo H, Zhang R (2022a) Labile carbon input and temperature effects on soil organic matter turnover in subtropical forests. Ecol Indic 145:109726. https://doi.org/10.1016/j.ecolind.2022.109726
Li JH, Han YW, Ye LF et al (2023a) Effects of nitrogen and phosphorus fertilization on soil organic matter priming and net carbon balance in alpine meadows. Land Degrad Dev 34:2681–2692. https://doi.org/10.1002/ldr.4642
Li K, Lu H, Nkoh JN et al (2022b) Aluminum mobilization as influenced by soil organic matter during soil and mineral acidification: a constant pH study. Geoderma 418:115853. https://doi.org/10.1016/j.geoderma.2022.115853
Li Q, Jia W, Wu J et al (2023b) Changes in soil organic carbon stability in association with microbial carbon use efficiency following 40 years of afforestation. Catena (Amst) 228:107179. https://doi.org/10.1016/j.catena.2023.107179
Li Y, Li J, Chang Y et al (2023c) Comparing bacterial dynamics for the conversion of organics and humus components during manure composting from different sources. Front Microbiol 14. https://doi.org/10.3389/fmicb.2023.1281633
Li Y, Li M, Liu H, Qin W (2021b) Influence of soil texture on the process of subsurface drainage in saturated-unsaturated zones. Int J Agric Biol Eng 14:82–89. https://doi.org/10.25165/j.ijabe.20211401.5699
Li Y, Zhang J, Chang SX et al (2013) Long-term intensive management effects on soil organic carbon pools and chemical composition in Moso bamboo (Phyllostachys pubescens) forests in subtropical China. For Ecol Manag 303:121–130. https://doi.org/10.1016/j.foreco.2013.04.021
Li Z, Song M, Li D et al (2020) Effect of long-term fertilization on decomposition of crop residues and their incorporation into microbial communities of 6-year stored soils. Biol Fertil Soils 56:25–37. https://doi.org/10.1007/s00374-019-01398-z
Liao C, Huang W, Wells J et al (2022) Microbe-iron interactions control lignin decomposition in soil. Soil Biol Biochem 173:108803. https://doi.org/10.1016/j.soilbio.2022.108803
Liao C-S, Xie Z-H, Jien S-H (2021) Decomposition and nutrient releasing of biochar compound materials in soil with different textures. Processes 9:1521. https://doi.org/10.3390/pr9091521
Lin D, Dou P, Yang G et al (2020) Home-field advantage of litter decomposition differs between leaves and fine roots. New Phytol 227:995–1000. https://doi.org/10.1111/nph.16517
Liptzin D, Norris CE, Cappellazzi SB et al (2022) An evaluation of carbon indicators of soil health in long-term agricultural experiments. Soil Biol Biochem 172:108708. https://doi.org/10.1016/j.soilbio.2022.108708
Liu C, Wang S, Zhao Y et al (2023) Enhanced microbial contribution to mineral-associated organic carbon accrual in drained wetlands: beyond direct lignin-iron interactions. Soil Biol Biochem 185:109152. https://doi.org/10.1016/j.soilbio.2023.109152

Liu X, Chen D, Yang T et al (2020) Changes in soil labile and recalcitrant carbon pools after land-use change in a semi-arid agro-pastoral ecotone in Central Asia. Ecol Indic 110:105925. https://doi.org/10.1016/j.ecolind.2019.105925

Liu X, Herbert SJ, Hashemi AM et al (2006) Effects of agricultural management on soil organic matter and carbon transformation – a review. Plant Soil Environ 52:531–543. https://doi.org/10.17221/3544-PSE

Liu Z, Han Y, Lai F et al (2022) Contribution of arbuscular mycorrhizal fungal communities to soil carbon accumulation during the development of Cunninghamia lanceolata plantations. Forests 13:2099. https://doi.org/10.3390/f13122099

Lorenz K, Lal R (2015) Managing soil carbon stocks to enhance the resilience of urban ecosystems. Carbon Manag 6:35–50. https://doi.org/10.1080/17583004.2015.1071182

Lorenz K, Lal R (2018) Soil Carbon Stock. In: Carbon sequestration in agricultural ecosystems. Springer International Publishing, Cham, pp 39–136

Lorenz K, Lal R, Ehlers K (2019) Soil organic carbon stock as an indicator for monitoring land and soil degradation in relation to United Nations' Sustainable Development Goals. Land Degrad Dev 30:824–838. https://doi.org/10.1002/ldr.3270

Lorenz K, Lal R, Preston CM, Nierop KGJ (2007) Strengthening the soil organic carbon pool by increasing contributions from recalcitrant aliphatic bio(macro)molecules. Geoderma 142:1–10. https://doi.org/10.1016/j.geoderma.2007.07.013

Lu C, Kotze DJ, Setälä HM (2021) Evergreen trees stimulate carbon accumulation in urban soils via high root production and slow litter decomposition. Sci Total Environ 774:145129. https://doi.org/10.1016/j.scitotenv.2021.145129

Lucas M (2022) Perspectives from the Fritz-Scheffer awardee 2020—the mutual interactions between roots and soil structure and how these affect rhizosphere processes #. J Plant Nutr Soil Sci 185:8–18. https://doi.org/10.1002/jpln.202100385

Luo Y, Ma L, Seibold S et al (2023) The diversity of mycorrhiza-associated fungi and trees shapes subtropical mountain forest ecosystem functioning. J Biogeogr 50:715–729. https://doi.org/10.1111/jbi.14563

Luo Z, Feng W, Luo Y et al (2017) Soil organic carbon dynamics jointly controlled by climate, carbon inputs, soil properties and soil carbon fractions. Glob Chang Biol 23:4430–4439. https://doi.org/10.1111/gcb.13767

Ma C, Tu Q, Zheng S et al (2022) Soil acidification induced by intensive agricultural use depending on climate. J Soils Sediments 22:2604–2607. https://doi.org/10.1007/s11368-022-03265-1

Magdoff F (2007) Ecological agriculture: principles, practices, and constraints. Renew Agric Food Syst 22:109–117. https://doi.org/10.1017/S1742170507001846

Malamoud K, AlexB MB, Minasny B, Field DJ (2009) Modelling how carbon affects soil structure. Geoderma 149:19–26. https://doi.org/10.1016/j.geoderma.2008.10.018

Manns HR, Parkin GW, Martin RC (2016) Evidence of a union between organic carbon and water content in soil. Can J Soil Sci 96:305–316. https://doi.org/10.1139/cjss-2015-0084

Marin M, Hallett PD, Feeney DS et al (2022) Impact of root hairs on microscale soil physical properties in the field. Plant Soil 476:491–509. https://doi.org/10.1007/s11104-022-05530-1

Marschner B, Brodowski S, Dreves A et al (2008) How relevant is recalcitrance for the stabilization of organic matter in soils? J Plant Nutr Soil Sci 171:91–110. https://doi.org/10.1002/jpln.200700049

Masoudi M, Centeri C, Karlik M, Jakab G (2023) Extracted samples and in situ soil investigations to assess the effects of different land use and tillage management practices on soil organic matter composition. Land Degrad Dev 34:2560–2572. https://doi.org/10.1002/ldr.4629

Matzek V, Lewis D, O'Geen A et al (2020) Increases in soil and woody biomass carbon stocks as a result of rangeland riparian restoration. Carbon Balance Manag 15:16. https://doi.org/10.1186/s13021-020-00150-7

McBratney AB, Morgan CLS, Jarrett LE (2017) The value of soil's contributions to ecosystem services. In: Field DJ, Morgan CLS, McBratney AB (eds) Global soil security. Springer, Cham, pp 227–235

McGeary R, West DJ, Roberton SD, Bennett JM (2022) The soil-water cost of heavy machinery traffic on a Queensland Vertisol, Australia. Geoderma Regional 29:e00506. https://doi.org/10.1016/j.geodrs.2022.e00506
Mikutta R, Kleber M, Torn MS, Jahn R (2006) Stabilization of soil organic matter: association with minerals or chemical recalcitrance? Biogeochemistry 77:25–56. https://doi.org/10.1007/s10533-005-0712-6
Mills RTE, Tipping E, Bryant CL, Emmett BA (2014) Long-term organic carbon turnover rates in natural and semi-natural topsoils. Biogeochemistry 118:257–272. https://doi.org/10.1007/s10533-013-9928-z
Miner GL, Delgado JA, Ippolito JA, Stewart CE (2020) Soil health management practices and crop productivity. Agric Environ Lett 5. https://doi.org/10.1002/ael2.20023
Miyajima S, Uoi N, Murata T et al (2015) Effect of structural modification on heat transfer through man-made soils in urban green areas. Soil Sci Plant Nutr 61:70–87. https://doi.org/10.1080/00380768.2015.1051929
Momma N, Yamamoto K, Simandi P, Shishido M (2006) Role of organic acids in the mechanisms of biological soil disinfestation (BSD). J Gen Plant Pathol 72:247–252. https://doi.org/10.1007/s10327-006-0274-z
Monhonval A, Mauclet E, Hirst C et al (2023) Mineral organic carbon interactions in dry versus wet tundra soils. Geoderma 436:116552. https://doi.org/10.1016/j.geoderma.2023.116552
Moravcová A, Barbi F, Brabcová V et al (2023) Climate-driven shifts in plant and fungal communities can lead to topsoil carbon loss in alpine ecosystems. FEMS Microbiol Ecol 99. https://doi.org/10.1093/femsec/fiad041
Mulimbi W, Brye KR, Nalley LL, Birindwa DR (2023) Conservation agriculture assists smallholder farmers and their agroecosystem in The Democratic Republic of the Congo. Agric Ecosyst Environ 355:108597. https://doi.org/10.1016/j.agee.2023.108597
Murphy BW (2015) Impact of soil organic matter on soil properties—a review with emphasis on Australian soils. Soil Research 53:605. https://doi.org/10.1071/SR14246
Mutio JM, Kebeney S, Njoroge R et al (2023) Effect of land rehabilitation measures on soil organic carbon fractions in semi-arid environment. Front Sustain Food Syst 7. https://doi.org/10.3389/fsufs.2023.1095865
Naorem A, Jayaraman S, Dalal RC et al (2022) Soil inorganic carbon as a potential sink in carbon storage in dryland soils—a review. Agriculture 12:1256. https://doi.org/10.3390/agriculture12081256
Ndzana GM, Zhang Y, Yao S et al (2022) The adsorption capacity of root exudate organic carbon onto clay mineral surface changes depending on clay mineral types and organic carbon composition. Rhizosphere 23:100545. https://doi.org/10.1016/j.rhisph.2022.100545
Nickels MCL, Prescott CE (2021) Soil carbon stabilization under coniferous, deciduous and grass vegetation in Post-mining reclaimed ecosystems. Front For Glob Change 4:689594. https://doi.org/10.3389/ffgc.2021.689594
Nolan RH, Collins L, Gibson RK et al (2022) The carbon cost of the 2019–20 Australian fires varies with fire severity and forest type. Glob Ecol Biogeogr 31:2131–2146. https://doi.org/10.1111/geb.13548
van Noordwijk M, Goverse T, Ballabio C et al (2015) Soil carbon transition curves: reversal of land degradation through management of soil organic matter for multiple benefits. In: Soil carbon: science, management and policy for multiple benefits. CABI, UK, pp 26–46
O'Rourke SM, Angers DA, Holden NM, McBratney AB (2015) Soil organic carbon across scales. Glob Chang Biol 21:3561–3574. https://doi.org/10.1111/gcb.12959
Obalum SE, Chibuike GU, Peth S, Ouyang Y (2017) Soil organic matter as sole indicator of soil degradation. Environ Monit Assess 189:176. https://doi.org/10.1007/s10661-017-5881-y
Oorts K, Vanlauwe B, Merckx R (2003) Cation exchange capacities of soil organic matter fractions in a ferric Lixisol with different organic matter inputs. Agric Ecosyst Environ 100:161–171. https://doi.org/10.1016/S0167-8809(03)00190-7

Ortner M, Seidel M, Semella S et al (2022) Content of soil organic carbon and labile fractions depend on local combinations of mineral-phase characteristics. Soil 8:113–131. https://doi.org/10.5194/soil-8-113-2022

Orwin KH, Kirschbaum MUF, St John MG, Dickie IA (2011) Organic nutrient uptake by mycorrhizal fungi enhances ecosystem carbon storage: a model-based assessment. Ecol Lett 14:493–502. https://doi.org/10.1111/j.1461-0248.2011.01611.x

Osanai Y, Janes JK, Newton PCD, Hovenden MJ (2015) Warming and elevated CO2 combine to increase microbial mineralisation of soil organic matter. Soil Biol Biochem 85:110–118. https://doi.org/10.1016/j.soilbio.2015.02.032

Osborne SL, Chim BK, Riedell WE, Schumacher TE (2020) Root length density of cereal and grain legume crops grown in diverse rotations. Crop Sci 60:2611–2620. https://doi.org/10.1002/csc2.20164

Osseni AA, Gbesso GHF, Nansi KM, Tente ABH (2020) Phytodiversité et services écosystémiques associés aux plantations d'alignement des rues aménagées de la ville de Grand-Popo au Bénin. Bois & Forets Des Tropiques 345:87–99. https://doi.org/10.19182/bft2020.345.a31808

Paul EA (2016) The nature and dynamics of soil organic matter: plant inputs, microbial transformations, and organic matter stabilization. Soil Biol Biochem 98:109–126. https://doi.org/10.1016/j.soilbio.2016.04.001

Pauwels R, Graefe J, Bitterlich M (2023) An arbuscular mycorrhizal fungus alters soil water retention and hydraulic conductivity in a soil texture specific way. Mycorrhiza 33:165–179. https://doi.org/10.1007/s00572-023-01106-8

Phil-Eze PO (2010) Variability of soil properties related to vegetation cover in a tropical rainforest landscape. J Geogr Reg Plann 3:177–184

Phillips DL, Johnson MG, Tingey DT et al (2000) Minirhizotron installation in sandy, rocky soils with minimal soil disturbance. Soil Sci Soc Am J 64:761–764. https://doi.org/10.2136/sssaj2000.642761x

Picchio R, Neri F, Petrini E et al (2012) Machinery-induced soil compaction in thinning two pine stands in Central Italy. For Ecol Manag 285:38–43. https://doi.org/10.1016/j.foreco.2012.08.008

Pintaldi E, D'Amico ME, Stanchi S et al (2018) Humus forms affect soil susceptibility to water erosion in the Western Italian Alps. Appl Soil Ecol 123:478–483. https://doi.org/10.1016/j.apsoil.2017.04.007

Pisani O, Hills KM, Courtier-Murias D et al (2014) Accumulation of aliphatic compounds in soil with increasing mean annual temperature. Org Geochem 76:118–127. https://doi.org/10.1016/j.orggeochem.2014.07.009

Pocknee S, Sumner ME (1997) Cation and nitrogen contents of organic matter determine its soil liming potential. Soil Sci Soc Am J 61:86–92. https://doi.org/10.2136/sssaj1997.03615995006100010014x

Poeplau C, Kätterer T, Leblans NIW, Sigurdsson BD (2017) Sensitivity of soil carbon fractions and their specific stabilization mechanisms to extreme soil warming in a subarctic grassland. Glob Chang Biol 23:1316–1327. https://doi.org/10.1111/gcb.13491

Pold G, Grandy AS, Melillo JM, DeAngelis KM (2017) Changes in substrate availability drive carbon cycle response to chronic warming. Soil Biol Biochem 110:68–78. https://doi.org/10.1016/j.soilbio.2017.03.002

Polyakov VO, Lal R (2004) SOIL EROSION AND CARBON DYNAMICS UNDER SIMULATED RAINFALL. Soil Sci 169:590–599. https://doi.org/10.1097/01.ss.0000138414.84427.40

Ponge J-F (2003) Humus forms in terrestrial ecosystems: a framework to biodiversity. Soil Biol Biochem 35:935–945. https://doi.org/10.1016/S0038-0717(03)00149-4

Ponge J-F (2013) Plant–soil feedbacks mediated by humus forms: a review. Soil Biol Biochem 57:1048–1060. https://doi.org/10.1016/j.soilbio.2012.07.019

Possignolo-Vitti NV, Bertoncini EI, Vitti AC (2017) Decomposition of the organic matter of natural and concentrated vinasse in sandy and clayey soils. Water Sci Technol 76:728–738. https://doi.org/10.2166/wst.2017.239

Pouyat R, Groffman P, Yesilonis I, Hernandez L (2002) Soil carbon pools and fluxes in urban ecosystems. Environ Pollut 116:S107–S118. https://doi.org/10.1016/S0269-7491(01)00263-9
Poveda K, Gómez MI, Martínez E (2008) Diversification practices: their effect on pest regulation and production. Rev Colomb Entomol 34:131–144. https://doi.org/10.25100/socolen.v34i2.9269
Powlson D, Cai ZuCong CZ, Lemanceau P (2015) Soil carbon dynamics and nutrient cycling. In: Soil carbon: science, management and policy for multiple benefits. CABI, UK, pp 98–107
Pribyl DW (2010) A critical review of the conventional SOC to SOM conversion factor. Geoderma 156:75–83. https://doi.org/10.1016/j.geoderma.2010.02.003
Puissant J, Cécillon L, Mills RTE et al (2015) Seasonal influence of climate manipulation on microbial community structure and function in mountain soils. Soil Biol Biochem 80:296–305. https://doi.org/10.1016/j.soilbio.2014.10.013
Qi Y, Li J, Deng S et al (2022) Long-term irrigation reduces soil carbon sequestration by affecting soil microbial communities in agricultural ecosystems of northern China. Eur J Soil Sci 73. https://doi.org/10.1111/ejss.13143
Quartucci F, Gocke M, Denich M et al (2023) Deep soil carbon loss offsets rapid aboveground carbon accumulation after reforestation. For Ecol Manag 548:121403. https://doi.org/10.1016/j.foreco.2023.121403
Quijas S, Schmid B, Balvanera P (2010) Plant diversity enhances provision of ecosystem services: a new synthesis. Basic Appl Ecol 11:582–593. https://doi.org/10.1016/j.baae.2010.06.009
Rakesh S, Sinha AK, Sarkar D et al (2023) Active and passive carbon fractions in contrasting cropping systems, tillage practices, and soil types. Land (Basel) 12:365. https://doi.org/10.3390/land12020365
Ramadan A, Sahar M, Abdou M, Allam A (2021) The efficiency of organic mulching for improving the water productivity under dry regions. Egypt J Chem. https://doi.org/10.21608/ejchem.2021.61497.3324
Ramos FT, Dores EFDC, Weber OLS et al (2018) Soil organic matter doubles the cation exchange capacity of tropical soil under no-till farming in Brazil. J Sci Food Agric 98:3595–3602. https://doi.org/10.1002/jsfa.8881
Rawls WJ, Pachepsky YA, Ritchie JC et al (2003) Effect of soil organic carbon on soil water retention. Geoderma 116:61–76. https://doi.org/10.1016/S0016-7061(03)00094-6
Redlich S, Martin EA, Steffan-Dewenter I (2021) Sustainable landscape, soil and crop management practices enhance biodiversity and yield in conventional cereal systems. J Appl Ecol 58:507–517. https://doi.org/10.1111/1365-2664.13821
Ritchie G, Dolling P (1985) The role of organic matter in soil acidification. Soil Res 23:569. https://doi.org/10.1071/SR9850569
Rivers EN, Heitman JL, McLaughlin RA, Howard AM (2021) Reducing roadside runoff: tillage and compost improve stormwater mitigation in urban soils. J Environ Manag 280:111732. https://doi.org/10.1016/j.jenvman.2020.111732
Rovira P, Jorba M, Romanyà J (2010) Active and passive organic matter fractions in Mediterranean forest soils. Biol Fertil Soils 46:355–369. https://doi.org/10.1007/s00374-009-0437-0
Rovira P, Vallejo VR (2002) Labile and recalcitrant pools of carbon and nitrogen in organic matter decomposing at different depths in soil: an acid hydrolysis approach. Geoderma 107:109–141. https://doi.org/10.1016/S0016-7061(01)00143-4
Rubenstein MA, Crowther TW, Maynard DS et al (2017) Decoupling direct and indirect effects of temperature on decomposition. Soil Biol Biochem 112:110–116. https://doi.org/10.1016/j.soilbio.2017.05.005
Rukshana F, Butterly CR, Xu J-M et al (2014) Organic anion-to-acid ratio influences pH change of soils differing in initial pH. J Soils Sediments 14:407–414. https://doi.org/10.1007/s11368-013-0682-6
Russell EW (1977) The role of organic matter in soil fertility. Philos Trans R Soc Lond B 281:209–219

Saleem M, Fetzer I, Harms H, Chatzinotas A (2013) Diversity of protists and bacteria determines predation performance and stability. ISME J 7:1912–1921. https://doi.org/10.1038/ismej.2013.95

Sanderman J, Amundson R (2014) Biogeochemistry of decomposition and detrital processing. In: Treatise on geochemistry. Elsevier, pp 217–272

Savy D, Mercl F, Cozzolino V et al (2020) Soil amendments with lignocellulosic residues of biorefinery processes affect soil organic matter accumulation and microbial growth. ACS Sustain Chem Eng 8:3381–3391. https://doi.org/10.1021/acssuschemeng.9b07474

Schäffer B, Stauber M, Müller R, Schulin R (2007) Changes in the macro-pore structure of restored soil caused by compaction beneath heavy agricultural machinery: a morphometric study. Eur J Soil Sci 58:1062–1073. https://doi.org/10.1111/j.1365-2389.2007.00886.x

Schrumpf M, Kaiser K, Guggenberger G et al (2013) Storage and stability of organic carbon in soils as related to depth, occlusion within aggregates, and attachment to minerals. Biogeosciences 10:1675–1691. https://doi.org/10.5194/bg-10-1675-2013

Selvarani M, Alagumani T (2018) Crop diversification in Ariyalur District of Tamil Nadu-an empirical study#. Indian J Econ Dev 14:564. https://doi.org/10.5958/2322-0430.2018.00172.5

Shaghaleh H, Zhu Y, Shi X et al (2023) Co-effects of nitrogen fertilizer and straw-decomposing microbial inoculant on decomposition and transformation of field composted wheat straw. Life 13:1993. https://doi.org/10.3390/life13101993

Shahgholi G, Moinfar A (2021) The effect of crop residue percentages and their chopped size on soil compactibility. Acta Technologica Agriculturae 24:181–186. https://doi.org/10.2478/ata-2021-0030

Shao P, Liang C, Lynch L et al (2019) Reforestation accelerates soil organic carbon accumulation: evidence from microbial biomarkers. Soil Biol Biochem 131:182–190. https://doi.org/10.1016/j.soilbio.2019.01.012

Shawver CJ, Ippolito JA, Brummer JE et al (2021) Soil health changes following transition from an annual cropping to perennial management-intensive grazing agroecosystem. Agrosyst Geosci Environ 4:e20181. https://doi.org/10.1002/agg2.20181

Shein EV, Pachepsky YA (1995) Influence of root density on the critical soil water potential. Plant Soil 171:351–357. https://doi.org/10.1007/BF00010291

Shi P, Schulin R (2018) Erosion-induced losses of carbon, nitrogen, phosphorus and heavy metals from agricultural soils of contrasting organic matter management. Sci Total Environ 618:210–218. https://doi.org/10.1016/j.scitotenv.2017.11.060

Shoba VN, Chudnenko KV (2014) Ion exchange properties of humus acids. Eurasian Soil Sci 47:761–771. https://doi.org/10.1134/S1064229314080110

Siahaan H, Kunarso A, Sumadi A et al (2020) Carbon loss affected by fires on various forests and land types in South Sumatera. Indones J For Res 7:15–25. https://doi.org/10.20886/ijfr.2020.7.1.15-25

Siebielec S, Siebielec G, Klimkowicz-Pawlas A et al (2020) Impact of water stress on microbial community and activity in sandy and loamy soils. Agronomy 10:1429. https://doi.org/10.3390/agronomy10091429

Simpson AJ, Kingery WL, Hayes MH et al (2002) Molecular structures and associations of humic substances in the terrestrial environment. Naturwissenschaften 89:84–88. https://doi.org/10.1007/s00114-001-0293-8

Singh M, Sarkar B, Bolan NS et al (2019) Decomposition of soil organic matter as affected by clay types, pedogenic oxides and plant residue addition rates. J Hazard Mater 374:11–19. https://doi.org/10.1016/j.jhazmat.2019.03.135

Singh S, Sharma S (2020) Temporal changes in rhizosphere biological soil quality indicators of wheat in response to nitrogen and straw incorporation. Trop Ecol 61:328–344. https://doi.org/10.1007/s42965-020-00092-8

Six J, Conant RT, Paul EA, Paustian K (2002) Stabilization mechanisms of soil organic matter: implications for C-saturation of soils. Plant Soil 241:155–176. https://doi.org/10.1023/A:1016125726789

Six J, Jastrow JD (2002) Organic matter turnover. In: Lal R (ed) Encyclopedia of soil science. CRC Press, Boca Taton, FL, pp 936–942

Smith KA (1980) A model of the extent of anaerobic zones in aggregated soils, and its potential application to estimates of denitrification. J Soil Sci 31:263–277. https://doi.org/10.1111/j.1365-2389.1980.tb02080.x

Sohrabi H, Bakhtiarvand-Bakhtiari S, Ahmadi K (2016) Above- and below-ground biomass and carbon stocks of different tree plantations in Central Iran. J Arid Land 8:138–145. https://doi.org/10.1007/s40333-015-0087-z

Solly EF, Schöning I, Boch S et al (2014) Factors controlling decomposition rates of fine root litter in temperate forests and grasslands. Plant Soil 382:203–218. https://doi.org/10.1007/s11104-014-2151-4

Song X-D, Yang F, Wu H-Y et al (2022) Significant loss of soil inorganic carbon at the continental scale. Natl Sci Rev 9. https://doi.org/10.1093/nsr/nwab120

Soudzilovskaia NA, van Bodegom PM, Terrer C et al (2019) Global mycorrhizal plant distribution linked to terrestrial carbon stocks. Nat Commun 10:5077. https://doi.org/10.1038/s41467-019-13019-2

Spoor G (2006) Alleviation of soil compaction: requirements, equipment and techniques. Soil Use Manag 22:113–122. https://doi.org/10.1111/j.1475-2743.2006.00015.x

Staton T, Walters RJ, Smith J, Girling RD (2019) Evaluating the effects of integrating trees into temperate arable systems on pest control and pollination. Agric Syst 176:102676. https://doi.org/10.1016/j.agsy.2019.102676

Stevenson FJ (1994) Humus chemistry. Genesis, composition, reactions, Second Edition. Wiley, New York

Stevenson FJ (1995) Humus chemistry: genesis, composition, reactions, second edition (Stevenson, F. J.). J Chem Educ 72:A93. https://doi.org/10.1021/ed072pA93.6

Stewart RD, Bhaskar AS, Parolari AJ et al (2019) An analytical approach to ascertain saturation-excess versus infiltration-excess overland flow in urban and reference landscapes. Hydrol Process 33:3349–3363. https://doi.org/10.1002/hyp.13562

Sutton R, Sposito G (2005) Molecular structure in soil humic substances: the new view. Environ Sci Technol 39:9009–9015. https://doi.org/10.1021/es050778q

Talbot JM, Allison SD, Treseder KK (2008) Decomposers in disguise: mycorrhizal fungi as regulators of soil C dynamics in ecosystems under global change. Funct Ecol 22:955–963. https://doi.org/10.1111/j.1365-2435.2008.01402.x

Thaymuang W, Kheoruenromne I, Suddhipraharn A, Sparks DL (2013) The role of mineralogy in organic matter stabilization in tropical soils. Soil Sci 178:308–315. https://doi.org/10.1097/SS.0b013e3182a4dac4

Theng BKG (2012) Humic substances. In: Developments in Clay Science. Elsevier, pp 391–456

Tiunov AV, Scheu S (1999) Microbial respiration, biomass, biovolume and nutrient status in burrow walls of Lumbricus terrestris L. (Lumbricidae). Soil Biol Biochem 31:2039–2048. https://doi.org/10.1016/S0038-0717(99)00127-3

Tiwari J, Ramanathan A, Bauddh K, Korstad J (2023) Humic substances: structure, function and benefits for agroecosystems—a review. Pedosphere 33:237–249. https://doi.org/10.1016/j.pedsph.2022.07.008

TNC (2016) reThink Soil: a Roadmap for U.S. Soil Health. In: The Nature Conservancy. https://www.nature.org/content/dam/tnc/nature/en/documents/rethink-soil-external-paper-103116.pdf. Accessed 31 May 2025

Tomar U, Baishya R (2020) Moisture regime influence on soil carbon stock and carbon sequestration rates in semi-arid forests of the National Capital Region, India. J For Res (Harbin) 31:2323–2332. https://doi.org/10.1007/s11676-019-01032-6

Tombácz E, Szekeres M, Baranyi L, Michéli E (1998) Surface modification of clay minerals by organic polyions. Colloids Surf A Physicochem Eng Asp 141:379–384. https://doi.org/10.1016/S0927-7757(98)00241-6

Tresch S, Frey D, Le Bayon R-C et al (2019) Litter decomposition driven by soil fauna, plant diversity and soil management in urban gardens. Sci Total Environ 658:1614–1629. https://doi.org/10.1016/j.scitotenv.2018.12.235

Treseder KK (2016) Model behavior of arbuscular mycorrhizal fungi: predicting soil carbon dynamics under climate change. Botany 94:417–423. https://doi.org/10.1139/cjb-2015-0245

Tu C, Liu C, Quine TA et al (2018) Dynamics of soil organic carbon following land-use change: insights from stable C-isotope analysis in black soil of Northeast China. Acta Geochimica 37:746–757. https://doi.org/10.1007/s11631-018-0282-7

Tumer AR, Karacaoglu E, Namli A et al (2013) Effects of different types of soil on decomposition: an experimental study. Legal Med 15:149–156. https://doi.org/10.1016/j.legalmed.2012.11.003

Twerdoff DA, Chanasyk DS, Mapfumo E et al (1999) Impacts of forage grazing and cultivation on near-surface relative compaction. Can J Soil Sci 79:465–471. https://doi.org/10.4141/S98-076

Védère C, Vieublé Gonod L, Pouteau V et al (2020) Spatial and temporal evolution of detritusphere hotspots at different soil moistures. Soil Biol Biochem 150:107975. https://doi.org/10.1016/j.soilbio.2020.107975

Veenstra JJ, Horwath WR, Mitchell JP (2007) Tillage and cover cropping effects on aggregate-protected carbon in cotton and tomato. Soil Sci Soc Am J 71:362–371. https://doi.org/10.2136/sssaj2006.0229

Veloso MG, Angers DA, Chantigny MH, Bayer C (2022) Mineral–organic associations are enriched in both microbial metabolites and plant residues in a subtropical soil profile under no-tillage and legume cover cropping. Soil Research 60:590–600. https://doi.org/10.1071/SR21151

Vogel C, Heister K, Buegger F et al (2015) Clay mineral composition modifies decomposition and sequestration of organic carbon and nitrogen in fine soil fractions. Biol Fertil Soils 51:427–442. https://doi.org/10.1007/s00374-014-0987-7

von Lützow M, Kögel-Knabner I, Ekschmitt K et al (2007) SOM fractionation methods: relevance to functional pools and to stabilization mechanisms. Soil Biol Biochem 39:2183–2207. https://doi.org/10.1016/j.soilbio.2007.03.007

von Lützow M, Kögel-Knabner I, Ludwig B et al (2008) Stabilization mechanisms of organic matter in four temperate soils: development and application of a conceptual model. J Plant Nutr Soil Sci 171:111–124. https://doi.org/10.1002/jpln.200700047

Vukicevich E, Lowery T, Bowen P et al (2016) Cover crops to increase soil microbial diversity and mitigate decline in perennial agriculture. A review. Agron Sustain Dev 36:48. https://doi.org/10.1007/s13593-016-0385-7

Wander MM, Cihacek LJ, Coyne M et al (2019) Developments in agricultural soil quality and health: reflections by the research committee on soil organic matter management. Front Environ Sci 7. https://doi.org/10.3389/fenvs.2019.00109

Wang G, Li X, Xi X, Cong W-F (2022a) Crop diversification reinforces soil microbiome functions and soil health. Plant Soil 476:375–383. https://doi.org/10.1007/s11104-022-05436-y

Wang H, Wu J, Li G, Yan L (2020) Changes in soil carbon fractions and enzyme activities under different vegetation types of the northern Loess Plateau. Ecol Evol 10:12211–12223. https://doi.org/10.1002/ece3.6852

Wang Q, Huang Q, Wang J et al (2021) Dissolved organic carbon drives nutrient cycling via microbial community in paddy soil. Chemosphere 285:131472. https://doi.org/10.1016/j.chemosphere.2021.131472

Wang X, Jelinski NA, Toner B, Yoo K (2019) Long-term agricultural management and erosion change soil organic matter chemistry and association with minerals. Sci Total Environ 648:1500–1510. https://doi.org/10.1016/j.scitotenv.2018.08.110

Wang X, Tian L, Li Y et al (2022b) Effects of exogenous cellulose-degrading bacteria on humus formation and bacterial community stability during composting. Bioresour Technol 359:127458. https://doi.org/10.1016/j.biortech.2022.127458

Wang Y, Tang C, Wu J et al (2013) Impact of organic matter addition on pH change of paddy soils. J Soils Sediments 13:12–23. https://doi.org/10.1007/s11368-012-0578-x

Wei X, Li Q, Liu Y et al (2013) Restoring ecosystem carbon sequestration through afforestation: a sub-tropic restoration case study. For Ecol Manag 300:60–67. https://doi.org/10.1016/j.foreco.2012.06.018
Han W (2011) A simulation test of the impact on soil moisture by agricultural machinery. Afr J Biotechnol 10. https://doi.org/10.5897/AJB11.2237
Wiesmeier M, Steffens M, Mueller CW et al (2012) Aggregate stability and physical protection of soil organic carbon in semi-arid steppe soils. Eur J Soil Sci 63:22–31. https://doi.org/10.1111/j.1365-2389.2011.01418.x
Windsor FM, Tavella J, Rother DC et al (2021) Identifying plant mixes for multiple ecosystem service provision in agricultural systems using ecological networks. J Appl Ecol 58:2770–2782. https://doi.org/10.1111/1365-2664.14007
Wiseman CLS, Püttmann W (2006) Interactions between mineral phases in the preservation of soil organic matter. Geoderma 134:109–118. https://doi.org/10.1016/j.geoderma.2005.09.001
Wolińska A, Banach A, Szafranek-Nakonieczna A et al (2018) Easily degradable carbon – an indicator of microbial hotspots and soil degradation. Int Agrophys 32:123–131. https://doi.org/10.1515/intag-2016-0098
Wu Q, Wu F, Zhu J, Ni X (2023) Leaf and root inputs additively contribute to soil organic carbon formation in various forest types. J Soils Sediments 23:1135–1145. https://doi.org/10.1007/s11368-022-03403-9
Xiong P, Zhang Z, Peng X (2022) Root and root-derived biopore interactions in soils: a review. J Plant Nutr Soil Sci 185:643–655. https://doi.org/10.1002/jpln.202200003
Xu S, Su F, Sayer EJ et al (2023) Fine root litter quality regulates soil carbon storage efficiency in subtropical forest soils. Soil Ecol Lett 5:230182. https://doi.org/10.1007/s42832-023-0182-6
Xue B, Huang L, Huang Y et al (2019) Roles of soil organic carbon and iron oxides on aggregate formation and stability in two paddy soils. Soil Tillage Res 187:161–171. https://doi.org/10.1016/j.still.2018.12.010
Xue B, Huang L, Huang Y et al (2020) Straw management influences the stabilization of organic carbon by Fe (oxyhydr)oxides in soil aggregates. Geoderma 358:113987. https://doi.org/10.1016/j.geoderma.2019.113987
Yadav GS, Babu S, Das A et al (2019) Soil carbon dynamics and productivity of rice–rice system under conservation tillage in submerged and unsubmerged ecologies of Eastern Indian Himalaya. Carbon Manag 10:51–62. https://doi.org/10.1080/17583004.2018.1545518
Yan M, Zhang X, Liu K et al (2022) Particle size primarily shifts chemical composition of organic matter under long-term fertilization in paddy soil. Eur J Soil Sci 73. https://doi.org/10.1111/ejss.13170
Yang W, Li Z, Cai C et al (2013) Mechanical properties and soil stability affected by fertilizer treatments for an Ultisol in subtropical China. Plant Soil 363:157–174. https://doi.org/10.1007/s11104-012-1266-8
Yu H, Song W (2023) Research Progress on the impact of land use change on soil carbon sequestration. Land (Basel) 12:213. https://doi.org/10.3390/land12010213
Yu L, Tian R, Zhang R et al (2020) Coupling effects of humus and 2:1 type electrolyte on soil water movement. Geoderma 375:114482. https://doi.org/10.1016/j.geoderma.2020.114482
Yu M, Tariq SM, Yang H (2022) Engineering clay minerals to manage the functions of soils. Clay Miner 57:51–69. https://doi.org/10.1180/clm.2022.19
Zacháry D, Filep T, Jakab G et al (2023) The effect of mineral composition on soil organic matter turnover in temperate forest soils. J Soils Sediments 23:1389–1402. https://doi.org/10.1007/s11368-022-03393-8
Zak DR, Pellitier PT, Argiroff WA et al (2019) Exploring the role of ectomycorrhizal fungi in soil carbon dynamics. New Phytol 223:33–39. https://doi.org/10.1111/nph.15679
Zamanian K, Pustovoytov K, Kuzyakov Y (2016) Pedogenic carbonates: forms and formation processes. Earth Sci Rev 157:1–17. https://doi.org/10.1016/j.earscirev.2016.03.003
Zavarzina AG, Danchenko NN, Demin VV et al (2021) Humic substances: hypotheses and reality (a review). Eurasian Soil Sci 54:1826–1854. https://doi.org/10.1134/S1064229321120164

Zhang H, Zhou Z (2018) Recalcitrant carbon controls the magnitude of soil organic matter mineralization in temperate forests of northern China. For Ecosyst 5:17. https://doi.org/10.1186/s40663-018-0137-z

Zhang X, Liu Z, Liu X, Liang X (2016) Relationship between soil humus dissimilation, soil biological and chemical properties, and leaf litter characteristics in pure forests. Emir J Food Agric 28:616. https://doi.org/10.9755/ejfa.2016-04-355

Zhang X, Song J, Wang Y et al (2022) Threshold effects of vegetation coverage on runoff and soil loss in the loess plateau of China: a meta-analysis. Geoderma 412:115720. https://doi.org/10.1016/j.geoderma.2022.115720

Zhang Y-Y, Wu W, Liu H (2019) Factors affecting variations of soil pH in different horizons in hilly regions. PLoS One 14:e0218563. https://doi.org/10.1371/journal.pone.0218563

Zhao D, Cheng C, Lyu D et al (2017) Effects of residue coverage on the characteristics of soil carbon pools in orchards. Arch Agron Soil Sci 63:771–783. https://doi.org/10.1080/03650340.2016.1241390

Zhao J, Wang X, Shao Y et al (2011) Effects of vegetation removal on soil properties and decomposer organisms. Soil Biol Biochem 43:954–960. https://doi.org/10.1016/j.soilbio.2011.01.010

Zhao Q, Thompson AM, Callister SJ et al (2022) Dynamics of organic matter molecular composition under aerobic decomposition and their response to the nitrogen addition in grassland soils. Sci Total Environ 806:150514. https://doi.org/10.1016/j.scitotenv.2021.150514

Zhao X, Zhang L, Liu D (2012) Biomass recalcitrance. Part I: the chemical compositions and physical structures affecting the enzymatic hydrolysis of lignocellulose. Biofuels Bioprod Biorefin 6:465–482. https://doi.org/10.1002/bbb.1331

Zhou Y, Boutton TW, Ben WX (2019) A three-dimensional assessment of soil δ13C in a subtropical savanna: implications for vegetation change and soil carbon dynamics. Soil Syst 3:73. https://doi.org/10.3390/soilsystems3040073

Zhu Y, Guo B, Liu C et al (2021) Soil fertility, enzyme activity, and microbial community structure diversity among different soil textures under different land use types in coastal saline soil. J Soils Sediments 21:2240–2252. https://doi.org/10.1007/s11368-021-02916-z

Zoghlami A, Paës G (2019) Lignocellulosic biomass: understanding recalcitrance and predicting hydrolysis. Front Chem 7. https://doi.org/10.3389/fchem.2019.00874

Chapter 4
Carbon Farming Techniques

Abstract This chapter introduces key carbon (C) farming techniques that help capture and store C in agricultural systems while supporting soil health and productivity. It covers agroforestry, cover cropping, conservation agriculture, crop rotation, and biochar application. For each technique, the chapter explains how practice works, followed by a discussion of its potential to sequester C. By adopting these approaches, farmers can contribute to climate change mitigation while improving the long-term sustainability and resilience of their land.

4.1 Agroforestry

4.1.1 Introduction

Agroforestry, as defined by Lundgren and Raintree (1983), is a comprehensive term encompassing land-use systems and technologies where woody perennials are deliberately integrated into the same land management units as agricultural crops and/or animals. Distinct from a mere practice, agroforestry is more appropriately considered a system, allowing for the flexibility to group various practices such as alley cropping, home gardens, improved fallows, silvopasture, shelterbelts/windbreaks, living fences, and forest farming etc. (Lorenz and Lal 2018; Elevitch et al. 2018) (Table 4.1). This multifaceted approach provides an array of ecosystem services (ESs) and delivers socio-economic and environmental benefits (Jose 2009). These include biodiversity conservation (Acharya 2006; Mayorga et al. 2022), improved soil fertility (Octavia et al. 2023), carbon sequestration (CS) (Lorenz and Lal 2014a), microclimate modification (Lin 2007; Wang et al. 2022), pollination and pest control (Staton et al. 2019), climate change mitigation, restoration of degraded lands (Sarvade et al. 2019; Celentano et al. 2020), increased water use efficiency (Loria and Bhardwaj 2021), and effective control of soil, water, and wind erosion (König 1992). Additionally, agroforestry contributes to income diversification for farmers (Antle et al. 2007; Takimoto et al. 2008; Donaghy et al. 2010;

N. Loria, R. Lal, *Carbon Farming*, https://doi.org/10.1007/978-3-032-00842-8_4

Table 4.1 Agroforestry practices

Agroforestry Practice	Description	Benefits	Geographical Suitability	References
Shelterbelt/ Windbreak	Single or multiple rows of trees or shrubs planted along agricultural fields or livestock facilities to protect crops, animals, and soil from wind damage.	Wind protection, reduced soil erosion, microclimate modification, reduce GHG emissions, improve crop yield	Varies, suitable in regions with strong winds and potential soil erosion.	Van Thuyet et al. (2014), Dhillon and Van Rees (2017), Ballesteros-Possu et al. (2017)
Alley Cropping	Widely spaced rows of trees or shrubs with agricultural or pasture crops grown in between	Improved soil fertility, enhanced biodiversity, diversified income streams, reduce wind and water erosion, reduced nutrient leaching, and minimized incidence of pests and disease	Widely adaptable, depending on tree and crop species chosen.	Böhm et al. (2014), van Ramshorst et al. (2022), Louhaichi and Hassan (2023)
Forest Farming	The cultivation of crops beneath the canopy of managed forests, combining agricultural and forestry practices.	Sustainable timber production, increased biodiversity, improved soil health.	Forested regions with suitable tree species and climatic conditions.	Lorenz and Lal (2018), Trozzo et al. (2021)
Taungya	A system where crops are cultivated alongside young forest plantations, allowing shared land use and management.	Efficient land use, diversified income sources, potential for sustainable timber production.	Commonly used in tropical regions with suitable tree species for afforestation.	Imo (2009), Hemida et al. (2023)
Home Gardens	Small-scale gardens around homes incorporating a variety of crops, trees, and sometimes livestock.	Diverse food production, increased resilience, enhanced microclimate.	Globally adaptable, especially in areas with limited land for agriculture.	Gebru et al. (2019), Nyong et al. (2020)
Improved Fallows	Allowing land to naturally regenerate or intentionally planting cover crops to restore soil fertility after cultivation.	Soil fertility improvement, weed suppression, erosion control.	Adaptable to various climates and regions, commonly used in tropical agroecosystems.	Sanchez (1999), Tassin et al. (2012), Joslin et al. (2019)

(continued)

Table 4.1 (continued)

Agroforestry Practice	Description	Benefits	Geographical Suitability	References
Silvopasture	Integrating trees or shrubs with pasture or forage to combine livestock grazing with forestry or agroforestry.	Improved animal welfare, enhanced forage quality, microclimate modification, carbon sequestration.	Variable, suitable in regions where livestock grazing, and tree cultivation coexist.	Kallenbach et al. (2006), Castillo et al. (2020), Kumar et al. (2021), López-Hernández et al. (2024)
Living Fences	The use of living plants, such as trees or shrubs, to create a fence, providing boundaries and additional benefits like wind protection.	Erosion control, windbreak, habitat for beneficial organisms, protection of new plantations from livestock trampling effects.	Adaptable to various climates and regions, especially in areas prone to wind erosion or snow.	Love et al. (2009), Pulido-Santacruz and Renjifo (2011), Ogdahl et al. (2018)

Waldén et al. 2020). The selection of tree species within agroforestry systems (AFS) is a critical aspect, emphasizing the importance of avoiding competition with adjacent crops while meeting farmers' needs. The principle of complementarity, both temporal and spatial, plays a pivotal role in optimizing the combination of tree and crop species. This strategy involves the thoughtful pairing of species that exhibit diverse resource utilization patterns. Each species is strategically placed to occupy a specific niche, allowing them to access and thrive on particular resources (Ong et al. 2007). This approach significantly contributes to enhancing the overall productivity and sustainability of AFS. For instance, integrating short-duration leguminous and oilseed crops such as cowpea (*Vigna unguiculata*) and mung bean (*Vigna radiata*) with fruit trees like Karonda (*Carissa carandas*) and Drumstick (*Moringa oleifera*) has proven to be a beneficial agricultural practice, especially in semi-arid regions (Rathore et al. 2022). Similarly, pairing maize (*Zea mays*) with *Leucaena leucocephala* (Lal 1989) and *Gliricidia sepium* in alley cropping systems (Kang et al. 1981) has also demonstrated significant benefits in such conditions. However, it is crucial to avoid agroforestry species exhibiting negative allelopathic effects on agricultural and fodder crops, while promoting those with no or positive allelopathic interactions to enhance the success of AFS (Rizvi 1999).

Climate change mitigation and CS stands out as the key ESs provided by agroforestry (Lorenz and Lal 2014a; Patel et al. 2023). Trees act as carbon (C) sinks, effectively trapping and storing C in their biomass—trunks, branches, leaves, and roots. The long-life span of trees, compared to annual crops, ensures sustained C storage over extended periods (Pardon et al. 2017; Besar et al. 2020). The choice of tree species within AFS further influences CS potential. Fast growing trees, such as

Salix and *Populus* spp. enhance soil organic C (SOC), thereby improving soil structure and influencing mycorrhizal associations within a short time (Kahle et al. 2005). Notably, in the Indo-Gangetic plains of India, poplar and eucalyptus exhibit considerable CS potential, with net rates of 10.3 Mg C ha^{-1} yr^{-1} and 12.7 Mg C ha^{-1} yr^{-1} respectively (Chavan et al. 2023). Furthermore, *Bambusa vulgaris* in the Central Himalayas demonstrates substantial SOC sequestration and lability enhancement (Melkani et al. 2024). The varying tree density and spatial arrangements among different tree species can significantly influence CS outcomes (Rasoarinaivo et al. 2022). The impact of tree species diversity on soil CS involves various mechanisms, including diverse root structures (Cadotte 2013; Verma et al. 2014; Bieluczyk et al. 2021), leaf litter qualities (Yadav et al. 2008; Akinyele and Donald-Amaeshi 2021; Sari et al. 2022), and interactions with soil microbes (Dobo et al. 2018; Chaiya et al. 2021). While considering soil depth and specific C fractions is crucial, certain AFS may not significantly affect soil C stocks at specific depths despite variations in fine root C (Upson and Burgess 2013). For instance, agroforestry practices such as those observed in Senegal's parklands with Faidherbia albida trees enhance crop biomass and feature extensive root exploration but contributed modestly to overall soil C stocks (Siegwart et al. 2023). Nonetheless, agroforestry remains vital for preventing soil erosion and providing crop protection against wind damage in vegetation-deprived areas (Jinger et al. 2022; Pan et al. 2022). This approach holds significant promise in the restoration of such lands, creating opportunities for increased C storage and contributing to overall environmental resilience (Schroeder 1994; Nyong et al. 2020; Siarudin et al. 2021; Fahad et al. 2022).

4.1.2 Carbon Sequestration Potential of Agroforestry

Agroforestry is widely acknowledged for its CS potential (Ramachandran Nair et al. 2009; Nair 2012; Lal et al. 2018), with studies highlighting its pivotal role in improving soil health (Schroeder 1994; Montagnini and Nair 2004; Amare et al. 2019). While obtaining a precise global estimate of agroforestry CS potential remains challenging, numerous efforts have been made to approximate its extent (Lorenz and Lal 2014a). Notably, from 2000 to 2010, a 3.7% increase in tree cover led to a significant rise of over 2 Pg (1 petagram = 10^{15} g = 1 Gt) in biomass C, emphasizing the crucial contribution of agroforestry to CS endeavors (Zomer et al. 2016). Young trees in agroforestry and top soils may have a higher CS potential compared to those of mature trees and soils deeper than 30 cm, respectively (Shi et al. 2018; Kumara et al. 2023). Globally, the net annual rate of change in C stocks through improved agroforestry management and land-use change to agroforestry is reported as 0.3 Mg C ha^{-1} yr^{-1} and 3.1 Mg C ha^{-1} yr^{-1} respectively (IPCC 2000). India's National Agroforestry Policy estimates the average CS potential of agroforestry at 0.21 Mg C ha^{-1} yr^{-1}, potentially mitigating millions of tons (~109 Tg) of

CO_2 annually (Ajit et al. 2017). Similarly, in the United States, AFS have the potential to sequester approximately 0.55 Pg C ha^{-1} yr^{-1}, offsetting a considerable portion of current emission rates (~34%) from fossil fuel combustion (Udawatta and Jose 2011). Research conducted across different regions of India provides valuable insights into the diverse impacts of AFS on SOC dynamics. Panwar et al. (2022) observed SOC contents ranging from 12.3 to 125.3 Mg ha^{-1} across AFS spanning 3 to 50 years. Among these, the highest levels were observed in coffee agroforests in southern India, while the lowest levels were recorded in agrisilviculture systems in western India. In contrast, Melkani et al. (2024) reported significant enhancements in SOC stocks of 19.82–46.33% under various AFS in the Central Himalayas, with SOC sequestration rates varying from 0.111 to 0.697 Mg C ha^{-1} yr^{-1}. While these numbers highlight the substantial CS potential of agroforestry, challenges remain in measurement due to methodological inconsistencies (Nair 2011; Ghale et al. 2022) and limited financial resources (Holderieath et al. 2012). Site-specific choice of species and spatial and temporal arrangements of trees and crops is critical to optimize symbiotic relations (Lal 1991). Nevertheless, agroforestry holds promise as a sustainable land-use practice for CS, contributing to climate change mitigation efforts while providing additional environmental and socioeconomic benefits.

4.2 Cover Cropping

4.2.1 Introduction

Cover cropping, a fundamental component of conservation agriculture (CA), is a potent tool in C farming practices. This practice involves planting specific species on agricultural land to provide soil health benefits during periods when the primary cash crops are not growing, typically during fallow periods or between planting and harvesting seasons. Such benefits include adding soil organic matter (SOM) (Lal 2015a; Toom et al. 2019a), suppressing weed (Uchino et al. 2011; Büchi et al. 2020), minimizing soil erosion (Himmelbauer et al. 2013; Márquez-García et al. 2024), improving soil aggregation(Lal 2015b), reducing nutrient leaching (Wyland 1996; Nguyen et al. 2022; Kumar et al. 2023), increasing soil biodiversity (Vukicevich et al. 2016; Verzeaux et al. 2016; Zhang et al. 2022a), and reducing insect pest incidence (Sáenz-Romo et al. 2019; Martinez et al. 2020). In discussions surrounding cover crops, terms such as catch crops, living mulch, intercrops, and trap crops may arise. Although these terms are sometimes used interchangeably in agricultural discourse, it is important to recognize that each have distinct meanings and serves distinct functions within farming systems (Table 4.2). Cover crops have a neutral to positive effect on the main crop yield (Marcillo and Miguez 2017; Toom et al. 2019b; Fan et al. 2021; Waring et al. 2023). Yield benefits mostly depend on environmental factors such as decreased evapotranspiration and edaphic factors like

Table 4.2 Cover crop and related terms

Term	Definition	Purpose	Example	References
Cover Crops	Crops intentionally planted to cover and protect the soil during periods of fallow or between cash crops.	Soil improvement, erosion control, weed suppression	Clover, ryegrass planted in winter, oat as a spring cover crop	Teixeira et al. (2016), Meyer et al. (2020)
Catch Crops	Quick-growing crops planted between regular crop cycles to absorb excess nutrients and prevent soil leaching.	Nutrient capture, soil protection during fallow periods	Spring barley undersown with ryegrass or a mixture of ryegrass and red clover	Pawłowski et al. (2021), Vogeler et al. (2023)
Living mulch	Cover crop or low-growing plant species intentionally planted alongside main crops to conserve soil moisture and regulate temperature	Moisture retention, control runoff and erosion, improve soil properties, weed control	Barley, white clover, ryegrass, crown vetch	Qian et al. (2015), Kader et al. (2019), Iqbal et al. (2020)
Intercrops	Different crops cultivated simultaneously in the same field to maximize resource use efficiency and enhance biodiversity.	Resource optimization, pest and disease management through diversity	Barley and vetch mixture	Lithourgidis et al. (2011), Ramirez-Garcia et al. (2015)
Green Manure	Cover crops grown to be incorporated into the soil while still green or shortly after flowering, adding SOM and nutrients.	Increase SOM; Soil fertility enhancement, soil structure improvement, improve crop yield	Mustard greens plowed into soil	Sincik et al. (2008), Couëdel et al. (2018), Sun and Shahrajabian (2022)
Trap Crops	Specific plants strategically planted to attract pests away from primary crops, reducing pest damage.	Insect Pest management, reduce cash crop damage, less dependence on pesticides	Crimson clover, mustard	Parker et al. (2016), Harbach et al. (2021), Panwar et al. (2021)

soil pH and soil salinity within the same land management farming system (Leuthold et al. 2021; Wang et al. 2021; Nordblom et al. 2023). However, negative effects on cash crop yield are also reported in a few studies (Kaspar and Bakker 2015; Abdalla et al. 2019; Mennan et al. 2020; Qin et al. 2021). These negative effects are mainly due to allelopathic effects, nutrient deficiency, and decreased water availability in dry regions (Kaspar and Bakker 2015; Qin et al. 2021; Reginatto et al. 2023). Employing the 4Rs approach in cover crop management mitigates negative impacts on the main crop by ensuring the selection of the **r**ight cover crop or cover crop mixture, **r**ight timing for planting, harvesting, or termination, and implementation of **r**ight cover crop management practices at the **r**ight location (Delgado and Gantzer

2015). This comprehensive strategy optimizes the benefits of cover crops while minimizing potential drawbacks.

Cover crops are typically categorized into legumes and non-legumes (Fig. 4.1), though other crops also play crucial roles. Legumes like pea (*Pisum sativum*), soybean (*Glycine max*), hairy vetch (*Vicia villosa*), and clover (*Trifolium repens*) have the unique ability to fix nitrogen, enriching the soil and reducing the need for synthetic fertilizers (Hartwig and Ammon 2002; Gebhard et al. 2013; De Notaris et al. 2019). On the other hand, non-legumes such as wheat (*Triticum aestivum*), oats (*Avina sativa*), and ryegrass (*Lolium perenne*) contribute to soil health by improving structure, suppressing weeds, and preventing erosion (Moyer et al. 2000; Kaspar et al. 2001; Sun and Shahrajabian 2022). Furthermore, specialty cover crops like mustard greens and sunflowers offer specific benefits such as biofumigation or rapid biomass production (Ahmed et al. 2022; Motisi et al. 2009; Waisen et al. 2022). For instance, wild flax (*Camelina sativa*) demonstrates promising allelopathic and agronomic benefits, making it a viable option for sustainable weed management and integrated pest control in agriculture, highlighting its potential as a versatile cover crop (Ghidoli et al. 2023). Strategic combinations of cover crops enable farmers to improve soil fertility, combat erosion, and enhance agricultural sustainability (Table 4.3). Achieving this involves recognizing the limitations of individual species and leveraging the benefits of diversity (Ramírez-García et al. 2015). The adoption of multi-species cover crop mixtures is increasingly favored, driven by considerations such as grower objectives, crop rotation compatibility, ecosystem functions, adaptability to the local environment, and economic viability (Chapagain et al. 2020). However, challenges such as seed cost, labor, and time constraints exist, necessitating a multifaceted approach, including financial incentives, education, technology, and community support, to facilitate successful cover crop adoption (Clay et al. 2020; Golden et al. 2023). Nevertheless, cover crops offer rapid financial benefits in addressing various agricultural challenges, including herbicide-resistant weeds, soil compaction, transitioning to no-till practices, grazing, soil moisture deficits, high fertilizer costs, and incentivized usage (Myers et al. 2019).

4.2.2 Carbon Sequestration Potential of Cover Crops

Cover crops significantly contribute to soil health and CS by enriching SOM content and fostering microbial diversity, with their residues enhancing SOC mineralization, thus promoting soil C cycling within agricultural systems (Ghimire et al. 2017; Amsili and Kaye 2021; Liang et al. 2023). Factors influencing the impact of cover crops on soil C accumulation include biomass input, duration of use, initial soil C levels, soil type, cover crop species, tillage practices, and climate conditions (Blanco-Canqui et al. 2015; Meena et al. 2020; Jian et al. 2020; Liang et al. 2023). Studies indicate that cover cropping leads to significant annual SOC accumulation rates, contributing to global CS efforts. For instance, Poeplau and Don (2015) demonstrate an annual SOC accumulation rate of 0.32 ± 0.08 Mg ha^{-1} yr^{-1} in 0.22 m

Fig. 4.1 (**a**) Soybean (legume) cover crop, (**b**) Wheat (non-legume) cover crop (Photo taken by the author, Dr. Nancy Loria)

Table 4.3 Diversity in cover crop mixtures

Cover Crop Mixture	Example	Benefits	References
Grass-legume	Clover + Ryegrass; cowpea + sorghum	Nitrogen fixation from legumes. Soil stabilization and erosion control from grasses; weed suppression	Brainard et al. (2011), Kabelka et al. (2019), Seefeldt et al. (2023)
Crucifer-legume	Radish/Mustard + faba bean/crimson clover/hairy vetch	Crucifers reduce sulphate leaching; suppress fungal pathogens; bio fumigation potential; nitrogen fixation from legumes	Blanco-Canqui et al. (2015), Couëdel et al. (2018)
Cereal-Legume Mix	Pea + wheat	Soil improvement, higher yield; Pathogen control; improved economic efficiency	Fernández-Aparicio et al. (2010), Pelzer et al. (2012), Monti et al. (2019), Księżak et al. (2023)
Diverse Multi-Species Mix	Wheat + Clover + vetch + Rye	Improved soil properties; weed suppression; erosion control, pest control, improve nitrogen use efficiency	Ramírez-García et al. (2015), Chapagain et al. (2020), Franco et al. (2021), Larkin (2024)
Summer Cover Crop Mix	Sun hemp + Sorghum	Rapid biomass production; Drought tolerance and erosion control from sorghum	Wang et al. (2012), Parvin et al. (2022), Myrans et al. (2024)

deep soil, suggesting a potential global SOC sequestration of 0.12 ± 0.03 Pg C yr^{-1}, equivalent to approximately 8% of direct annual greenhouse gas (GHG) emissions from agriculture. In the United States, cover crops exhibit a potential to sequester SOC, averaging 0.41 Mg C ha^{-1} yr^{-1} in the upper 0.3 m soil depth across 29% of comparisons, suggesting a need for refined management strategies to optimize SOC accumulation (Blanco-Canqui 2022). In France, cover crops have been shown to enhance the mean direct GHG balance by 315 kg CO_2 eq. ha^{-1} yr^{-1}, offering a substantial opportunity to mitigate climate change by reducing annual CO_2 emissions in the agricultural sector by 4.5–9% (Tribouillois et al. 2018). In temperate climates, cover cropping significantly increases SOC stocks by an average of 1.11 Mg C/ha compared to no cover crop control, with key factors including planting and termination dates, annual cover crop biomass production, and soil clay content (McClelland et al. 2021). Integrating cover crops into agricultural rotations results in a notable rise in SOC levels. Jian et al. (2020) reported an average increase of 15.5% in SOC levels, potentially sequestering around 0.16 +/− 0.06 Pg of C annually if implemented on 15% of global cropland, comparable to 1–2% of current fossil fuel emissions.

Additionally, it's observed that cover crop mixtures tend to result in greater SOC increases compared to mono-species cover crops (Sainju et al. 2017; Amsili and Kaye 2021; Zhu et al. 2024), with reported enhancements of up to 14.94% in total particulate organic C (POC) and 5.56% in mineral-associated organic C fractions (Wooliver and Jagadamma 2023). However, Blanco-Canqui (2024) noted inconsistent improvements in soil physical properties between multispecies cover crop mixtures and monoculture cover crops, underscoring the necessity for site-specific assessments and long-term data to draw definitive conclusions. Topps et al. (2021) highlighted the potential of cover crop diversity to influence SOC contents within soil aggregates, particularly macro-aggregates. Cover cropping substantially increases soil CS, increasing large macroaggregate proportion by 18–22% and enhancing C fractions, particularly microaggregate-protected and mineral-associated C, compared to controls (Zhu et al. 2024). Rye cover crops, when incorporated over a nine-year period in Kanto, Japan, significantly increased SOC by 13.4 Mg ha^{-1}, underscoring their pivotal role in CS (Higashi et al. 2014). Overall, cover crops hold significant potential for enhancing soil CS and improving soil health in agricultural systems.

4.3 Conservation Agriculture

4.3.1 Introduction

Conservation agriculture (CA) is built upon three interconnected pillars, adapted to suit local conditions: minimal mechanical soil disturbance (e.g., zero or reduced tillage), permanent soil cover, and species diversification (FAO 2024). Lal (2015c) adds another crucial dimension with the inclusion of Integrated Nutrient Management (INM) as an additional pillar. CA encompasses tillage methods aimed at minimizing soil disturbance, preserving soil structure, and maintaining at least 30% soil surface coverage with crop residues, with the goal of mitigating soil erosion and conserving soil C to reduce CO_2 emissions (Fowler and Rockstrom 2001; Manhas et al. 2015; Singh et al. 2018). In this approach, traditional plowing methods involving heavy plows, such as moldboard plows, are replaced with specialized tillers designed to minimize soil disturbance (Masek et al. 2008). Minimum tillage practices present cost-effective alternatives to conventional moldboard plowing, rendering them a financially feasible choice for farmers (Sijtsma et al. 1998). There are two main types of tillage practices in CA: No-till/Zero tillage (NT) and reduced / minimized tillage, with additional variations including mulch tillage, strip tillage, vertical tillage, ridge tillage, contour tillage, rotational tillage, and subsoiling (Table 4.4). CA provides numerous benefits to soil physical, chemical, and biological properties, including improved soil structure, increased crop water productivity, reduced erosion, enhanced SOC, and boosted enzyme activity (Sharratt 1996; Uri 1998a; Navarro-Noya et al. 2013; Manhas et al. 2015; Busari et al. 2015; Li et al. 2020; Huang et al. 2021; Wen et al. 2023). Madden et al. (2008) demonstrate that adopting CA substantially decreases PM10 emissions compared to conventional tillage, primarily due to fewer tillage operations and higher soil moisture levels. However, challenges arise with reduced tillage and NT practices, including weed and pest management, stubble-borne disease control, nutrient distribution, nitrogen

Table 4.4 Conservation agriculture practices

Tillage type	Description	Area of applicability	References
No-till/Zero tillage	Involves planting crops without prior soil cultivation, with minimal or no surface disturbance and significant plant residue coverage	Applicable across diverse climates and soil types.	Soane et al. (2012)
Minimized tillage	Intermediate disturbance between conventional and no-tillage achieved by limiting ploughing with primary tillage equipment.	While transitioning from conventional to no-tillage	Busari et al. (2015)
Mulch tillage	Incorporation of organic mulch into the soil surface	Horticulture, organic farming systems	Hashimi et al. (2020)
Strip tillage	Soil is tilled only in narrow strips where crops will be planted	Precision agriculture, row-crop production	Morrison (2002)
Vertical tillage	Utilizing vertical tillage tools to break up soil without inverting it	Heavy residue areas, conservation-minded farming	Zeng and Chen (2018)
Ridge tillage	Ridges are cultivated and maintained for planting subsequent crops, minimizing soil disturbance.	Sloping terrain, areas with heavy rainfall	Thapa et al. (1999)
Contour tillage	Tillage parallel to the land's contour lines to minimize soil erosion	Hilly or sloping terrain, erosion-prone landscapes	Jia et al. (2020)
Rotational tillage	Rotating between tillage and no-tillage practices	Regions experiencing soil compaction issues	Cavalaris et al. (2023)
Subsoiling	Breaking up compacted soil layers found beneath the surface without disturbing the overall structure of the soil profile	Regions with heavy clay soils or where heavy machinery has caused soil compaction.	Lou et al. (2021)

availability, soil warming, and soil compaction (Dang et al. 2020). These challenges may necessitate increased fertilizer and pesticide applications (Uri 1998b; Abdalla et al. 2013; Hydbom et al. 2020), leading to higher N_2O emissions (Mei et al. 2018; Slayden et al. 2022). While some studies have shown improved crop yields with CA (Hou et al. 2012; Nunes et al. 2018; Huang et al. 2018), others have reported yield losses(Rieger et al. 2008; Zhang et al. 2017; Achankeng and Cornelis 2023), such as lower rice yields under no-tillage with straw mulch compared to plow tillage with straw return, attributed to declining soil nitrogen and phosphorus levels over time (Zhang et al. 2023b). Some studies have even found no significant effect on crop yield (Franchini et al. 2012). These inconclusive findings underscore the necessity for further research to comprehensively understand the intricate interactions between CA practices and crop yields.

The benefits of CA are contingent upon factors such as crop type (Martins et al. 2009), tillage duration (Díaz-Zorita and Grove 2002; Huang et al. 2015; Kan et al. 2020), soil texture (Nunes et al. 2018), and climatic conditions (Kladivko 2001), with profitability varying across regions based on weed and pest management efficacy (Uri 1999; Li et al. 2023d). CA adoption is prevalent on larger farms with clay soil predominance, with crop rotation often cited as a key factor due to crop-specific requirements (Hydbom et al. 2020). Effective adoption of CA requires site-specific technology (Huang et al. 2015; Kovács et al. 2023). Factors influencing the success of CA adoption in black soil areas of Northeast China include government subsidies, agricultural machinery adaptability, technology input, policy frameworks, tillage quality, farmer income, practical application capacity, and government outreach (Qu et al. 2021). Shen et al. (2022) identified three distinct farmer types, noting that factors such as land fragmentation and social relations play different roles in adoption behavior among economically, resource, and socially dominant farmers. Awareness among farmers regarding CA technology, proximity to agricultural markets, and the size of the active labor force within the family were found to have considerable adverse effects on the adoption of CA practices (Han et al. 2018).

4.3.2 Carbon Sequestration Potential of Conservation Agriculture

CA promotes CS by preserving and enhancing SOC levels (Lal and Kimble 1997; Lal 2004). When tillage is minimized or eliminated, the decomposition of SOM is slowed down, allowing more SOC to accumulate in the soil (Stockfisch et al. 1999). Crop residues left on the soil surface act as a source of SOM (Fig. 4.2), which gradually decomposes and contributes to SOC formation (Lehtinen et al. 2014).

Fig. 4.2 Residual corn stubble supporting soil cover in conservation agriculture (Photo taken by the author, Dr. Nancy Loria)

Transitioning from plow to no-till cultivation with proper crop residue management could significantly enhance SOC levels, potentially serving as a significant sink for atmospheric CO_2 (Duiker and Lal 1999). Additionally, reduced soil disturbance preserves soil aggregates, which provide a physical framework for stabilizing SOM and protecting it from decomposition (Yang et al. 2022a; Chen et al. 2022). No-till with residue retention emerges as a promising approach for enhancing soil C accumulation while minimizing temporal fluctuations (Qi et al. 2019). Research consistently demonstrates substantial SOC accumulation in soils under no-till or reduced tillage systems, with reported increases ranging from 25% to 73% compared to conventionally tilled soils (Zhu et al. 2022; Liu et al. 2023; Zhang et al. 2023a). CA practices, such as residue retention and reduced tillage, significantly enhance SOC sequestration, as evidenced by notable increases in SOC content under mulch ripping and tied ridging treatments compared to conventional tillage (Chivenge et al. 2007). The benefits of CS through CT extend beyond soil health to climate change mitigation by effectively removing CO_2 from the atmosphere and helping to offset GHG emissions (Allmaras et al. 2000). The potential annual CS in African agricultural soils through CA was estimated at 0.14 Pg of C, equivalent to 0.52 Pg of CO_2 (Gonzalez-Sanchez et al. 2019). CA, improved cropping systems, optimized nutrient management, and increased land cover offer substantial potential for CS in the US, with estimates suggesting a range of 30–105 million metric tons of C sequestered annually, emphasizing their significance for soil health and climate mitigation (Follett 2001). Assessing soil CS under CA practices requires accounting for various factors including soil erosion, initial C levels, and advancements in field management and crop varieties (Yang et al. 2013; Xiao et al. 2020). The adoption of CA should be customized to suit varying soil types and climates, given their impact on GHG emissions, soil quality, and CS (Page et al. 2013; Abdalla et al. 2013).

4.4 Crop Rotation

4.4.1 Introduction

Crop rotation involves strategically alternating different crop species over time to optimize land use and minimize risks associated with soil degradation, single crop failure, nutrient availability, yield variability, and high fertilizer and pesticide costs. This practice capitalizes on the unique requirements and behaviors of each crop, ensuring more resilient and sustainable agricultural systems (Bowles et al. 2020; Liu et al. 2022). Guided by four main rules, including minimum return periods between the same crop to break pest cycles, benefits or risks of immediate crop successions, within-year cycles influenced by climate and harvest timings, and overall proportions of crops to balance workload and minimize risk of crop failure (Castellazzi et al. 2008), crop rotation offers various benefits. These benefits include enhanced soil productivity, improved nutrient cycling efficiency, decreased dependence on pesticides and herbicides, and higher crop yields achieved through strategic crop alternation (Honeycutt et al. 1995; Kelley et al. 2003; Zegada-Lizarazu and Monti 2011; Franchini et al. 2012; Peralta et al. 2018; Bowles et al. 2020). Continuous monoculture of the same crop can lead to increased pest pressure and disease outbreaks (Li et al. 2022), which crop rotation disrupts by introducing different plant species that may be less susceptible to common pests and diseases or may actively suppress them through allelopathy or other mechanisms (Tesio and Ferrero 2010; Peralta et al. 2018). Over time, crop rotation significantly impacts microbial communities, enhancing soil biological composition through interactions with host plants, soil characteristics, and climatic conditions (Sun et al. 2016; Thapa et al. 2021; Singh and Kumar 2021). Additionally, diverse crop rotations can help maintain soil structure, moisture retention, and SOM content, contributing to long-term soil health and resilience (Thierfelder et al. 2013; Man et al. 2021; Iheshiulo et al. 2023). Other potential benefits include increasing SOC stocks and reducing GHG emissions (Behnke and Villamil 2019; Yang et al. 2023, 2024). Crop rotation significantly impacts microbial communities over time, enhancing soil biological composition through interactions with host plants, soil characteristics, and climatic conditions (Sun et al. 2016; Thapa et al. 2021; Singh and Kumar 2021). Additionally, diverse crop rotations can help maintain soil structure, moisture retention, and SOM content, contributing to long-term soil health and resilience (Thierfelder et al. 2013; Man et al. 2021; Iheshiulo et al. 2023). Factors influencing the benefits of crop rotation include the diversity and duration of the crop sequence, timing and frequency of rotations, soil and climatic conditions, crop selection, and management practices such as tillage, fertilization, and irrigation (Tanveer et al. 2019; Sun et al. 2023). While soil health is recognized as a primary benefit of adopting crop rotations, barriers to adoption include soil limitations, production costs, alternative crop productivity levels, water and equipment requirements, market access, and regional differences (Rosenberg et al. 2022). A well-designed crop rotation plan considers these factors to maximize benefits while minimizing potential drawbacks (Dias et al. 2015). Careful planning and coordination of agronomic activities are essential to maximize benefits and minimize potential drawbacks associated with crop rotation (Kusumastuti et al. 2016; Esteso et al. 2022).

4.4.2 Carbon Sequestration Potential of Crop Rotation

Crop rotations play a crucial role in shaping SOC dynamics and overall soil health, with significant impacts on farm profitability and sustainability. King and Blesh (2018) emphasize that increasing functional diversity within crop rotations, especially through the inclusion of cover crops and perennial crops, can significantly enhance SOC concentrations. Conversely, Rubio et al. (2021) found that crop rotations with higher proportions of annual crops may adversely affect SOC levels and associated soil properties, leading to rotation-induced SOC variations ranging from 1.2% to 2.6%. However, King et al. (2020) provides contrasting insights, indicating that while diverse crop rotations may not always lead to increased SOC concentrations or aggregate mean weight diameter (MWD), perennialized rotations incorporating crops like alfalfa demonstrate higher SOC levels. Studies, such as that by Ni et al. (2004), suggest that integrating green manure into crop rotations can effectively enhance SOM levels and microbial activity, thereby promoting sustainable agricultural practices (SAPs). Additionally, research by Zhou et al. (2023) highlights the benefits of intensifying winter crop rotations, including various winter crops, in double rice fields. This intensification not only boosts crop yields by 2.5–4.6% but also enhances soil health and sustainability, leading to a notable increase in SOC sequestration by 21% and fostering greater microbial diversity. However, contrasting findings regarding European cropland SOC dynamics, as indicated by Poyda et al. (2019), underscore the destabilizing impact of silage maize within rotations. These findings collectively highlight the importance of considering diverse crop rotation strategies, particularly those incorporating perennial phases, as a means to enhance CS and promote long-term soil health and sustainability.

Tillage practices, when combined with crop rotation, significantly influence SOC and nitrogen (N) cycling, accelerating the decomposition of residues and SOM (Zuber et al. 2017). A study on maize cultivation's impact on SOC dynamics in various tillage and wheat rotation systems found increased SOC stocks, particularly in subsoil under minimum tillage and wheat rotation, possibly due to stable C fractions and dissolved organic C (DOC) movement (Osanai et al. 2020). Improving cropping systems through rotation, cover cropping, and enhanced agronomic practices presents a promising approach to restoring SOC stocks in agricultural lands facing long-term depletion (Semmartin et al. 2023). Adoption of crop rotation, CA, and supplementation with crop residues and chemical fertilizers substantially increased SOC and N in topsoil layers, suggesting it as the most effective management strategy for enhancing soil productivity (Liu et al. 2005). Crop rotation significantly enhances soil aggregation and SOC content, leading to notable increases in macroaggregates, stability, and SOC across different aggregate sizes, typically rising by 7–8% (Zheng et al. 2023). However, its effectiveness varies depending on regional climate and edaphic factors, with additional benefits observed from practices such as no-till and straw retention. Longer-term monitoring is essential to fully capture any potential alterations in SOC content and aggregation patterns (Canalli et al. 2020). Moreover, the long-term adoption of diverse crop rotations, particularly the 4-year rotation, has been shown to significantly enhance SOC and soil aggregation compared to conventional 2-year corn-soybean rotations, indicating the potential of diversification alongside CA soil management for improving soil health and productivity (Maiga et al. 2019).

Crop rotation has emerged as an effective strategy for reducing CO_2 emissions, as highlighted by several studies (Campbell et al. 2014; Behnke et al. 2018; Li et al. 2023a). For instance, research has shown that crop rotation systems such as corn-rice (RC) and sweet sorghum-rice (RS) can significantly decrease net CO_2

equivalent emissions by 68–78% compared to rice-rice (RR) rotations. Additionally, RC and RS rotations have been found to preserve SOC stocks, thereby positively contributing to soil CS, whereas RR rotations show a decline in SOC levels (Cha-un et al. 2017). Changes in crop rotations in Finland between 2009 and 2018 have resulted in estimated annual mitigation of SOC loss by 1336 Mg C yr^{-1} in mineral soils and a reduction of CO_2 emissions from organic soils by 10,475 Mg C yr^{-1}, leading to a combined effect of 11,811 Mg C yr^{-1} (Kostensalo et al. 2024). Olesen et al. (2004) study underscores the importance of including crop rotations with cereals and catch crops to mitigate GHG emissions and promote sustainable agriculture. Furthermore, utilizing a crop rotation of legumes and non-legumes, along with reduced nitrogen fertilization, has been found to effectively decrease soil CO_2 emissions with minimal influence on soil methane emissions, as indicated by recent research findings (Sainju et al. 2021).

The rice-wheat (RW) rotation system practiced on about 24 Mha in South Asia of which 13.5 Mha in the Indo-Gangetic Plains including 10Mha in India (Mahajan and Gupta 2009). However, the conventional RW rotation in Indo-Gangetic plains has posed challenges, being resource-intensive and exacerbated by crop residue burning, leading to significant GHG emissions and environmental pollution (Kakraliya et al. 2021), along with potential indications of declining crop productivity (Fujisaka et al. 1994; Kumar et al. 2019). Conversely, CA-based RW rotation has shown promise, yielding increased crop productivity (Jat et al. 2014), reduced irrigation requirements (Devkota et al. 2015), improved soil physical properties (Kumar et al. 2018) and increased CS (Samal et al. 2017; Nandan et al. 2019; Mahala et al. 2023; Bana et al. 2023). Mungbean incorporation into the basmati RW system, within a CA framework, resulted in around 40% and 150% improvements in soil chemical and biological properties, respectively (Jat et al. 2019). Chaki et al.'s study suggests that the impact of CA on RW systems varies depending on the environment, with notable benefits observed in terms of crop yields, nitrogen use efficiency, water productivity in fine-textured soils, and significant improvements in soil properties (Chaki et al. 2021). However, further research is needed for full adoption in coarse-textured soils (Chaki et al. 2021).

4.5 Biochar

4.5.1 Introduction

Biochar is a C-enriched material derived from the thermal degradation of biomass such as agricultural residues, forestry waste, or animal manure through a process called pyrolysis. Pyrolysis involves heating organic materials in the absence of oxygen (Demirbas and Arin 2002), resulting in the decomposition of volatile compounds (López-Aguilar et al. 2022) and the formation of a stable C-rich residue that

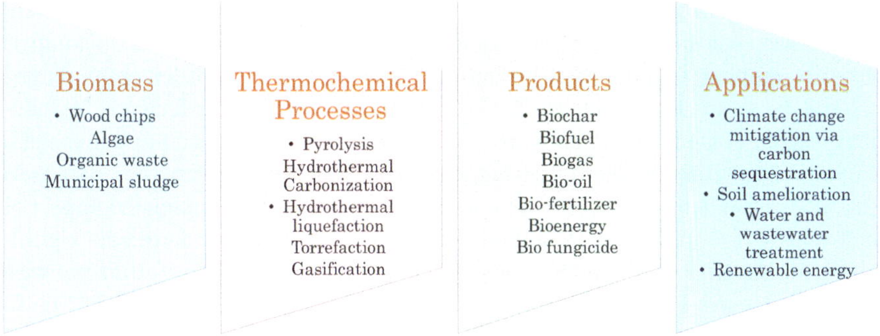

Fig. 4.3 Overview of biochar production: From biomass to applications

can persist in the soil for hundreds to thousands of years (Lobzenko et al. 2022), effectively sequestering C and mitigating climate change. The process of biochar production from biomass involves various thermochemical processes that yield different bioproducts suitable for diverse applications (Fig. 4.3). Biochar can be derived from various feedstocks, leading to the production of different types of biochar with varying properties (Kumari et al. 2023). Common feedstocks include wood chips, crop residues, manure, and even sewage sludge. Biochar possesses a complex and heterogeneous structure characterized by its porosity, surface area, and chemical composition, which are influenced by various factors including feedstock type, pyrolysis conditions, and post-treatment processes (Kavitha et al. 2018; Tomczyk et al. 2020; Singh Yadav et al. 2023). Studies underscore that feedstock selection predominantly determines biochar properties, with wood-based biochar exhibiting greater surface area and crop-based biochar showing higher nutrient retention capacities (Ippolito et al. 2020; Das et al. 2021). Additionally, variations in pyrolysis temperature notably impact biochar's physical, chemical, and structural traits, resulting in enhanced C stability and aromaticity alongside reduced yield and volatile components (Zhang et al. 2015). Interestingly, while feedstock types exert some influence, the duration of heating time shows minimal effect on biochar properties. These findings highlight the importance of tailored biochar production for environmental solutions and improved soil quality.

When added as a soil amendment it provides a range of benefits that include CS, soil amelioration, improved soil physicochemical properties, enhanced water retention, increased microbial activity, and higher crop yields (Hussain et al. 2017; Kavitha et al. 2018; Pankaj 2020; Murtaza et al. 2021; Campion et al. 2023; Bo et al. 2023; Wu et al. 2024). The properties of biochar vary depending on feedstock, temperature, pressure, and oxygen concentration during the pyrolysis process. Its effectiveness in enhancing crop yields varies across soil types, with greater benefits observed in soils with low CEC and SOC, suggesting targeted application strategies for climate change mitigation (Crane-Droesch et al. 2013). However, biochar may not consistently improve water retention, saturated hydraulic conductivity, and hydrophobicity in sandy soils (Wiersma et al. 2020). In tropical soils, high-nutrient biochar has been shown to enhance yields more significantly than in fertile temperate soils, emphasizing the need for tailored biochar management for optimal agricultural benefits (Jeffery et al. 2017). Additionally, biochar effectively reduces heavy metal content in contaminated soil, with its efficacy depending on specific properties and composition (Gholizadeh and Hu 2021). It also serves as a sustainable alternative to traditional soil amendments like lime for mitigating the impact of acid rains (Li et al. 2023c). Combining biochar with inorganic fertilizers significantly boosts crop yields, underscoring its efficacy comparable to fertilizers, with long-term studies revealing additional benefits beyond liming (Ye et al. 2020). To maximize the benefits of biochar in soil improvement, researchers have explored various modification techniques aimed at enhancing its effectiveness in different soil applications (Cheng et al. 2017; Wang and Wang 2019; Shafawi et al. 2021). These modifications involve altering the physical, chemical, and biological properties of biochar to better suit specific soil types, environmental conditions, and desired outcomes (Hafeez et al. 2022; Bao et al. 2022; Tian et al. 2023). For instance, Fe-modified biochar has been shown to improve soil structure stability and SOC content, thereby enhancing C stability in saline-alkaline soils (Chen et al. 2024). Thorough exploration of the interactions, modifications, and binding mechanisms of biochar with soil is essential for maximizing its efficacy and applicability as a sustainable soil amendment.

Despite its potential benefits, the widespread adoption of biochar faces several challenges that need to be addressed (Campion et al. 2023; Li and Tasnady 2023). Firstly, the cost associated with biochar production and application can be prohibitive, especially for small-scale farmers or those operating in resource-limited regions (Galinato et al. 2011; Campbell et al. 2018). Additionally, there is a lack of knowledge and awareness among farmers regarding biochar and its potential benefits, which impedes its uptake (Niemmanee et al. 2019; Sachdeva et al. 2023). Regulatory barriers related to biochar production, certification, and application also vary across regions, further complicating its adoption (Shackley et al. 2011; Verheijen et al. 2012; Gao et al. 2020). Moreover, the availability of suitable feedstocks for biochar production varies geographically, posing challenges for consistent supply (Zilberman et al. 2023). Furthermore, research gaps exist concerning the long-term effects of biochar on soil health, crop productivity, and environmental sustainability, which hinders informed decision-making and adoption efforts (Agegnehu et al. 2017; Wei et al. 2018; Nguyen et al. 2023). Assuming biochar as a panacea or one-size-fits-all without a strong science may aggravate controversies and dilemmas (Lal 2015d). Addressing these challenges through targeted policies, education initiatives, and research endeavors is crucial to realizing the full potential of biochar.

4.5.2 Carbon Sequestration Potential of Biochar

The CS potential of biochar relies on its ability to remain stable in soil over time and its impact on the decomposition of native SOC (Weng et al. 2017). Essentially, biochar can store C in the soil by resisting decomposition itself while also influencing the decomposition rates of existing SOC through a process known as priming (Wang et al. 2016). This interaction between biochar stability and its priming effect determines its overall effectiveness as a CS agent. Soil properties such as texture, pH, SOM content, and microbial activity can affect the adsorption, retention, and degradation of biochar C (Ameloot et al. 2015; Yang et al. 2022b; Li et al. 2023b; Wei et al. 2023). Soil pH significantly influences biochar's CS potential, with acidic soils exhibiting increased CO_2 release and accelerated degradation of both native SOC and biochar (Sheng et al. 2016). Additionally, the physicochemical properties of biochar, such as surface area, pore structure, surface chemistry, and degree of carbonization, play a crucial role in determining its stability in soil (Aller 2016). For example, biochar derived from woody plant species, known for its higher surface area and pore volume, exhibits enhanced adsorption capacity for organic compounds and nutrients, enhancing its stability in soil (Leng et al. 2021; Ceriani et al. 2024). Moreover, biochar produced at higher pyrolysis temperatures generally exhibit greater thermal stability and resistance to microbial degradation, leading to prolonged persistence of biochar C in soil (Zhang et al. 2020). Higher pyrolysis temperatures result in lower mineralization of biochar C, indicating greater stability and CS potential, particularly in clayey soils (Yang et al. 2022b). Understanding the complex interactions between soil properties and biochar characteristics is essential for predicting the long-term fate of biochar C in soil and optimizing its potential for CS and soil improvement.

Biochar can enhance SOC stocks, contributing to long-term C storage and climate change mitigation efforts (Lorenz and Lal 2014b; Criscuoli et al. 2021). Acting as a protective barrier against decomposition, biochar improves soil aggregation, increases nutrient and water retention, influences soil microbial activity, and stabilizes labile C, all of which contribute to CS (Li and Tasnady 2023). Additionally, biochar has resulted in reductions of methane and nitrous oxide emissions by 25 and 20 Mt. CO_2-eq yr^{-1}, respectively, while also increasing crop yields by 19% (Bo et al. 2023). Moreover, biochar application over different durations has shown promising results in increasing SOC by 37–76% across various particle-size fractions (El-Naggar et al. 2018). Investigations into the effects of biochar on different C fractions, including DOC, POC, and microbial biomass C (MBC), have underscored its significant role in CS (Demisie et al. 2014; Sandhu et al. 2017; Shi et al. 2021; Das et al. 2023). For instance, Yang et al. (2018) found that biochar addition over 3 years notably increased SOC, POC, easily oxidizable C, and light fraction organic C, except for DOC which decreased post-biochar addition while Bakshi et al. (2018) emphasized the importance of assessing the stability of biochar in soil based on the C:N ratios of its labile fractions. Additionally, straw biochar application in a 3-year field experiment showcased its potential to enhance micro-aggregate C, improve soil nutrient availability, and facilitate C storage and nutrient retention in vegetable rotations, with an optimal dosage of 20 Mg ha^{-1} (Zhang et al. 2022a, b).

Furthermore, biochar's positive effects on soil inorganic C (SIC) and its composition further strengthen its role in CS (Dong et al. 2019; Shi et al. 2021). The production and soil storage of biochar offer a promising solution for mitigating climate change, with the potential to significantly reduce annual emissions by up to 1.8 Pg CO_2-eq yr^{-1} and total emissions by 130 Pg CO_2-eq over a century without compromising soil health (Woolf et al. 2010). However, the impact of biochar on GHG emissions may vary across its life cycle stages and feedstock types, as highlighted by Xia et al. (2024). While biochar application effectively curtails soil CO_2 emissions and enhances SOM levels and aggregate stability (Hua et al. 2014), its effects on GHG emissions during application stages warrant further investigation. Overall, biochar presents a multifaceted approach to CS, with significant potential for mitigating climate change and improving soil health.

4.6 Conclusion

C farming techniques offer promising strategies for enhancing CS in agricultural systems. Agroforestry, for instance, involves integrating trees and crops, providing multiple environmental benefits including C storage in both above and belowground biomass. This technique requires careful planning and management to optimize interactions between trees and crops while maximizing CS potential. Similarly, cover cropping entails planting of non-harvested crops during fallow periods, contributing to soil CS through the addition of SOM and the prevention of erosion. While cover cropping provides short-term C inputs and erosion control benefits, it may require additional considerations for water and nutrient management to ensure its effectiveness. CA practices, such as no-till or reduced tillage, help preserve SOC by minimizing soil disturbance and reducing decomposition rates. However, adjustments in weed and pest management may be necessary to fully realize the benefits of CA. Crop rotation, a practice involving the sequential planting of different crops in the same field, enhances CS by diversifying root systems and organic inputs. Careful selection and rotation of crops are essential to maximize the benefits of crop rotation for C storage. Furthermore, biochar application has shown significant potential for long-term C storage and soil fertility. Produced through biomass pyrolysis, biochar can effectively sequester C in soil while improving soil structure and nutrient retention. However, careful consideration of feedstock sources, pyrolysis conditions, application rates, and long-term effects on soil health is necessary to ensure the sustainable use of biochar. Given that each C farming technique has distinct strengths and considerations, integrating a combination of these practices into agricultural systems presents opportunities to enhance CS, improve soil health, and increase resilience to climate variability. Continued research and investment in these practices are essential to realizing their full potential in mitigating climate change and promoting sustainable agriculture for future generations.

References

Abdalla M, Osborne B, Lanigan G et al (2013) Conservation tillage systems: a review of its consequences for greenhouse gas emissions. Soil Use Manag 29:199–209. https://doi.org/10.1111/sum.12030

Abdalla M, Hastings A, Cheng K et al (2019) A critical review of the impacts of cover crops on nitrogen leaching, net greenhouse gas balance and crop productivity. Glob Chang Biol 25:2530–2543. https://doi.org/10.1111/gcb.14644

Achankeng E, Cornelis W (2023) Conservation tillage effects on European crop yields: a meta-analysis. Field Crop Res 298:108967. https://doi.org/10.1016/j.fcr.2023.108967

Acharya KP (2006) Linking trees on farms with biodiversity conservation in subsistence farming systems in Nepal. Biodivers Conserv 15:631–646. https://doi.org/10.1007/s10531-005-2091-7

Agegnehu G, Srivastava AK, Bird MI (2017) The role of biochar and biochar-compost in improving soil quality and crop performance: a review. Appl Soil Ecol 119:156–170. https://doi.org/10.1016/j.apsoil.2017.06.008

Ahmed N, Galaup B, Desplanques J et al (2022) Ecosystem services provided by cover crops and biofumigation in sunflower cultivation. Agronomy 12:120. https://doi.org/10.3390/agronomy12010120

Ajit DSK, Handa AK et al (2017) Estimating carbon sequestration potential of existing agroforestry systems in India. Agrofor Syst 91:1101–1118. https://doi.org/10.1007/s10457-016-9986-z

Akinyele AO, Donald-Amaeshi U (2021) Leaf litter decomposition and nutrient release of three selected agroforestry tree species. Agrofor Syst 95:559–570. https://doi.org/10.1007/s10457-021-00606-w

Aller MF (2016) Biochar properties: transport, fate, and impact. Crit Rev Environ Sci Technol 46:1183–1296. https://doi.org/10.1080/10643389.2016.1212368

Allmaras R, Schomberg H, Douglas C, Dao T (2000) Soil organic carbon sequestration potential of adopting conservation tillage in U.S. croplands. J Soil Water Conserv 55:365–373

Amare D, Wondie M, Mekuria W, Darr D (2019) Agroforestry of smallholder farmers in Ethiopia: practices and benefits. Small-scale Forest 18:39–56. https://doi.org/10.1007/s11842-018-9405-6

Ameloot N, Sleutel S, Das KC et al (2015) Biochar amendment to soils with contrasting organic matter level: effects on N mineralization and biological soil properties. GCB Bioenergy 7:135–144. https://doi.org/10.1111/gcbb.12119

Amsili JP, Kaye JP (2021) Root traits of cover crops and carbon inputs in an organic grain rotation. Renew Agric Food Syst 36:182–191. https://doi.org/10.1017/S1742170520000216

Antle JM, Stoorvogel JJ, Valdivia RO (2007) Assessing the economic impacts of agricultural carbon sequestration: terraces and agroforestry in the Peruvian Andes. Agric Ecosyst Environ 122:435–445. https://doi.org/10.1016/j.agee.2007.02.003

Bakshi S, Banik C, Laird DA (2018) Quantification and characterization of chemically-and thermally-labile and recalcitrant biochar fractions. Chemosphere 194:247–255. https://doi.org/10.1016/j.chemosphere.2017.11.151

Ballesteros-Possu W, Brandle J, Schoeneberger M (2017) Potential of windbreak trees to reduce carbon emissions by agricultural operations in the US. Forests 8:138. https://doi.org/10.3390/f8050138

Bana RS, Bamboriya SD, Kumar V et al (2023) Identifying optimum residue levels for stable crop and water productivity and carbon sequestration under a conservation agriculture based rice-wheat system. Soil Tillage Res 232:105745. https://doi.org/10.1016/j.still.2023.105745

Bao Z, Shi C, Tu W et al (2022) Recent developments in modification of biochar and its application in soil pollution control and ecoregulation. Environ Pollut 313:120184. https://doi.org/10.1016/j.envpol.2022.120184

Behnke GD, Villamil MB (2019) Cover crop rotations affect greenhouse gas emissions and crop production in Illinois, USA. Field Crop Res 241:107580. https://doi.org/10.1016/j.fcr.2019.107580

Behnke GD, Zuber SM, Pittelkow CM et al (2018) Long-term crop rotation and tillage effects on soil greenhouse gas emissions and crop production in Illinois, USA. Agric Ecosyst Environ 261:62–70. https://doi.org/10.1016/j.agee.2018.03.007

Besar NA, Suardi H, Phua M-H et al (2020) Carbon stock and sequestration potential of an agroforestry system in Sabah. Malaysia Forests 11:210. https://doi.org/10.3390/f11020210

Bieluczyk W, de Piccolo MC, Pereira MG et al (2021) Eucalyptus tree influence on spatial and temporal dynamics of fine-root growth in an integrated crop-livestock-forestry system in southeastern Brazil. Rhizosphere 19:100415. https://doi.org/10.1016/j.rhisph.2021.100415

Blanco-Canqui H (2022) Cover crops and carbon sequestration: lessons from U.S. studies. Soil Sci Soc Am J 86:501–519. https://doi.org/10.1002/saj2.20378

Blanco-Canqui H (2024) Do cover crop mixtures improve soil physical health more than monocultures? Plant Soil 495:99–112. https://doi.org/10.1007/s11104-023-06086-4

Blanco-Canqui H, Shaver TM, Lindquist JL et al (2015) Cover crops and ecosystem services: insights from studies in temperate soils. Agron J 107:2449–2474. https://doi.org/10.2134/agronj15.0086

Bo X, Zhang Z, Wang J et al (2023) Benefits and limitations of biochar for climate-smart agriculture: a review and case study from China. Biochar 5:77. https://doi.org/10.1007/s42773-023-00279-x

Böhm C, Kanzler M, Freese D (2014) Wind speed reductions as influenced by woody hedgerows grown for biomass in short rotation alley cropping systems in Germany. Agrofor Syst 88:579–591. https://doi.org/10.1007/s10457-014-9700-y

Bowles TM, Mooshammer M, Socolar Y et al (2020) Long-term evidence shows that crop-rotation diversification increases agricultural resilience to adverse growing conditions in North America. One Earth 2:284–293. https://doi.org/10.1016/j.oneear.2020.02.007

Brainard DC, Bellinder RR, Kumar V (2011) Grass–Legume mixtures and soil fertility affect cover crop performance and weed seed production. Weed Technol 25:473–479. https://doi.org/10.1614/WT-D-10-00134.1

Büchi L, Wendling M, Amossé C et al (2020) Cover crops to secure weed control strategies in a maize crop with reduced tillage. Field Crop Res 247:107583. https://doi.org/10.1016/j.fcr.2019.107583

Busari MA, Kukal SS, Kaur A et al (2015) Conservation tillage impacts on soil, crop and the environment. Int Soil Water Conserv Res 3:119–129. https://doi.org/10.1016/j.iswcr.2015.05.002

Cadotte MW (2013) Experimental evidence that evolutionarily diverse assemblages result in higher productivity. Proc Natl Acad Sci 110:8996–9000

Campbell B, Chen L, Dygert C, Dick W (2014) Tillage and crop rotation impacts on greenhouse gas fluxes from soil at two long-term agronomic experimental sites in Ohio. J Soil Water Conserv 69:543–552. https://doi.org/10.2489/jswc.69.6.543

Campbell RM, Anderson NM, Daugaard DE, Naughton HT (2018) Financial viability of biofuel and biochar production from forest biomass in the face of market price volatility and uncertainty. Appl Energy 230:330–343. https://doi.org/10.1016/j.apenergy.2018.08.085

Campion L, Bekchanova M, Malina R, Kuppens T (2023) The costs and benefits of biochar production and use: a systematic review. J Clean Prod 408:137138. https://doi.org/10.1016/j.jclepro.2023.137138

dos Canalli LBS, dos Santos JB, Benassi DA et al (2020) Soil carbon and structural quality in crop rotations under no-tillage system. Braz Arch Biol Technol 63. https://doi.org/10.1590/1678-4324-solo-2020190603

Castellazzi MS, Wood GA, Burgess PJ et al (2008) A systematic representation of crop rotations. Agric Syst 97:26–33. https://doi.org/10.1016/j.agsy.2007.10.006

Castillo MS, Tiezzi F, Franzluebbers AJ (2020) Tree species effects on understory forage productivity and microclimate in a silvopasture of the Southeastern USA. Agric Ecosyst Environ 295:106917. https://doi.org/10.1016/j.agee.2020.106917

Cavalaris C, Gemtos T, Karamoutis C (2023) Rotational tillage practices to deal with soil compaction in carbon farming. Soil Syst 7:90. https://doi.org/10.3390/soilsystems7040090

Celentano D, Rousseau GX, Paixão LS et al (2020) Carbon sequestration and nutrient cycling in agroforestry systems on degraded soils of Eastern Amazon, Brazil. Agrofor Syst 94:1781–1792. https://doi.org/10.1007/s10457-020-00496-4

Ceriani A, Dalle Fratte M, Agosto G et al (2024) Woody and herbaceous invasive alien plant species-derived biochars are potentially optimal for soil amendment, soil remediation, and carbon storage. GCB Bioenergy 16. https://doi.org/10.1111/gcbb.13117

Chaiya L, Gavinlertvatana P, Teaumroong N et al (2021) Enhancing teak (Tectona grandis) seedling growth by rhizosphere microbes: a sustainable way to optimize agroforestry. Microorganisms 9:1990. https://doi.org/10.3390/microorganisms9091990

Chaki AK, Gaydon DS, Dalal RC et al (2021) Conservation agriculture enhances the rice-wheat system of the Eastern Gangetic Plains in some environments, but not in others. Field Crop Res 265:108109. https://doi.org/10.1016/j.fcr.2021.108109

Chapagain T, Lee EA, Raizada MN (2020) The potential of multi-species mixtures to diversify cover crop benefits. Sustainability 12:2058. https://doi.org/10.3390/su12052058

Cha-un N, Chidthaisong A, Yagi K et al (2017) Greenhouse gas emissions, soil carbon sequestration and crop yields in a rain-fed rice field with crop rotation management. Agric Ecosyst Environ 237:109–120. https://doi.org/10.1016/j.agee.2016.12.025

Chavan SB, Dhillon RS, Sirohi C et al (2023) Carbon sequestration potential of commercial agroforestry systems in Indo-Gangetic plains of India: poplar and eucalyptus-based agroforestry systems. Forests 14:559. https://doi.org/10.3390/f14030559

Chen S, Cao Y, Zhang T et al (2022) Improvement of soil aggregate-associated carbon sequestration capacity after 14 years of conservation tillage. Exp Agric 58:e55. https://doi.org/10.1017/S0014479722000370

Chen M, Wu L, Ding X et al (2024) Fe-modified biochar improved the stability of soil aggregates and organic carbon: evidence from enzymatic activity and microbial composition. Land Degrad Dev 35:732–743. https://doi.org/10.1002/ldr.4948

Cheng B-H, Zeng RJ, Jiang H (2017) Recent developments of post-modification of biochar for electrochemical energy storage. Bioresour Technol 246:224–233. https://doi.org/10.1016/j.biortech.2017.07.060

Chivenge P, Murwira H, Giller K et al (2007) Long-term impact of reduced tillage and residue management on soil carbon stabilization: implications for conservation agriculture on contrasting soils. Soil Tillage Res 94:328–337. https://doi.org/10.1016/j.still.2006.08.006

Clay L, Perkins K, Motallebi M et al (2020) The perceived benefits, challenges, and environmental effects of cover crop implementation in South Carolina. Agriculture 10:372. https://doi.org/10.3390/agriculture10090372

Couëdel A, Alletto L, Justes É (2018) Crucifer-legume cover crop mixtures provide effective sulphate catch crop and sulphur green manure services. Plant Soil 426:61–76. https://doi.org/10.1007/s11104-018-3615-8

Crane-Droesch A, Abiven S, Jeffery S, Torn MS (2013) Heterogeneous global crop yield response to biochar: a meta-regression analysis. Environ Res Lett 8:044049. https://doi.org/10.1088/1748-9326/8/4/044049

Criscuoli I, Ventura M, Wiedner K et al (2021) Stability of woodchips biochar and impact on soil carbon stocks: results from a two-year field experiment. Forests 12:1350. https://doi.org/10.3390/f12101350

Dang YP, Page KL, Dalal RC, Menzies NW (2020) No-till farming systems for sustainable agriculture: an overview. In: No-till farming systems for sustainable agriculture. Springer International Publishing, Cham, pp 3–20

Das SK, Ghosh GK, Avasthe RK, Sinha K (2021) Compositional heterogeneity of different biochar: effect of pyrolysis temperature and feedstocks. J Environ Manag 278:111501. https://doi.org/10.1016/j.jenvman.2020.111501

Das SK, Choudhury BU, Hazarika S et al (2023) Long-term effect of organic fertilizer and biochar on soil carbon fractions and sequestration in maize-black gram system. Biomass Convers Biorefinery. https://doi.org/10.1007/s13399-023-04165-1

De Notaris C, Rasmussen J, Sørensen P et al (2019) Manipulating cover crop growth by adjusting sowing time and cereal inter-row spacing to enhance residual nitrogen effects. Field Crop Res 234:15–25. https://doi.org/10.1016/j.fcr.2019.02.008

Delgado JA, Gantzer CJ (2015) The 4Rs for cover crops and other advances in cover crop management for environmental quality. J Soil Water Conserv 70:142A–145A. https://doi.org/10.2489/jswc.70.6.142A

Demirbas A, Arin G (2002) An overview of biomass pyrolysis. Energy Sources 24:471–482. https://doi.org/10.1080/00908310252889979

Demisie W, Liu Z, Zhang M (2014) Effect of biochar on carbon fractions and enzyme activity of red soil. Catena (Amst) 121:214–221. https://doi.org/10.1016/j.catena.2014.05.020

Devkota KP, Hoogenboom G, Boote KJ et al (2015) Simulating the impact of water saving irrigation and conservation agriculture practices for rice–wheat systems in the irrigated semi-arid drylands of Central Asia. Agric For Meteorol 214–215:266–280. https://doi.org/10.1016/j.agrformet.2015.08.264

Dhillon GS, Van Rees KCJ (2017) Soil organic carbon sequestration by shelterbelt agroforestry systems in Saskatchewan. Can J Soil Sci 93:1–16. https://doi.org/10.1139/cjss-2016-0094

Dias T, Dukes A, Antunes PM (2015) Accounting for soil biotic effects on soil health and crop productivity in the design of crop rotations. J Sci Food Agric 95:447–454. https://doi.org/10.1002/jsfa.6565

Diaz-Zorita M, Grove JH (2002) Duration of tillage management affects carbon and phosphorus stratification in phosphatic Paleudalfs. Soil Tillage Res 66:165–174. https://doi.org/10.1016/S0167-1987(02)00024-7

Dobo B, Asefa F, Asfaw Z (2018) Diversity and abundance of arbuscular mycorrhizal fungi under different plant and soil properties in Sidama, Southern Ethiopia. Agrofor Syst 92:91–101. https://doi.org/10.1007/s10457-016-0017-x

Donaghy P, Bray S, Gowen R et al (2010) The bioeconomic potential for agroforestry in Australia's northern grazing systems. Small-scale Forestry 9:463–484. https://doi.org/10.1007/s11842-010-9126-y

Dong X, Singh BP, Li G et al (2019) Biochar increased field soil inorganic carbon content five years after application. Soil Tillage Res 186:36–41. https://doi.org/10.1016/j.still.2018.09.013

Duiker SW, Lal R (1999) Crop residue and tillage effects on carbon sequestration in a Luvisol in central Ohio. Soil Tillage Res 52:73–81. https://doi.org/10.1016/S0167-1987(99)00059-8

Elevitch C, Mazaroli D, Ragone D (2018) Agroforestry standards for regenerative agriculture. Sustainability 10:3337. https://doi.org/10.3390/su10093337

El-Naggar A, Awad YM, Tang X et al (2018) Biochar influences soil carbon pools and facilitates interactions with soil: a field investigation. Land Degrad Dev 29:2162–2171. https://doi.org/10.1002/ldr.2896

Esteso A, Alemany M, Ortiz Á, Iannacone R (2022) Crop planting and harvesting planning: conceptual framework and sustainable multi-objective optimization for plants with variable molecule concentrations and minimum time between harvests. Appl Math Model 112:136–155. https://doi.org/10.1016/j.apm.2022.07.023

Fahad S, Chavan SB, Chichaghare AR et al (2022) Agroforestry systems for soil health improvement and maintenance. Sustain For 14:14877. https://doi.org/10.3390/su142214877

Fan F, van der Werf W, Makowski D et al (2021) Cover crops promote primary crop yield in China: a meta-regression of factors affecting yield gain. Field Crop Res 271:108237. https://doi.org/10.1016/j.fcr.2021.108237

FAO (2024) Conservation agriculture. Food and Agriculture Organization of the United Nations. https://www.fao.org/conservation-agriculture/en/. Accessed 23 May 2024

Fernández-Aparicio M, Amri M, Kharrat M, Rubiales D (2010) Intercropping reduces Mycosphaerella pinodes severity and delays upward progress on the pea plant. Crop Prot 29:744–750. https://doi.org/10.1016/j.cropro.2010.02.013

Follett RF (2001) Soil management concepts and carbon sequestration in cropland soils. Soil Tillage Res 61:77–92. https://doi.org/10.1016/S0167-1987(01)00180-5

Fowler R, Rockstrom J (2001) Conservation tillage for sustainable agriculture. Soil Tillage Res 61:93–108. https://doi.org/10.1016/S0167-1987(01)00181-7

Franchini JC, Debiasi H, Balbinot Junior AA et al (2012) Evolution of crop yields in different tillage and cropping systems over two decades in southern Brazil. Field Crop Res 137:178–185. https://doi.org/10.1016/j.fcr.2012.09.003

Franco JG, Gramig GG, Beamer KP, Hendrickson JR (2021) Cover crop mixtures enhance stability but not productivity in a semi-arid climate. Agron J 113:2664–2680. https://doi.org/10.1002/agj2.20695

Fujisaka S, Harrington L, Hobbs P (1994) Rice-Wheat in South Asia: systems and long-term priorities established through diagnostic research. Agric Syst 46:169–187. https://doi.org/10.1016/0308-521X(94)90096-X

Galinato SP, Yoder JK, Granatstein D (2011) The economic value of biochar in crop production and carbon sequestration. Energy Policy 39:6344–6350. https://doi.org/10.1016/j.enpol.2011.07.035

Gao Y, Li T, Fu Q et al (2020) Biochar application for the improvement of water-soil environments and carbon emissions under freeze-thaw conditions: an in-situ field trial. Sci Total Environ 723:138007. https://doi.org/10.1016/j.scitotenv.2020.138007

Gebhard CA, Büchi L, Liebisch F et al (2013) Screening of legumes as cover crops: nitrogen and weeds. Agrarforsch Schweiz 4:384–393

Gebru BM, Wang SW, Kim SJ, Lee W-K (2019) Socio-ecological niche and factors affecting agroforestry practice adoption in different agroecologies of Southern Tigray, Ethiopia. Sustainability 11:3729. https://doi.org/10.3390/su11133729

Ghale B, Mitra E, Sodhi HS et al (2022) Carbon sequestration potential of agroforestry systems and its potential in climate change mitigation. Water Air Soil Pollut 233:228. https://doi.org/10.1007/s11270-022-05689-4

Ghidoli M, Pesenti M, Colombo F et al (2023) Camelina sativa (L.) crantz as a promising cover crop species with allelopathic potential. Agronomy 13:2187. https://doi.org/10.3390/agronomy13082187

Ghimire B, Ghimire R, VanLeeuwen D, Mesbah A (2017) Cover crop residue amount and quality effects on soil organic carbon mineralization. Sustain For 9:2316. https://doi.org/10.3390/su9122316

Gholizadeh M, Hu X (2021) Removal of heavy metals from soil with biochar composite: a critical review of the mechanism. J Environ Chem Eng 9:105830. https://doi.org/10.1016/j.jece.2021.105830

Golden LA, Hubbard ML, Utych S et al (2023) Benefits, barriers, and use of cover crops in the western United States: regional survey results. J Soil Water Conserv 78:260–271. https://doi.org/10.2489/jswc.2023.00170

Gonzalez-Sanchez EJ, Veroz-Gonzalez O, Conway G et al (2019) Meta-analysis on carbon sequestration through conservation agriculture in Africa. Soil Tillage Res 190:22–30. https://doi.org/10.1016/j.still.2019.02.020

Hafeez A, Pan T, Tian J, Cai K (2022) Modified biochars and their effects on soil quality: a review. Environments 9:60. https://doi.org/10.3390/environments9050060

Han Q, Siddique K, Li F (2018) Adoption of conservation tillage on the semi-arid loess plateau of Northwest China. Sustain For 10:2621. https://doi.org/10.3390/su10082621

Harbach CJ, Wlezien E, Tylka GL (2021) A mechanistic approach to assessing the potential for cover crops to serve as trap crops for the Soybean Cyst Nematode. Plant Dis 105:1136–1142. https://doi.org/10.1094/PDIS-05-20-0964-RE

Hartwig NL, Ammon HU (2002) Cover crops and living mulches. Weed Sci 50:688–699

Hashimi R, Kaneda S, Kaneko N et al (2020) Aggregate distribution and substrate-induced respiration under different tillage and mulching management systems in organic farming. Soil Sci Plant Nutr 66:878–888. https://doi.org/10.1080/00380768.2020.1824534

Hemida M, Vityi A, ZeinabM H (2023) Socio-economic traits and constraints associated with smallholder farmers in Taungya agroforestry program in Sudan. Agrofor Syst 97:1169–1184. https://doi.org/10.1007/s10457-023-00855-x

Higashi T, Yunghui M, Komatsuzaki M et al (2014) Tillage and cover crop species affect soil organic carbon in Andosol, Kanto, Japan. Soil Tillage Res 138:64–72. https://doi.org/10.1016/j.still.2013.12.010

Himmelbauer ML, Vateva V, Lozanova L et al (2013) Site effects on root characteristics and soil protection capability of two cover crops grown in South Bulgaria. J Hydrol Hydromech 61:30–38. https://doi.org/10.2478/johh-2013-0005

Holderieath J, Valdivia C, Godsey L, Barbieri C (2012) The potential for carbon offset trading to provide added incentive to adopt silvopasture and alley cropping in Missouri. Agrofor Syst 86:345–353. https://doi.org/10.1007/s10457-012-9543-3

Honeycutt CW, Clapham WM, Leach SS (1995) Influence of crop rotation on selected chemical and physical soil properties in potato cropping systems. Am Potato J 72:721–735. https://doi.org/10.1007/BF02849157

Hou XQ, Li R, Jia ZK et al (2012) Effects of rotational tillage practices on soil structure, organic carbon concentration and crop yields in semi-arid areas of northwest China. Soil Use Manag 28:551–558. https://doi.org/10.1111/j.1475-2743.2012.00429.x

Hua L, Lu Z, Ma H, Jin S (2014) Effect of biochar on carbon dioxide release, organic carbon accumulation, and aggregation of soil. Environ Prog Sustain Energy 33:941–946. https://doi.org/10.1002/ep.11867

Huang M, Zhou X, Cao F et al (2015) No-tillage effect on rice yield in China: a meta-analysis. Field Crop Res 183:126–137. https://doi.org/10.1016/j.fcr.2015.07.022

Huang Y, Ren W, Wang L et al (2018) Greenhouse gas emissions and crop yield in no-tillage systems: a meta-analysis. Agric Ecosyst Environ 268:144–153. https://doi.org/10.1016/j.agee.2018.09.002

Huang Y, Tao B, Xiaochen Z et al (2021) Conservation tillage increases corn and soybean water productivity across the Ohio River Basin. Agric Water Manag 254:106962. https://doi.org/10.1016/j.agwat.2021.106962

Hussain M, Farooq M, Nawaz A et al (2017) Biochar for crop production: potential benefits and risks. J Soils Sediments 17:685–716. https://doi.org/10.1007/s11368-016-1360-2

Hydbom S, Alkan Olsson J, Olsson PA (2020) The use of conservation tillage in an agro-intensive region: results from a survey of farmers in Scania, Sweden. Renew Agric Food Syst 35:59–68. https://doi.org/10.1017/S174217051800025X

Iheshiulo EM-A, Larney FJ, Hernandez-Ramirez G et al (2023) Do diversified crop rotations influence soil physical health? A meta-analysis. Soil Tillage Res 233:105781. https://doi.org/10.1016/j.still.2023.105781

Imo M (2009) Interactions amongst trees and crops in taungya systems of western Kenya. Agrofor Syst 76:265–273. https://doi.org/10.1007/s10457-008-9164-z

IPCC (2000) IPCC Special Report: land use, land-use change, and Forestry. Cambridge University Press

Ippolito JA, Cui L, Kammann C et al (2020) Feedstock choice, pyrolysis temperature and type influence biochar characteristics: a comprehensive meta-data analysis review. Biochar 2:421–438. https://doi.org/10.1007/s42773-020-00067-x

Iqbal R, Raza MAS, Valipour M et al (2020) Potential agricultural and environmental benefits of mulches—a review. Bull Natl Res Cent 44:75. https://doi.org/10.1186/s42269-020-00290-3

Jat RK, Sapkota TB, Singh RG et al (2014) Seven years of conservation agriculture in a rice–wheat rotation of Eastern Gangetic Plains of South Asia: yield trends and economic profitability. Field Crop Res 164:199–210. https://doi.org/10.1016/j.fcr.2014.04.015

Jat HS, Kumar P, Sutaliya JM et al (2019) Conservation agriculture based sustainable intensification of basmati rice-wheat system in North-West India. Arch Agron Soil Sci 65:1370–1386. https://doi.org/10.1080/03650340.2019.1566708

Jeffery S, Abalos D, Prodana M et al (2017) Biochar boosts tropical but not temperate crop yields. Environ Res Lett 12:053001. https://doi.org/10.1088/1748-9326/aa67bd

Jia L, Zhao W, Zhai R et al (2020) Quantifying the effects of contour tillage in controlling water erosion in China: a meta-analysis. Catena (Amst) 195:104829. https://doi.org/10.1016/j.catena.2020.104829

Jian J, Du X, Reiter MS, Stewart RD (2020) A meta-analysis of global cropland soil carbon changes due to cover cropping. Soil Biol Biochem 143:107735. https://doi.org/10.1016/j.soilbio.2020.107735

Jinger D, Kumar R, Kakade V et al (2022) Agroforestry for controlling soil erosion and enhancing system productivity in ravine lands of Western India under climate change scenario. Environ Monit Assess 194:267. https://doi.org/10.1007/s10661-022-09910-z

Jose S (2009) Agroforestry for ecosystem services and environmental benefits: an overview. Agrofor Syst 76:1–10. https://doi.org/10.1007/s10457-009-9229-7

Joslin AH, Vasconcelos SS, de Assis OF et al (2019) A slash-and-mulch improved-fallow agroforestry system: growth and nutrient budgets over two rotations. Forests 10:1125. https://doi.org/10.3390/f10121125

Kabelka D, Kincl D, Janeček M et al (2019) Reduction in soil organic matter loss caused by water erosion in inter-rows of hop gardens. Soil Water Res 14:172–182. https://doi.org/10.17221/135/2018-SWR

Kader MA, Singha A, Begum MA et al (2019) Mulching as water-saving technique in dryland agriculture: review article. Bull Natl Res Cent 43:147. https://doi.org/10.1186/s42269-019-0186-7

Kahle P, Baum C, Boelcke B (2005) Effect of afforestation on soil properties and mycorrhizal formation. Pedosphere 15:754–760

Kakraliya SK, Jat HS, Sapkota TB et al (2021) Effect of climate-smart agriculture practices on climate change adaptation, greenhouse gas mitigation and economic efficiency of rice-wheat system in India. Agriculture 11:1269. https://doi.org/10.3390/agriculture11121269

Kallenbach RL, Kerley MS, Bishop-Hurley GJ (2006) Cumulative forage production, forage quality and livestock performance from an annual ryegrass and cereal rye mixture in a pine walnut silvopasture. Agrofor Syst 66:43–53. https://doi.org/10.1007/s10457-005-6640-6

Kan Z-R, Virk AL, He C et al (2020) Characteristics of carbon mineralization and accumulation under long-term conservation tillage. Catena (Amst) 193:104636. https://doi.org/10.1016/j.catena.2020.104636

Kang BT, Wilson GF, Sipkens L (1981) Alley cropping maize (Zea mays L.) and leucaena (Leucaena leucocephala Lam) in southern Nigeria. Plant Soil 63:165–179. https://doi.org/10.1007/BF02374595

Kaspar TC, Bakker MG (2015) Biomass production of 12 winter cereal cover crop cultivars and their effect on subsequent no-till corn yield. J Soil Water Conserv 70:353–364. https://doi.org/10.2489/jswc.70.6.353

Kaspar TC, Radke JK, Laflen JM (2001) Small grain cover crops and wheel traffic effects on infiltration, runoff, and erosion. J Soil Water Conserv 56:160–164

Kavitha B, Reddy PVL, Kim B et al (2018) Benefits and limitations of biochar amendment in agricultural soils: a review. J Environ Manag 227:146–154. https://doi.org/10.1016/j.jenvman.2018.08.082

Kelley KW, Long JH, Todd TC (2003) Long-term crop rotations affect soybean yield, seed weight, and soil chemical properties. Field Crop Res 83:41–50. https://doi.org/10.1016/S0378-4290(03)00055-8

King AE, Blesh J (2018) Crop rotations for increased soil carbon: perenniality as a guiding principle. Ecol Appl 28:249–261. https://doi.org/10.1002/eap.1648

King AE, Congreves KA, Deen B et al (2020) Crop rotations differ in soil carbon stabilization efficiency, but the response to quality of structural plant inputs is ambiguous. Plant Soil 457:207–224. https://doi.org/10.1007/s11104-020-04728-5

Kladivko EJ (2001) Tillage systems and soil ecology. Soil Tillage Res 61:61–76. https://doi.org/10.1016/S0167-1987(01)00179-9

König D (1992) The potential of agroforestry methods for erosion control in Rwanda. Soil Technol 5:167–176. https://doi.org/10.1016/0933-3630(92)90017-U
Kostensalo J, Hyväluoma J, Jauhiainen L et al (2024) Diversification of crop rotations and soil carbon balance: impact assessment based on national-scale monitoring data. Carbon Manag 15. https://doi.org/10.1080/17583004.2023.2298373
Kovács GP, Simon B, Balla I et al (2023) Conservation tillage improves soil quality and crop yield in Hungary. Agronomy 13:894. https://doi.org/10.3390/agronomy13030894
Księżak J, Staniak M, Stalenga J (2023) Restoring the importance of cereal-grain legume mixtures in low-input farming systems. Agriculture 13:341. https://doi.org/10.3390/agriculture13020341
Kumar V, Kumar M, Singh SK, Jat RK (2018) Impact of conservation agriculture on soil physical properties in rice-wheat system of eastern indo-gangetic plains. J Anim Plant Sci 28:1432–1440
Kumar R, Saurabh K, Kumawat N et al (2019) Conservation agriculture: perspectives on soil and environmental management in Indo-Gangetic plains of South Asia. In: Sustainable management of soil and environment. Springer Singapore, Singapore, pp 123–168
Kumar S, Bijalwan A, Singh B et al (2021) Comparison of carbon sequestration potential of Quercus leucotrichophora–based agroforestry systems and natural forest in Central Himalaya, India. Water Air Soil Pollut 232:350. https://doi.org/10.1007/s11270-021-05294-x
Kumar U, Thomsen IK, Eriksen J et al (2023) Delaying sowing of cover crops decreases the ability to reduce nitrate leaching. Agric Ecosyst Environ 355:108598. https://doi.org/10.1016/j.agee.2023.108598
Kumara TMK, Pal S, Chand P, Kandpal A (2023) Carbon sequestration potential of agroforestry systems in Indian agricultural landscape: a Meta-Analysis. Ecosyst Serv 62:101537. https://doi.org/10.1016/j.ecoser.2023.101537
Kumari K, Kumar R, Bordoloi N et al (2023) Unravelling the recent developments in the production technology and efficient applications of biochar for agro-ecosystems. Agriculture 13:512. https://doi.org/10.3390/agriculture13030512
Kusumastuti RD, van DP D, Teunter R (2016) Crop-related harvesting and processing planning: a review. Int J Prod Econ 174:76–92. https://doi.org/10.1016/j.ijpe.2016.01.010
Lal R (1989) Agroforestry systems and soil surface management of a tropical alfisol. Agrofor Syst 8:97–111. https://doi.org/10.1007/BF00123115
Lal R (1991) Myths and scientific realities of agroforestry as a strategy for sustainable management for soils in the tropics. In: Stewart BA (ed) Advances in soil science. Springer, New York, pp 91–137
Lal R (2004) Soil carbon sequestration impacts on global climate change and food security. Science 304:1623–1627
Lal R (2015a) Cover cropping and the "4 per Thousand" proposal. J Soil Water Conserv 70:141A. https://doi.org/10.2489/jswc.70.6.141A
Lal R (2015b) Soil carbon sequestration and aggregation by cover cropping. J Soil Water Conserv 70:329–339. https://doi.org/10.2489/jswc.70.6.329
Lal R (2015c) A system approach to conservation agriculture. J Soil Water Conserv 70:82A–88A. https://doi.org/10.2489/jswc.70.4.82A
Lal R (2015d) Biochar and soil carbon sequestration. In: Guo M, He Z, Uchimiya SM (eds) Agricultural and environmental applications of biochar: advances and barriers. Soil Science Society of America, Inc, pp 175–197
Lal R, Kimble JM (1997) Conservation tillage for carbon sequestration. Nutr Cycl Agroecosyst 49:243–253. https://doi.org/10.1023/A:1009794514742
Lal R, Smith P, Jungkunst HF et al (2018) The carbon sequestration potential of terrestrial ecosystems. J Soil Water Conserv 73:145A–152A. https://doi.org/10.2489/jswc.73.6.145A
Larkin RP (2024) Effects of different cover crops and amendments on soil and crop properties in organic vegetable production. Agronomy 14:171. https://doi.org/10.3390/agronomy14010171
Lehtinen T, Schlatter N, Baumgarten A et al (2014) Effect of crop residue incorporation on soil organic carbon and greenhouse gas emissions in European agricultural soils. Soil Use Manag 30:524–538. https://doi.org/10.1111/sum.12151

Leng L, Xiong Q, Yang L et al (2021) An overview on engineering the surface area and porosity of biochar. Sci Total Environ 763:144204. https://doi.org/10.1016/j.scitotenv.2020.144204

Leuthold SJ, Salmerón M, Wendroth O, Poffenbarger H (2021) Cover crops decrease maize yield variability in sloping landscapes through increased water during reproductive stages. Field Crop Res 265:108111. https://doi.org/10.1016/j.fcr.2021.108111

Li S, Tasnady D (2023) Biochar for soil carbon sequestration: current knowledge, mechanisms, and future perspectives. C J Carbon Res 9:67. https://doi.org/10.3390/c9030067

Li Y, Song D, Liang S et al (2020) Effect of no-tillage on soil bacterial and fungal community diversity: a meta-analysis. Soil Tillage Res 204:104721. https://doi.org/10.1016/j.still.2020.104721

Li H, Li C, Song X et al (2022) Impacts of continuous and rotational cropping practices on soil chemical properties and microbial communities during peanut cultivation. Sci Rep 12:2758. https://doi.org/10.1038/s41598-022-06789-1

Li C, Li Z, Zhang F et al (2023a) Seasonal dynamics of carbon dioxide and water fluxes in a rice-wheat rotation system in the Yangtze-Huaihe region of China. Agric Water Manag 275:107992. https://doi.org/10.1016/j.agwat.2022.107992

Li F, Zhou J, Guo X et al (2023b) Effect of biochar and earthworms on mineralization of organic matter in top soil and deep soil. Environ Eng Sci 40:340–346. https://doi.org/10.1089/ees.2023.0069

Li Y, Abdo AI, Shi Z et al (2023c) Biochar derived from invasive plants improved the pH, macronutrient availability and biological properties better than liming for acid rain-affected soil. Biochar 5:59. https://doi.org/10.1007/s42773-023-00251-9

Li Y, Xue C, Chai C et al (2023d) Influencing factors and spatiotemporal heterogeneity of net carbon sink of conservation tillage: evidence from China. Environ Sci Pollut Res 30:110913–110930. https://doi.org/10.1007/s11356-023-29969-6

Liang Z, Rasmussen J, Poeplau C, Elsgaard L (2023) Priming effects decrease with the quantity of cover crop residues—potential implications for soil carbon sequestration. Soil Biol Biochem 184:109110. https://doi.org/10.1016/j.soilbio.2023.109110

Lin BB (2007) Agroforestry management as an adaptive strategy against potential microclimate extremes in coffee agriculture. Agric For Meteorol 144:85–94. https://doi.org/10.1016/j.agrformet.2006.12.009

Lithourgidis A, Dordas C, Damalas CA, Vlachostergios DN (2011) Annual intercrops: an alternative pathway for sustainable agriculture. Aust J Crop Sci 5:396–410

Liu X, Liu J, Xing B et al (2005) Effects of long-term continuous cropping, tillage, and fertilization on soil organic carbon and nitrogen of black soils in China. Commun Soil Sci Plant Anal 36:1229–1239. https://doi.org/10.1081/CSS-200056917

Liu C, Plaza-Bonilla D, Coulter JA et al (2022) Diversifying crop rotations enhances agroecosystem services and resilience. In: Advances in agronomy. Elsevier, pp 299–335

Liu Z, Nie L, Zhang M et al (2023) Long-term subsoiling and straw return increase soil organic carbon fractions and crop yield. J Soil Water Conserv 78:234–244. https://doi.org/10.2489/jswc.2023.00094

Lobzenko I, Burachevskaya M, Zamulina I et al (2022) Development of a unique technology for the pyrolysis of rice husk biochar for promising heavy metal remediation. Agriculture 12:1689. https://doi.org/10.3390/agriculture12101689

López-Aguilar HA, Quiroz-Cardoza D, Pérez-Hernández A (2022) Volatile compounds of algal biomass pyrolysis. J Mar Sci Eng 10:928. https://doi.org/10.3390/jmse10070928

López-Hernández JC, Aryal DR, Villanueva-López G et al (2024) Carbon storage and sequestration rates in Leucaena leucocephala-based silvopasture in Southern Mexico. Agrofor Syst 98:1105–1121. https://doi.org/10.1007/s10457-023-00922-3

Lorenz K, Lal R (2014a) Soil organic carbon sequestration in agroforestry systems. A review. Agron Sustain Dev 34:443–454. https://doi.org/10.1007/s13593-014-0212-y

Lorenz K, Lal R (2014b) Biochar application to soil for climate change mitigation by soil organic carbon sequestration. J Plant Nutr Soil Sci 177:651–670. https://doi.org/10.1002/jpln.201400058

Lorenz K, Lal R (2018) Agroforestry systems. In: Carbon sequestration in agricultural ecosystems. Springer International Publishing, Cham, pp 235–260

Loria N, Bhardwaj SK (2021) Climate change and water availability to plants in terrestrial ecosystem: effects and adaptation strategies. In: Gautam V, Kapoor D, Bhardwaj R (eds) Environmental contamination and climate change: effect on plants and remedial strategies. Nova Science Publishers, Inc, New York, pp 119–153

Lou S, He J, Lu C et al (2021) A tillage depth monitoring and control system for the independent adjustment of each subsoiling shovel. Actuators 10:250. https://doi.org/10.3390/act10100250

Louhaichi M, Hassan S (2023) Evaluation of selected fodder species for alley cropping systems. Range Manag Agrofor 44:241–247. https://doi.org/10.59515/rma.2023.v44.i2.04

Love BE, Bork EW, Spaner D (2009) Tree seedling establishment in living fences: a low-cost agroforestry management practice for the tropics. Agrofor Syst 77:1–8. https://doi.org/10.1007/s10457-009-9244-8

Lundgren BO, Raintree JB (1983) Sustained agroforestry. In: Nestel B (ed) Agricultural Research for Development: Potentials and Challenges in Asia. ISNAR, The Hague, Netherlands, pp 37–49

Madden NM, Southard RJ, Mitchell JP (2008) Conservation tillage reduces PM10 emissions in dairy forage rotations. Atmos Environ 42:3795–3808. https://doi.org/10.1016/j.atmosenv.2007.12.058

Mahajan A, Gupta RD (2009) The rice–wheat cropping system. In: Integrated Nutrient Management (INM) in a sustainable rice—wheat cropping system. Springer, Dordrecht, pp 109–117

Mahala DM, Meena MC, Dwivedi BS et al (2023) Changes in soil organic carbon pools after 15 years of Conservation Agriculture in rice (Oryza sativa)-wheat (Triticum aestivum) cropping system of eastern Indo-Gangetic plains. Indian J Agric Sci 93. https://doi.org/10.56093/ijas.v93i6.136045

Maiga A, Alhameid A, Singh S et al (2019) Responses of soil organic carbon, aggregate stability, carbon and nitrogen fractions to 15 and 24 years of no-till diversified crop rotations. Soil Res 57:149. https://doi.org/10.1071/SR18068

Man M, Wagner-Riddle C, Dunfield KE et al (2021) Long-term crop rotation and different tillage practices alter soil organic matter composition and degradation. Soil Tillage Res 209:104960. https://doi.org/10.1016/j.still.2021.104960

Manhas SS, Sidhu AS, Brar KS (2015) Conservation tillage for soil management and crop production. In: Sustainable agriculture reviews. Springer, pp 61–87

Marcillo GS, Miguez FE (2017) Corn yield response to winter cover crops: an updated meta-analysis. J Soil Water Conserv 72:226–239. https://doi.org/10.2489/jswc.72.3.226

Márquez-García F, Hayas A, Peña A et al (2024) Influence of cover crops and tillage on organic carbon loss in Mediterranean olive orchards. Soil Tillage Res 235:105905. https://doi.org/10.1016/j.still.2023.105905

Martinez L, Soti P, Kaur J et al (2020) Impact of cover crops on insect community dynamics in organic farming. Agriculture 10:209. https://doi.org/10.3390/agriculture10060209

dos Martins MR, Corá JE, Jorge RF, Marcelo AV (2009) Crop type influences soil aggregation and organic matter under no-tillage. Soil Tillage Res 104:22–29. https://doi.org/10.1016/j.still.2008.11.003

Masek J, Procházka P, Kvíz Z, Sindelár R (2008) Machines in conservation soil tillage technologies. In: 7th international scientific conference on engineering for rural development. Jelgava, pp 131–135

Mayorga I, Vargas de Mendonça JL, Hajian-Forooshani Z et al (2022) Tradeoffs and synergies among ecosystem services, biodiversity conservation, and food production in coffee agroforestry. Front Forest Global Change 5. https://doi.org/10.3389/ffgc.2022.690164

McClelland SC, Paustian K, Schipanski ME (2021) Management of cover crops in temperate climates influences soil organic carbon stocks: a meta-analysis. Ecol Appl 31. https://doi.org/10.1002/eap.2278

Meena RS, Kumar S, Yadav GS (2020) Soil carbon sequestration in crop production. In: Nutrient dynamics for sustainable crop production. Springer Singapore, Singapore, pp 1–39

Mei K, Wang Z, Huang H et al (2018) Stimulation of N2O emission by conservation tillage management in agricultural lands: a meta-analysis. Soil Tillage Res 182:86–93. https://doi.org/10.1016/j.still.2018.05.006

Melkani S, Singh V, Bhadha JH (2024) Quantifying the impact of different agroforestry systems on soil carbon fractions lability and long-term carbon sequestration in Central Himalayas. Soil Use Manag 40. https://doi.org/10.1111/sum.13012

Mennan H, Jabran K, Zandstra BH, Pala F (2020) Non-chemical weed management in vegetables by using cover crops: a review. Agronomy 10:257. https://doi.org/10.3390/agronomy10020257

Meyer N, Bergez J-E, Constantin J et al (2020) Cover crops reduce drainage but not always soil water content due to interactions between rainfall distribution and management. Agric Water Manag 231:105998. https://doi.org/10.1016/j.agwat.2019.105998

Montagnini F, Nair PKR (2004) Carbon sequestration: an underexploited environmental benefit of agroforestry systems. Agrofor Syst 61–62:281–295. https://doi.org/10.1023/B:AGFO.0000029005.92691.79

Monti M, Pellicanò A, Pristeri A et al (2019) Cereal/grain legume intercropping in rotation with durum wheat in crop/livestock production systems for Mediterranean farming system. Field Crop Res 240:23–33. https://doi.org/10.1016/j.fcr.2019.05.019

Morrison JE (2002) Strip tillage for "no-till" row crop production. Appl Eng Agric 18:277–284

Motisi N, Montfort F, Faloya V et al (2009) Growing Brassica juncea as a cover crop, then incorporating its residues provide complementary control of Rhizoctonia root rot of sugar beet. Field Crop Res 113:238–245. https://doi.org/10.1016/j.fcr.2009.05.011

Moyer JR, Blackshaw RE, Smith EG, McGinn SM (2000) Cereal cover crops for weed suppression in a summer fallow-wheat cropping sequence. Can J Plant Sci 80:441–449. https://doi.org/10.4141/P99-099

Murtaza G, Ahmed Z, Usman M et al (2021) Biochar induced modifications in soil properties and its impacts on crop growth and production. J Plant Nutr 44:1–15. https://doi.org/10.1080/01904167.2021.1871746

Myers R, Weber A, Tellatin S (2019) Cover crop economics: opportunities to improve your bottom line in row crops. SARE Tech. Bulletin. https://www.sare.org/resources/cover-crop-economics/. Accessed 20 Feb 2024

Myrans H, McCausland JR, Johnson SN, Gleadow RM (2024) Phenotypic and genotypic variation in Australian native Sorghum species along aridity clines. Aust J Bot 72. https://doi.org/10.1071/BT23062

Nair PKR (2011) Methodological challenges in estimating carbon sequestration potential of agroforestry systems. In: Carbon sequestration potential of agroforestry systems: opportunities and challenges. Springer, pp 3–16

Nair PKR (2012) Carbon sequestration studies in agroforestry systems: a reality-check. Agrofor Syst 86:243–253. https://doi.org/10.1007/s10457-011-9434-z

Nandan R, Singh V, Singh SS et al (2019) Impact of conservation tillage in rice–based cropping systems on soil aggregation, carbon pools and nutrients. Geoderma 340:104–114. https://doi.org/10.1016/j.geoderma.2019.01.001

Navarro-Noya YE, Gómez-Acata S, Montoya-Ciriaco N et al (2013) Relative impacts of tillage, residue management and crop-rotation on soil bacterial communities in a semi-arid agroecosystem. Soil Biol Biochem 65:86–95. https://doi.org/10.1016/j.soilbio.2013.05.009

Nguyen LTT, Ortner KA, Tiemann LK et al (2022) Soil properties after one year of interseeded cover cropping in topographically diverse agricultural landscape. Agric Ecosyst Environ 326:107803. https://doi.org/10.1016/j.agee.2021.107803

Nguyen T-B, Sherpa K, Bui X-T et al (2023) Biochar for soil remediation: a comprehensive review of current research on pollutant removal. Environ Pollut 337:122571. https://doi.org/10.1016/j.envpol.2023.122571
Ni J, Xu J, Xie Z, Wang D (2004) Changes of labile organic carbon fractions in soils under different rotation systems. Pedosphere 14:103–109
Niemmanee T, Borwornchokchai K, Nindam T (2019) Farmer's perception, knowledge and attitude toward the use of biochar for agricultural soil improvement in Amphawa district, Samut Songkhram province. In: Proceedings of the 8th International Conference on Informatics, Environment, Energy and Applications. ACM, New York, pp 39–43
Nordblom T, Gurusinghe S, Erbacher A, Weston LA (2023) Opportunities and challenges for cover cropping in sustainable agriculture systems in Southern Australia. Agriculture 13:688. https://doi.org/10.3390/agriculture13030688
Nunes MR, van Es HM, Schindelbeck R et al (2018) No-till and cropping system diversification improve soil health and crop yield. Geoderma 328:30–43. https://doi.org/10.1016/j.geoderma.2018.04.031
Nyong AP, Ngankam TM, Felicite TL (2020) Enhancement of resilience to climate variability and change through agroforestry practices in smallholder farming systems in Cameroon. Agrofor Syst 94:687–705. https://doi.org/10.1007/s10457-019-00435-y
Octavia D, Murniati SS et al (2023) Smart agroforestry for sustaining soil fertility and community livelihood. Forest Sci Technol 19:315–328. https://doi.org/10.1080/21580103.2023.2269970
Ogdahl E, Zamora D, Johnson G, Wyatt G (2018) Comparison of woody species for use in living snow fences in the midwestern United States. J Soil Water Conserv 73:321–328. https://doi.org/10.2489/jswc.73.3.321
Olesen JE, Rubæk GH, Heidmann T et al (2004) Effect of climate change on greenhouse gas emissions from arable crop rotations. Nutr Cycl Agroecosyst 70:147–160. https://doi.org/10.1023/B:FRES.0000048478.78669.33
Ong CK, Anyango S, Muthuri CW, Black CR (2007) Water use and water productivity of agroforestry systems in semi-arid tropics. Ann Arid Zone 46:255–284
Osanai Y, Knox O, Nachimuthu G, Wilson B (2020) Increasing soil organic carbon with maize in cotton-based cropping systems: mechanisms and potential. Agric Ecosyst Environ 299:106985. https://doi.org/10.1016/j.agee.2020.106985
Page KL, Dalal RC, Pringle MJ et al (2013) Organic carbon stocks in cropping soils of Queensland, Australia, as affected by tillage management, climate, and soil characteristics. Soil Res 51:596. https://doi.org/10.1071/SR12225
Pan J, Liu C, Li H et al (2022) Soil-resistant organic carbon improves soil erosion resistance under agroforestry in the Yellow River Flood Plain, of China. Agrofor Syst 96:997–1008. https://doi.org/10.1007/s10457-022-00757-4
Pankaj U (2020) Multifarious benefits of biochar application in different soil types. In: Biochar applications in agriculture and environment management. Springer International Publishing, Cham, pp 259–272
Panwar L, Devi S, Singh Y (2021) Insect pest management in vegetable crops through trap cropping: review. Indian J Agric Sci 91:1433–1437. https://doi.org/10.56093/ijas.v91i10.117421
Panwar P, Mahalingappa DG, Kaushal R et al (2022) Biomass production and carbon sequestration potential of different agroforestry systems in India: a critical review. Forests 13:1274. https://doi.org/10.3390/f13081274
Pardon P, Reubens B, Reheul D et al (2017) Trees increase soil organic carbon and nutrient availability in temperate agroforestry systems. Agric Ecosyst Environ 247:98–111. https://doi.org/10.1016/j.agee.2017.06.018
Parker JE, Crowder DW, Eigenbrode SD, Snyder WE (2016) Trap crop diversity enhances crop yield. Agric Ecosyst Environ 232:254–262. https://doi.org/10.1016/j.agee.2016.08.011
Parvin S, Condon J, Rose TJ (2022) Potential nitrogen contributions by tropical legume summer cover crops in mediterranean-type cropping systems. Nitrogen 3:592–599. https://doi.org/10.3390/nitrogen3040038

Patel R, Patel B, Mukherjee S et al (2023) Impact of agroforestry ecosystem on carbon sequestration potential and climate change. In: Agricultural soil sustainability and carbon management. Elsevier, pp 269–297

Pawłowski L, Kwiatkowski CA, Harasim E et al (2021) Environmental benefits of catch crops cultivation. Chem-Didact-Ecol-Metrol 26:109–121. https://doi.org/10.2478/cdem-2021-0009

Pelzer E, Bazot M, Makowski D et al (2012) Pea–wheat intercrops in low-input conditions combine high economic performances and low environmental impacts. Eur J Agron 40:39–53. https://doi.org/10.1016/j.eja.2012.01.010

Peralta AL, Sun Y, McDaniel MD, Lennon JT (2018) Crop rotational diversity increases disease suppressive capacity of soil microbiomes. Ecosphere 9. https://doi.org/10.1002/ecs2.2235

Poeplau C, Don A (2015) Carbon sequestration in agricultural soils via cultivation of cover crops—A meta-analysis. Agric Ecosyst Environ 200:33–41. https://doi.org/10.1016/j.agee.2014.10.024

Poyda A, Wizemann H-D, Ingwersen J et al (2019) Carbon fluxes and budgets of intensive crop rotations in two regional climates of southwest Germany. Agric Ecosyst Environ 276:31–46. https://doi.org/10.1016/j.agee.2019.02.011

Pulido-Santacruz P, Renjifo LM (2011) Live fences as tools for biodiversity conservation: a study case with birds and plants. Agrofor Syst 81:15–30. https://doi.org/10.1007/s10457-010-9331-x

Qi J, Wang X, Zhao X et al (2019) Temporal variability of soil organic carbon in paddies during 13-year conservation tillage. Land Degrad Dev 30:1840–1850. https://doi.org/10.1002/ldr.3384

Qian X, Gu J, Pan H et al (2015) Effects of living mulches on the soil nutrient contents, enzyme activities, and bacterial community diversities of apple orchard soils. Eur J Soil Biol 70:23–30. https://doi.org/10.1016/j.ejsobi.2015.06.005

Qin Z, Guan K, Zhou W et al (2021) Assessing the impacts of cover crops on maize and soybean yield in the U.S. Midwestern agroecosystems. Field Crop Res 273:108264. https://doi.org/10.1016/j.fcr.2021.108264

Qu Y, Pan C, Guo H (2021) Factors affecting the promotion of conservation tillage in black soil—the case of Northeast China. Sustain For 13:9563. https://doi.org/10.3390/su13179563

Ramachandran Nair PK, Mohan Kumar B, Nair VD (2009) Agroforestry as a strategy for carbon sequestration. J Plant Nutr Soil Sci 172:10–23. https://doi.org/10.1002/jpln.200800030

Ramírez-García J, Carrillo JM, Ruiz M et al (2015) Multicriteria decision analysis applied to cover crop species and cultivars selection. Field Crop Res 175:106–115. https://doi.org/10.1016/j.fcr.2015.02.008

Ramirez-Garcia J, Martens HJ, Quemada M, Thorup-Kristensen K (2015) Intercropping effect on root growth and nitrogen uptake at different nitrogen levels. J Plant Ecol 8:380–389. https://doi.org/10.1093/jpe/rtu024

van Ramshorst JGV, Siebicke L, Baumeister M et al (2022) Reducing wind erosion through agroforestry: a case study using large eddy simulations. Sustain For 14:13372. https://doi.org/10.3390/su142013372

Rasoarinaivo AR, Razafimbelo TM, Blanchart E et al (2022) Specific effects of tree species on soil carbon sequestration in a rice-tree association mesocosm experiment: evidence from natural 13C abundance. Rhizosphere 21:100485. https://doi.org/10.1016/j.rhisph.2022.100485

Rathore SS, Babu S, El-Sappah AH et al (2022) Integrated agroforestry systems improve soil carbon storage, water productivity, and economic returns in the marginal land of the semi-arid region. Saudi J Biol Sci 29:103427. https://doi.org/10.1016/j.sjbs.2022.103427

Reginatto M, da Bonome LTS, Giovanetti LK et al (2023) Chemical characteristics and phytotoxicity of root exudates from cover crops. Iheringia, Série Botânica 78:e2023009. https://doi.org/10.21826/2446-82312023v78e2023009

Rieger S, Richner W, Streit B et al (2008) Growth, yield, and yield components of winter wheat and the effects of tillage intensity, preceding crops, and N fertilisation. Eur J Agron 28:405–411. https://doi.org/10.1016/j.eja.2007.11.006

Rizvi S (1999) Allelopathic interactions in agroforestry systems. CRC Crit Rev Plant Sci 18:773–796. https://doi.org/10.1016/S0735-2689(99)00400-1

Rosenberg S, Crump A, Brim-DeForest W et al (2022) Crop rotations in california rice systems: assessment of barriers and opportunities. Front Agron 4. https://doi.org/10.3389/fagro.2022.806572
Rubio V, Diaz-Rossello R, Quincke JA, van Es HM (2021) Quantifying soil organic carbon's critical role in cereal productivity losses under annualized crop rotations. Agric Ecosyst Environ 321:107607. https://doi.org/10.1016/j.agee.2021.107607
Sachdeva S, Kumar R, Sahoo PK, Nadda AK (2023) Recent advances in biochar amendments for immobilization of heavy metals in an agricultural ecosystem: a systematic review. Environ Pollut 319:120937. https://doi.org/10.1016/j.envpol.2022.120937
Sáenz-Romo MG, Veas-Bernal A, Martínez-García H et al (2019) Effects of ground cover management on insect predators and pests in a mediterranean vineyard. Insects 10:421. https://doi.org/10.3390/insects10120421
Sainju UM, Singh HP, Singh BP (2017) Cover crops for enriching soil carbon and nitrogen under bioenergy sorghum. In: Global Soil Security. Springer, pp 181–192
Sainju UM, Ghimire R, Dangi S (2021) Soil carbon dioxide and methane emissions and carbon balance with crop rotation and nitrogen fertilization. Sci Total Environ 775:145902. https://doi.org/10.1016/j.scitotenv.2021.145902
Samal SK, Rao KK, Poonia SP et al (2017) Evaluation of long-term conservation agriculture and crop intensification in rice-wheat rotation of Indo-Gangetic Plains of South Asia: carbon dynamics and productivity. Eur J Agron 90:198–208. https://doi.org/10.1016/j.eja.2017.08.006
Sanchez PA (1999) Improved fallows come of age in the tropics. Agrofor Syst 47:3–12. https://doi.org/10.1023/A:1006287702265
Sandhu SS, Ussiri DAN, Kumar S et al (2017) Analyzing the impacts of three types of biochar on soil carbon fractions and physiochemical properties in a corn-soybean rotation. Chemosphere 184:473–481. https://doi.org/10.1016/j.chemosphere.2017.05.165
Sari RR, Rozendaal DMA, Saputra DD et al (2022) Balancing litterfall and decomposition in cacao agroforestry systems. Plant Soil 473:251–271. https://doi.org/10.1007/s11104-021-05279-z
Sarvade S, Loria N, Singh I et al (2019) Agroforestry role in sustainable management of degraded watersheds. In: Applied agricultural practices for mitigating climate change. Routledge, pp 133–149
Schroeder P (1994) Carbon storage benefits of agroforestry systems. Agrofor Syst 27:89–97. https://doi.org/10.1007/BF00704837
Seefeldt SS, Una TM, McMoran D et al (2023) Impacts of two years of autumn cover crops in northwestern Washington on winter annual weed populations. Weed Sci 71:150–159. https://doi.org/10.1017/wsc.2023.7
Semmartin M, Cosentino D, Poggio SL et al (2023) Soil carbon accumulation in continuous cropping systems of the rolling Pampa (Argentina): the role of crop sequence, cover cropping and agronomic technology. Agric Ecosyst Environ 347:108368. https://doi.org/10.1016/j.agee.2023.108368
Shackley S, Hammond J, Gaunt J, Ibarrola R (2011) The feasibility and costs of biochar deployment in the UK. Carbon Manag 2:335–356. https://doi.org/10.4155/cmt.11.22
Shafawi AN, Mohamed AR, Lahijani P, Mohammadi M (2021) Recent advances in developing engineered biochar for CO2 capture: an insight into the biochar modification approaches. J Environ Chem Eng 9:106869. https://doi.org/10.1016/j.jece.2021.106869
Sharratt BS (1996) Tillage and straw management for modifying physical properties of a subarctic soil. Soil Tillage Res 38:239–250. https://doi.org/10.1016/S0167-1987(96)01035-5
Shen Y, Kong W, Shi R et al (2022) Farmers' adoption behavior of conservation tillage technology: a multidimensional heterogeneity perspective. Environ Sci Pollut Res 30:37744–37761. https://doi.org/10.1007/s11356-022-24716-9
Sheng Y, Zhan Y, Zhu L (2016) Reduced carbon sequestration potential of biochar in acidic soil. Sci Total Environ 572:129–137. https://doi.org/10.1016/j.scitotenv.2016.07.140

Shi L, Feng W, Xu J, Kuzyakov Y (2018) Agroforestry systems: meta-analysis of soil carbon stocks, sequestration processes, and future potentials. Land Degrad Dev 29:3886–3897. https://doi.org/10.1002/ldr.3136

Shi S, Zhang Q, Lou Y et al (2021) Soil organic and inorganic carbon sequestration by consecutive biochar application: results from a decade field experiment. Soil Use Manag 37:95–103. https://doi.org/10.1111/sum.12655

Siarudin M, Rahman SA, Artati Y et al (2021) Carbon sequestration potential of agroforestry systems in degraded landscapes in West Java. Indonesia Forests 12:714. https://doi.org/10.3390/f12060714

Siegwart L, Bertrand I, Roupsard O, Jourdan C (2023) Contribution of tree and crop roots to soil carbon stocks in a Sub-Sahelian agroforestry parkland in Senegal. Agric Ecosyst Environ 352:108524. https://doi.org/10.1016/j.agee.2023.108524

Sijtsma CH, Campbell AJ, McLaughlin NB, Carter MR (1998) Comparative tillage costs for crop rotations utilizing minimum tillage on a farm scale. Soil Tillage Res 49:223–231. https://doi.org/10.1016/S0167-1987(98)00175-5

Sincik M, Turan ZM, Göksoy AT (2008) Responses of potato (Solanum tuberosum L.) to green manure cover crops and nitrogen fertilization rates. Am J Potato Res 85:150–158. https://doi.org/10.1007/s12230-008-9011-9

Singh J, Kumar S (2021) Responses of soil microbial community structure and greenhouse gas fluxes to crop rotations that include winter cover crops. Geoderma 385:114843. https://doi.org/10.1016/j.geoderma.2020.114843

Singh Yadav SP, Bhandari S, Bhatta D et al (2023) Biochar application: a sustainable approach to improve soil health. J Agric Food Res 11:100498. https://doi.org/10.1016/j.jafr.2023.100498

Singh BP, Setia R, Wiesmeier M, Kunhikrishnan A (2018) Agricultural management practices and soil organic carbon storage. In: Soil carbon storage : modulators, mechanisms and modeling. Elsevier, pp 207–244

Slayden JM, Brye KR, Della LD et al (2022) Site position and tillage treatment effects on nitrous oxide emissions from furrow-irrigated rice on a silt-loam Alfisol in the Mid-south, USA. Geoderma Reg 28:e00491. https://doi.org/10.1016/j.geodrs.2022.e00491

Soane BD, Ball BC, Arvidsson J et al (2012) No-till in northern, western and south-western Europe: a review of problems and opportunities for crop production and the environment. Soil Tillage Res 118:66–87. https://doi.org/10.1016/j.still.2011.10.015

Staton T, Walters RJ, Smith J, Girling RD (2019) Evaluating the effects of integrating trees into temperate arable systems on pest control and pollination. Agric Syst 176:102676. https://doi.org/10.1016/j.agsy.2019.102676

Stockfisch N, Forstreuter T, Ehlers W (1999) Ploughing effects on soil organic matter after twenty years of conservation tillage in Lower Saxony, Germany. Soil Tillage Res 52:91–101. https://doi.org/10.1016/S0167-1987(99)00063-X

Sun W, Shahrajabian MH (2022) The effectiveness of Rhizobium bacteria on soil fertility and sustainable crop production under cover and catch crops management and green manuring. Not Bot Horti Agrobot Cluj Napoca 50:12560. https://doi.org/10.15835/nbha50212560

Sun B, Jia S, Zhang S et al (2016) No tillage combined with crop rotation improves soil microbial community composition and metabolic activity. Environ Sci Pollut Res 23:6472–6482. https://doi.org/10.1007/s11356-015-5812-9

Sun L, Wang S, Narsing Rao MP et al (2023) The shift of soil microbial community induced by cropping sequence affect soil properties and crop yield. Front Microbiol 14. https://doi.org/10.3389/fmicb.2023.1095688

Takimoto A, Nair PKR, Alavalapati JRR (2008) Socioeconomic potential of carbon sequestration through agroforestry in the West African Sahel. Mitig Adapt Strateg Glob Chang 13:745–761. https://doi.org/10.1007/s11027-007-9140-3

Tanveer A, Ikram RM, Ali HH (2019) Crop rotation: principles and practices. In: Agronomic crops. Springer Singapore, Singapore, pp 1–12

Tassin J, Rangan H, Kull CA (2012) Hybrid improved tree fallows: harnessing invasive woody legumes for agroforestry. Agrofor Syst 84:417–428. https://doi.org/10.1007/s10457-012-9493-9
Teixeira EI, Johnstone P, Chakwizira E et al (2016) Sources of variability in the effectiveness of winter cover crops for mitigating N leaching. Agric Ecosyst Environ 220:226–235. https://doi.org/10.1016/j.agee.2016.01.019
Tesio F, Ferrero A (2010) Allelopathy, a chance for sustainable weed management. Int J Sustain Dev World Ecol 17:377–389. https://doi.org/10.1080/13504509.2010.507402
Thapa BB, Cassel DK, Garrity DP (1999) Ridge tillage and contour natural grass barrier strips reduce tillage erosion. Soil Tillage Res 51:341–356. https://doi.org/10.1016/S0167-1987(99)00047-1
Thapa VR, Ghimire R, Acosta-Martínez V et al (2021) Cover crop biomass and species composition affect soil microbial community structure and enzyme activities in semiarid cropping systems. Appl Soil Ecol 157:103735. https://doi.org/10.1016/j.apsoil.2020.103735
Thierfelder C, Cheesman S, Rusinamhodzi L (2013) Benefits and challenges of crop rotations in maize-based conservation agriculture (CA) cropping systems of southern Africa. Int J Agric Sustain 11:108–124. https://doi.org/10.1080/14735903.2012.703894
Tian L, Li H, Chang Z et al (2023) Biochar modification to enhance arsenic removal from water: a review. Environ Geochem Health 45:2763–2778. https://doi.org/10.1007/s10653-022-01462-y
Tomczyk A, Sokołowska Z, Boguta P (2020) Biochar physicochemical properties: pyrolysis temperature and feedstock kind effects. Rev Environ Sci Biotechnol 19:191–215. https://doi.org/10.1007/s11157-020-09523-3
Toom M, Talgre L, Mäe A et al (2019a) Selecting winter cover crop species for northern climatic conditions. Biol Agric Hortic 35:263–274. https://doi.org/10.1080/01448765.2019.1627908
Toom M, Tamm S, Talgre L et al (2019b) The effect of cover crops on the yield of spring barley in Estonia. Agriculture 9:172. https://doi.org/10.3390/agriculture9080172
Topps D, ul Khabir MI, Abdelmagid H et al (2021) Impact of cover crop monocultures and mixtures on organic carbon contents of soil aggregates. Soil Syst 5:43. https://doi.org/10.3390/soilsystems5030043
Tribouillois H, Constantin J, Justes E (2018) Cover crops mitigate direct greenhouse gases balance but reduce drainage under climate change scenarios in temperate climate with dry summers. Glob Chang Biol 24:2513–2529. https://doi.org/10.1111/gcb.14091
Trozzo KE, Munsell JF, Chamberlain JL et al (2021) Forest farming: who wants in? J For 119:478–492. https://doi.org/10.1093/jofore/fvab023
Uchino H, Iwama K, Jitsuyama Y et al (2011) Stable characteristics of cover crops for weed suppression in organic farming systems. Plant Prod Sci 14:75–85. https://doi.org/10.1626/pps.14.75
Udawatta RP, Jose S (2011) Carbon sequestration potential of agroforestry practices in temperate North America. In: Carbon sequestration potential of agroforestry systems. Springer, pp 17–42
Upson MA, Burgess PJ (2013) Soil organic carbon and root distribution in a temperate arable agroforestry system. Plant Soil 373:43–58. https://doi.org/10.1007/s11104-013-1733-x
Uri ND (1998a) Trends in the use of conservation tillage in US agriculture. Soil Use Manag 14:111–116. https://doi.org/10.1111/j.1475-2743.1998.tb00626.x
Uri ND (1998b) The environmental consequences of the conservation tillage adoption decision in agriculture in the United States. Water Air Soil Pollut 103:9–34. https://doi.org/10.1023/A:1004968616527
Uri ND (1999) Factors affecting the use of conservation tillage in the United States. Water Air Soil Pollut 116:621–638. https://doi.org/10.1023/A:1005168928627
Van Thuyet D, Van Do T, Sato T, Thai Hung T (2014) Effects of species and shelterbelt structure on wind speed reduction in shelter. Agrofor Syst 88:237–244. https://doi.org/10.1007/s10457-013-9671-4
Verheijen FGA, Montanarella L, Bastos AC (2012) Sustainability, certification, and regulation of biochar. Pesq Agrop Brasileira 47:649–653. https://doi.org/10.1590/S0100-204X2012000500003
Verma KS, Kohli S, Kaushal R, Chaturvedi OP (2014) Root structure, distribution and biomass in five multipurpose tree species of Western Himalayas. J Mt Sci 11:519–525. https://doi.org/10.1007/s11629-013-2479-x

Verzeaux J, Alahmad A, Habbib H et al (2016) Cover crops prevent the deleterious effect of nitrogen fertilisation on bacterial diversity by maintaining the carbon content of ploughed soil. Geoderma 281:49–57. https://doi.org/10.1016/j.geoderma.2016.06.035

Vogeler I, Hansen EM, Thomsen IK (2023) The effect of catch crops in spring barley on nitrate leaching and their fertilizer replacement value. Agric Ecosyst Environ 343:108282. https://doi.org/10.1016/j.agee.2022.108282

Vukicevich E, Lowery T, Bowen P et al (2016) Cover crops to increase soil microbial diversity and mitigate decline in perennial agriculture. A review. Agron Sustain Dev 36:48. https://doi.org/10.1007/s13593-016-0385-7

Waisen P, Cheng Z, Sipes BS, Wang K-H (2022) Biofumigation effects of brassicaceous cover crops on soil health in cucurbit agroecosystems in Hawaii, USA. Pedosphere 32:521–531. https://doi.org/10.1016/S1002-0160(21)60054-1

Waldén P, Ollikainen M, Kahiluoto H (2020) Carbon revenue in the profitability of agroforestry relative to monocultures. Agrofor Syst 94:15–28. https://doi.org/10.1007/s10457-019-00355-x

Wang J, Wang S (2019) Preparation, modification and environmental application of biochar: a review. J Clean Prod 227:1002–1022. https://doi.org/10.1016/j.jclepro.2019.04.282

Wang Q, Li Y, Alva A (2012) Cover crops in mono- and biculture for accumulation of biomass and soil organic carbon. J Sustain Agric 36:423–439. https://doi.org/10.1080/10440046.2011.627991

Wang J, Xiong Z, Kuzyakov Y (2016) Biochar stability in soil: meta-analysis of decomposition and priming effects. GCB Bioenergy 8:512–523. https://doi.org/10.1111/gcbb.12266

Wang J, Zhang S, Sainju UM et al (2021) A meta-analysis on cover crop impact on soil water storage, succeeding crop yield, and water-use efficiency. Agric Water Manag 256:107085. https://doi.org/10.1016/j.agwat.2021.107085

Wang X, Shen L, Liu T et al (2022) Microclimate, yield, and income of a jujube–cotton agroforestry system in Xinjiang, China. Ind Crop Prod 182:114941. https://doi.org/10.1016/j.indcrop.2022.114941

Waring E, Licht M, Ripley E et al (2023) Cover crop mixtures versus single species: water quality and cash crop yield. J Soil Water Conserv 78:1–15. https://doi.org/10.2489/jswc.2023.00174

Wei D, Li B, Huang H et al (2018) Biochar-based functional materials in the purification of agricultural wastewater: fabrication, application and future research needs. Chemosphere 197:165–180. https://doi.org/10.1016/j.chemosphere.2017.12.193

Wei B, Peng Y, Lin L et al (2023) Drivers of biochar-mediated improvement of soil water retention capacity based on soil texture: a meta-analysis. Geoderma 437:116591. https://doi.org/10.1016/j.geoderma.2023.116591

Wen L, Peng Y, Zhou Y et al (2023) Effects of conservation tillage on soil enzyme activities of global cultivated land: a meta-analysis. J Environ Manag 345:118904. https://doi.org/10.1016/j.jenvman.2023.118904

Weng Z, Van Zwieten L, Singh BP et al (2017) Biochar built soil carbon over a decade by stabilizing rhizodeposits. Nat Clim Chang 7:371–376. https://doi.org/10.1038/nclimate3276

Wiersma W, van der Ploeg MJ, Sauren IJMH, Stoof CR (2020) No effect of pyrolysis temperature and feedstock type on hydraulic properties of biochar and amended sandy soil. Geoderma 364:114209. https://doi.org/10.1016/j.geoderma.2020.114209

Woolf D, Amonette JE, Street-Perrott FA et al (2010) Sustainable biochar to mitigate global climate change. Nat Commun 1:56. https://doi.org/10.1038/ncomms1053

Wooliver R, Jagadamma S (2023) Response of soil organic carbon fractions to cover cropping: a meta-analysis of agroecosystems. Agric Ecosyst Environ 351:108497. https://doi.org/10.1016/j.agee.2023.108497

Wu D, Zhang Y, Gu W et al (2024) Long term co-application of biochar and fertilizer could increase soybean yield under continuous cropping: insights from photosynthetic physiology. J Sci Food Agric. https://doi.org/10.1002/jsfa.13202

Wyland L (1996) Winter cover crops in a vegetable cropping system: impacts on nitrate leaching, soil water, crop yield, pests and management costs. Agric Ecosyst Environ 59:1–17. https://doi.org/10.1016/0167-8809(96)01048-1

Xia F, Zhang Z, Zhang Q et al (2024) Life cycle assessment of greenhouse gas emissions for various feedstocks-based biochars as soil amendment. Sci Total Environ 911:168734. https://doi.org/10.1016/j.scitotenv.2023.168734
Xiao L, Zhou S, Zhao R et al (2020) Evaluating soil organic carbon stock changes induced by no-tillage based on fixed depth and equivalent soil mass approaches. Agric Ecosyst Environ 300:106982. https://doi.org/10.1016/j.agee.2020.106982
Yadav RS, Yadav BL, Chhipa BR (2008) Litter dynamics and soil properties under different tree species in a semi-arid region of Rajasthan, India. Agrofor Syst 73:1–12. https://doi.org/10.1007/s10457-008-9106-9
Yang X, Drury CF, Wander MM (2013) A wide view of no-tillage practices and soil organic carbon sequestration. Acta Agric Scand B Soil Plant Sci 63:523–530. https://doi.org/10.1080/09064710.2013.816363
Yang X, Wang D, Lan Y et al (2018) Labile organic carbon fractions and carbon pool management index in a 3-year field study with biochar amendment. J Soils Sediments 18:1569–1578. https://doi.org/10.1007/s11368-017-1874-2
Yang S, Wang Y, Wang Z et al (2022a) Interactive effects of conservation tillage on the aggregate stability and soil organic carbon. J Plant Nutr Soil Sci 185:505–512. https://doi.org/10.1002/jpln.202200044
Yang Y, Sun K, Han L et al (2022b) Biochar stability and impact on soil organic carbon mineralization depend on biochar processing, aging and soil clay content. Soil Biol Biochem 169:108657. https://doi.org/10.1016/j.soilbio.2022.108657
Yang Y, Ti J, Zou J et al (2023) Optimizing crop rotation increases soil carbon and reduces GHG emissions without sacrificing yields. Agric Ecosyst Environ 342:108220. https://doi.org/10.1016/j.agee.2022.108220
Yang X, Xiong J, Du T et al (2024) Diversifying crop rotation increases food production, reduces net greenhouse gas emissions and improves soil health. Nat Commun 15:198. https://doi.org/10.1038/s41467-023-44464-9
Ye L, Camps-Arbestain M, Shen Q et al (2020) Biochar effects on crop yields with and without fertilizer: a meta-analysis of field studies using separate controls. Soil Use Manag 36:2–18. https://doi.org/10.1111/sum.12546
Zegada-Lizarazu W, Monti A (2011) Energy crops in rotation. A review. Biomass Bioenergy 35:12–25. https://doi.org/10.1016/j.biombioe.2010.08.001
Zeng Z, Chen Y (2018) The performance of a fluted coulter for vertical tillage as affected by working speed. Soil Tillage Res 175:112–118. https://doi.org/10.1016/j.still.2017.08.006
Zhang J, Liu J, Liu R (2015) Effects of pyrolysis temperature and heating time on biochar obtained from the pyrolysis of straw and lignosulfonate. Bioresour Technol 176:288–291. https://doi.org/10.1016/j.biortech.2014.11.011
Zhang Y, Wang R, Wang S et al (2017) Effects of different sub-soiling frequencies incorporated into no-tillage systems on soil properties and crop yield in dryland wheat-maize rotation system. Field Crop Res 209:151–158. https://doi.org/10.1016/j.fcr.2017.05.002
Zhang X, Zhang P, Yuan X et al (2020) Effect of pyrolysis temperature and correlation analysis on the yield and physicochemical properties of crop residue biochar. Bioresour Technol 296:122318. https://doi.org/10.1016/j.biortech.2019.122318
Zhang C, Xue J, Li N et al (2022a) Afterlife effect of cover crops on soil nematode food web: Implications from the plant ecological strategy. Biol Fertil Soils 58:937–947. https://doi.org/10.1007/s00374-022-01676-3
Zhang J, Zhang X, Sun H et al (2022b) Carbon sequestration and nutrients improvement meditated by biochar in a 3-year vegetable rotation system. J Soils Sediments 22:1385–1396. https://doi.org/10.1007/s11368-022-03175-2
Zhang X, Guo L, Hu Y et al (2023a) Soil carbon sequestration and aggregate stability improvement by tillage methods in China's Loess Plateau. Arch Agron Soil Sci 69:1718–1733. https://doi.org/10.1080/03650340.2022.2108804

Zhang X, Wang J, Feng X et al (2023b) Effects of tillage on soil organic carbon and crop yield under straw return. Agric Ecosyst Environ 354:108543. https://doi.org/10.1016/j.agee.2023.108543

Zheng F, Liu X, Ding W et al (2023) Positive effects of crop rotation on soil aggregation and associated organic carbon are mainly controlled by climate and initial soil carbon content: a meta-analysis. Agric Ecosyst Environ 355:108600. https://doi.org/10.1016/j.agee.2023.108600

Zhou Q, Zhang P, Wang Z et al (2023) Winter crop rotation intensification to increase rice yield, soil carbon, and microbial diversity. Heliyon 9:e12903. https://doi.org/10.1016/j.heliyon.2023.e12903

Zhu K, Ran H, Wang F et al (2022) Conservation tillage facilitated soil carbon sequestration through diversified carbon conversions. Agric Ecosyst Environ 337:108080. https://doi.org/10.1016/j.agee.2022.108080

Zhu S, Sainju UM, Zhang S et al (2024) Cover cropping promotes soil carbon sequestration by enhancing microaggregate-protected and mineral-associated carbon. Sci Total Environ 908:168330. https://doi.org/10.1016/j.scitotenv.2023.168330

Zilberman D, Laird D, Rainey C et al (2023) Biochar <scp>supply-chain</scp> and challenges to commercialization. GCB Bioenergy 15:7–23. https://doi.org/10.1111/gcbb.12952

Zomer RJ, Neufeldt H, Xu J et al (2016) Global tree cover and biomass carbon on agricultural land: the contribution of agroforestry to global and national carbon budgets. Sci Rep 6:29987. https://doi.org/10.1038/srep29987

Zuber SM, Behnke GD, Nafziger ED, Villamil MB (2017) Multivariate assessment of soil quality indicators for crop rotation and tillage in Illinois. Soil Tillage Res 174:147–155. https://doi.org/10.1016/j.still.2017.07.007

Chapter 5
Carbon Farming Among Diverse Agricultural Systems

Abstract This chapter discusses how different agricultural systems can contribute to carbon (C) farming. It covers row cropping, plantation farming, livestock farming, horticulture and specialty cropping, and wetland agriculture, highlighting practices that improve soil health and C sequestration. The chapter also identifies knowledge gaps and future research needs to strengthen C farming across these diverse systems.

5.1 Introduction

Agriculture, often overshadowed by sectors like energy and forestry in discussions of carbon (C) sequestration, has significant potential to adapt and mitigate climate change through site-specific management practices which focus on efficient use of natural resources such as soil, water, nutrients, and plants. Despite its association with greenhouse gas (GHG) emissions, agriculture can play a crucial role in enhancing crop productivity and contributing to climate resilience. Sustainable approaches that prioritize environmental impact and utilize resources effectively can transform agricultural systems into vital players in C sequestration (CS) efforts. C farming, a sustainable approach discussed in previous chapters, involves storing C in soils over the long term and offers farmers the opportunity to trade stored C for credits in C markets, receive payments for ecosystem services (ESs) while promoting both climate change mitigation and economic incentives for adopting sustainable agriculture. This strategy is highly applicable across various agricultural sectors, including row cropping, plantation farming, livestock farming, horticultural and specialty cropping, and wetland agriculture. Each sector offers unique opportunities for CS at scale by optimizing soil and vegetation management practices. By examining specific management strategies within each agricultural practice, discussion in this chapter aims to underscore the significant role that agriculture can play in global C mitigation efforts. Emphasizing the importance of understanding C dynamics in diverse agricultural contexts, this chapter seeks to promote the adoption of C farming practices as integral components of sustainable agriculture. Ultimately,

N. Loria, R. Lal, *Carbon Farming*, https://doi.org/10.1007/978-3-032-00842-8_5

integrating these practices into mainstream agricultural systems not only enhances C storage capacities but also supports food and nutritional security, ecosystem health, and resilience in the face of anthropogenic climate change.

5.2 Row Cropping

5.2.1 Introduction

Row cropping is a strategic agricultural practice where crops are arranged in parallel rows, either equidistant single rows or in multiple rows forming strips (Fig. 5.1). This practice has been widely practiced worldwide, renowned for its ability to meet the growing demand for food and fiber, hence contributing to global food security (Blösch et al. 2023; Smith et al. 2008; Testa et al. 2016; Zalles et al. 2019). Optimizing plant density and inter-row spacing through row cropping is crucial for agricultural productivity, as it shapes plant morphology and influences yield outcomes (Testa et al. 2016). Likewise, east-west row crop orientation is important, as it significantly reduces weed biomass and increases grain yield in wheat (*Triticum aestivum*) and barley (*Hordeum vulgare*) by enhancing light interception (Borger et al. 2010; van der Meulen and Chauhan 2017). Benefits of row cropping are manifold: it promotes efficient photosynthesis by enhancing light distribution, regulating humidity levels by improving gas exchange between rows, thereby enhancing crop yield potential (Bareja 2022). Additionally, it facilitates the mechanization of agricultural operations like planting, cultivation, spraying, and harvesting, leading to reduced labor costs (Bautista and Regalado 2023). Row cropping enables the modern technologies like precision agriculture to be efficiently applied for optimizing the on-fam management such as utilizing Unmanned Aerial Vehicles (UAVs) managing for irrigation (Ronchetti et al. 2020); detecting crops with high weed pressure through image segmentation approaches based on vegetation indices (Montalvo

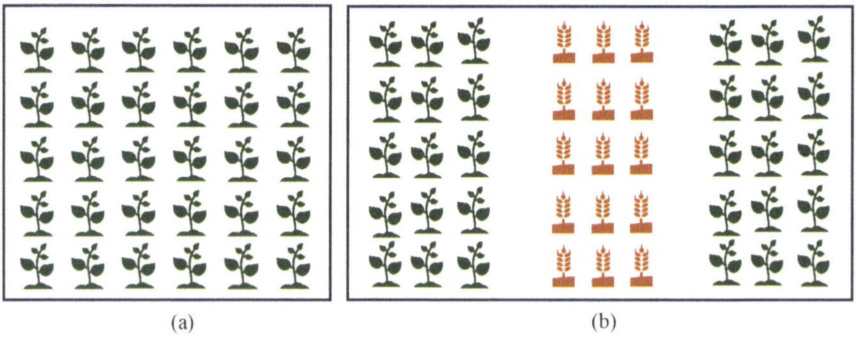

Fig. 5.1 Row cropping patterns (**a**) Equidistant single row cropping pattern, and (**b**) Multiple row cropping forming strips, with each strip composed of three rows

et al. 2012), and managing diseases and pests using GIS approaches (Willers and Riggins 2010). It involves growing of single (monoculture) or multiple crops (multiple cropping or intercropping) in rows on the same piece of land. Monoculture practices are predominantly used for industrialized agriculture for field and plantation crops (Javad Faraji 2011; Elouattassi et al. 2023).

While monoculture farming offers certain advantages in terms of efficiency and mechanization, it also poses challenges such as soil degradation (Xiong et al. 2015; Zhao et al. 2018), pest and disease outbreaks (Bai et al. 2015), reduced biodiversity (Strom et al. 2020), and environmental pollution resulting from heavy reliance on chemical and pesticides to sustain crop production (Capa et al. 2015; Yan et al. 2016). As a result, there is an increasing interest in multiple cropping practices that promote crop diversity and soil health. Transition from monoculture to multiple cropping not only improves crop yields but also increases economical resilience by better withstanding price fluctuations (Singh and Joshi 1994; Clément et al. 2023). Multiple cropping/ intercropping, as compared to solo cropping/ monoculture, efficiently utilizes resources, improves soil fertility and conservation, reduces pests, enhances forage quality, offers insurance against crop failure and market instability, and decreases input needs (Lehrsch et al. 1994; Lithourgidis et al. 2011). However, challenges arise in selecting appropriate species, determining optimal sowing densities, managing increased workload for preparation and maintenance, and navigating the complexity of system selection based on environmental factors and crop interactions (Lithourgidis et al. 2011; Kantola et al. 2017). Inter-row management significantly influences soil water dynamics and plant growth, especially amidst changing environmental conditions (Horel and Zsigmond 2023). Among different intercropping practices (Glaze-Corcoran et al. 2020), row arrangement is common, except for mixed intercropping, which lacks any spatial arrangement (Table 5.1).

5.2.2 Row Cropping for Soil Health

Row cropping has been a longstanding agricultural practice, known for its agronomic, ecological, and economic benefits, crucial in addressing global food security (Bareja 2022) and environmental concerns (Robertson et al. 2018). However, the historical emphasis on maximizing yields and profits often led to adverse environmental and health impacts (Zaidelman et al. 2013). As the era of climate change dawned since 1990s, recognition of agriculture's contribution to environmental degradation spurred a shift towards sustainable practices within row cropping systems (Kantola et al. 2017). Efforts to improve soil health and sequester C have gained traction, reflecting a commitment to balancing productivity with environmental stewardship. Integration of conservation agriculture (CA) techniques, such as no-till or reduced tillage, cover cropping, and diversified crop rotations, has become central to enhancing biodiversity, optimizing nutrient cycling, and reducing reliance on synthetic inputs in row cropping systems (Fageria et al. 2005; Sharma et al. 2018b; Schmidt et al. 2019b). Innovative methods like Soil Functional Zone Management

Table 5.1 Intercropping types within row cropping practices

Intercropping type	Strip intercropping	Row intercropping	Relay intercropping
Description	Crops are planted in parallel strips consisting of multiple rows, with each strip containing either different species of crops or various cultivars of the same species.	Different crops are planted in rows within the same area, with at least one companion crop grown in single or double rows.	Different species or cultivars are cultivated together in the same field, organized in either strip or row arrangements, with planting and harvesting done at different times.
Strengths	Efficient use of space; reduced soil erosion; potential for complementary relationship between crops; improves crop yields and forage quality	Facilitates weed and disease suppression; easy to manage and harvest; allows for mechanized farming practices	Maximizes use of resources and space over time; enhances biodiversity and ecosystem services; extends growing season and spreads workload
Challenges	Requires precision planting; interspecific above and below ground competition from root mingling, shade effects and wind disruption; management complexity increases with more crop types; may require specialized equipment for harvest	Crop compatibility; competition for water and nutrients; potential for allelopathic effects between rows	Initial establishment can be labor-intensive; requires careful timing of planting and harvesting; monitoring and managing multiple crops simultaneously
Examples	Maize-pea (Tan et al. 2020); Corn and Alfalfa (Smith and Carter 1998); Wheat-Alfalfa (Ma et al. 2024a)	Barley-peas (Jaskulska et al. 2022); Sugarcane-soybean (Li et al. 2013)	Maize-soybean (Yang et al. 2017); Tobacco-peanut (Gao et al. 2019); Sorghum with alfalfa, winter wheat, and cotton as intercrops (Phoofolo et al. 2010)

(SFZM), exemplified by ridge tillage, introduces spatial variation between crop rows and inter-rows, facilitating a delicate balance between nutrient replenishment and soil building processes (Williams et al. 2017). Research across four US states demonstrated that ridge tillage significantly improved soil organic C (SOC) content and nutrient provisioning processes, particularly within crop rows during critical growth stages such as maize cultivation (Williams et al. 2017). Hedge row intercropping, particularly with Gliricidia, enhances soil health by reducing run-off, soil loss, and nutrient loss while increasing soil moisture storage and productivity on sloping agriculture lands (Hombegowda et al. 2020). Incorporating energy crops like switchgrass and miscanthus into buffer strips adjacent to agricultural fields offers a sustainable solution to utilize nutrients from runoff and leachate for biomass production, benefiting both the environment and agricultural productivity (Gopalakrishnan et al. 2012). Integrating living mulch into row cropping systems offers a dual benefit by enhancing soil infiltration capacity and crop productivity while simultaneously reducing soil loss, ultimately improving soil health and water quality (Baker et al. 2022).

Row cropping, particularly intercrops, significantly increases SOC levels through residue addition, with lower cumulative soil CO_2 emissions, indicating potential for climate change mitigation (Bichel et al. 2016; Oelbermann et al. 2017). The role of cover crop mixtures in enhancing SOC contents is underscored by the fact that certain species combinations (e.g., triticale, crimson clover, canola, Australian winter pea, and red clover) improve root biomass production and distribution compared to monocultures, leading to greater cumulative C inputs and soil health benefits (Amsili and Kaye 2021). Perennial grain crops, emerges as a significant determinant in enhancing SOC dynamics and soil biological activity, as evidenced by the considerable increase in root biomass and permanganate oxidizable C (POXC) compared to annual wheat (Sprunger et al. 2019). Perennial legumes further enhance SOC and nitrogen (N) pools, with fine root N significantly higher in corn-forage-forage no-till systems, positively influencing soil health indicators (Martin et al. 2022). Pitahaya (dragon fruit) cultivation study reveals that SOC concentration is higher below the plants compared to the rows, and older plantations exhibit greater horizontal root extension and enhanced SOC sequestration (Guillen et al. 2023). Row cropping in agroforestry systems (AFS) significantly enhances SOC storage, particularly in the tree rows where herbaceous vegetation grew and tillage was absent, though most additional SOC is found in labile fractions (Cardinael et al. 2015). A 12-year study by Snapp et al. (2010) revealed that organic row cropping management significantly increased SOC by 36% and enhanced N retention, compared to minimal effects from crop diversity, emphasizing the importance of management practices over crop diversity in improving soil health.

5.3 Plantation Farming

5.3.1 Introduction

Plantation farming involves large-scale cultivation of cash crops like coffee (Velmourougane and Bhat 2017), tea(Liang et al. 2021), sugar cane(Hartemink 2008), rubber (Dove 1993), and the trees or perennial crops for commercial purposes, such as timber production, pulp and paper manufacturing, bioenergy production, and AFS (Hartemink 2005; Khamzina et al. 2012; van Eijck et al. 2012). It's characterized by monoculture, extensive land holdings, labor intensity, export orientation, and significant environmental and social implications (Hayami 2010). Primarily practiced in tropical regions, plantation farming presents environmental challenges such as biodiversity loss, soil erosion, and land conflicts (Hartemink 2005), yet it also contributes to employment opportunities and improved livelihood (Khamzina et al. 2012). Smallholding plantations, unlike large-scale agriculture, prioritize environmental benefits with crop diversification, minimal chemical use, higher ecological knowledge, and increased resilience to environmental changes (Bissonnette and De Koninck 2017). Amidst growing sustainability interests, mitigating environmental challenges in plantations entails adopting practices for sustainable resource management, biodiversity conservation, and soil fertility restoration (Velmourougane and Bhat 2017). Nadaraja et al. (2021) identified eight sustainability indicators of plantation agriculture focusing on the environmental aspect of land, including their associated potential issues (Table 5.2). These indicators are crucial for promoting sustainable land management, enhancing SOC levels, and thereby supporting C farming initiatives. For instance, cow manure addition in tea plantations enhances soil bacterial diversity and improves soil fertility, promoting healthier tea cultivation compared to synthetic fertilizers (Zhang et al. 2020). Increases in SOC stocks from manure management are influenced by management decisions (e.g., tillage intensity, amount, and duration of application), site properties (climate, initial SOC content, soil texture), and manure characteristics such as origin and combined application with synthetic fertilizer (Gross and Glaser 2021). The amount of fertilizer use can affect SOC levels by influencing microbial activity

Table 5.2 Land-based environmental sustainability indicators in plantation agriculture (Adapted from (Nadaraja et al. 2021))

Indicator	Benefit	Potential issue
Manure management	Reduces dependency on chemical fertilizers; improves soil health and quality.	Lack of legitimacy among agribusinesses; limited adoption due to lack of widespread prevalence.
Amount of fertilizer used	Relevant for preventing excessive use of chemical fertilizers; helps mitigate environmental impacts.	Agribusinesses may exceed recommended application guidelines due to global food demand pressures.
Intercropping	Increases crop yields; reduces need for chemical inputs and GHG emissions associated with N fixation.	Lack of guidelines on growing compatible crop species together; economic disincentives for farmers.
Tillage practices	Essential for crop production; affects soil erosion and water pollution.	Focuses on area cultivated or number of operations, neglecting impact of different tillage types; requires consideration of environmental impacts.
Crop rotation	Increases yield quantity; aids in pest management and soil fertility improvement.	Lack of knowledge on effective crop rotation methods; reluctance among agribusinesses due to economic reasons.
Soil nutrient content	Essential for plant growth; informs nutrient management strategies.	Costly and time-consuming lab analysis; variability in measurement techniques and accuracy.
Soil physical properties	Regulates soil structure, water movement, and aeration; crucial for plant root development.	Requires costly lab analysis or specialized equipment; variability in measurement techniques.
Soil chemical properties	Influences soil pH, nutrient availability, and microbial activity; guides fertilizer application.	High error rate in lab analysis; costly and time-consuming measurements; variability in measurement techniques.

(Dincă et al. 2022), nutrient balance, and organic matter (OM) decomposition rates (Goyal et al. 1999; Jiang et al. 2024). Excessive and indiscriminate use of chemical N fertilizers can accelerate OM breakdown and reduce SOC over time, while balanced and optimized fertilizer application supports SOC maintenance or increase (Li et al. 2022). Intercropping in rubber plantations with cash crops enhances ecosystem functions such as litterfall biomass, nutrient cycling efficiency, and soil fertility, thereby improving overall ESs compared to monoculture systems (Yuan et al. 2024). Furthermore, practices like incorporating post-harvest cotton residues into cotton plantations (Kaiyong et al. 2011) and implementing integrated pest management (IPM) using cotton seed oil in coconut plantations (Rodrigues et al. 2018) exemplify sustainable approaches in plantation agriculture. These practices not only promote environmental sustainability but also expand the scope of C farming initiatives.

5.3.2 Bioenergy Plantations

Energy crops, such as short-rotation woody and herbaceous species can help mitigate CO_2 emissions by sequestering SOC and offsetting fossil fuel use. Although data is limited, bioenergy cropping systems have the potential to reduce atmospheric CO_2 by 20% through enhanced SOC sequestration and biomass production (Sartori et al. 2006). Bioenergy plantations also play a crucial role in restoring degraded lands (Singh et al. 2020) but their impact on biodiversity and water resources, particularly in water-stressed regions like India and South Africa, requires careful consideration (Rahman et al. 2017). Key crops such as jatropha (*Jatropha curcas* L.), rapeseed (*Brassica napus* L.), castor bean (*Ricinus communis* L.), oil palm (*Elaeis guineensis*), and sweet sorghum are cultivated for biodiesel and biofuel production, aiming to reduce GHG emissions (Regassa and Wortmann 2014; Whitaker et al. 2018; Ferchaud et al. 2020). To address challenges like biodiversity loss and soil erosion associated with these crops, it is essential to manage plantations carefully, retaining native vegetation. This involves selecting suitable tree species, determining optimal rotation ages, and managing the understory to support floral and faunal communities (Flaspohler and Webster 2011). Compared to oil palm, rubber plantations are generally considered more sustainable due to their lower environmental impact, greater biodiversity support, and longer economic viability (Dove 1993). Large-scale bioenergy plantations aimed at C capture and storage require careful management to balance increased irrigation water demand. Optimizing water efficiency through appropriate crop selection and production region choices is crucial as biomass gains prominence in global energy strategies. For instance, bioelectricity from sugar beet, maize, and sugar cane demonstrates lower water footprints compared to other bioenergy crops, making them preferable for sustainable renewable energy solutions (Gerbens-Leenes et al. 2009). Achieving sustainable large-scale bioenergy production necessitates policies that optimize land use, minimize water consumption, and protect freshwater ecosystems through effective irrigation practices and ecosystem management strategies (Bonsch et al. 2016; Rahman et al. 2017). Encouraging perennial energy crops like switchgrass and miscanthus can enhance ESs by providing biodiversity habitats and improving soil health through deep root systems, while effectively sequestering C (Baumber 2017). Implementation of incentives such as tax credits and subsidies further promote sustainable bioenergy production (Baumber 2017). In response to land degradation on China's Loess Plateau, initiatives like Grain for Green and Western Development Action promote ecological restoration and economic growth, advocating for perennial grasses like switchgrass to enhance bioenergy production and ecosystem resilience (Cooney et al. 2017). Table 5.3 highlights diverse bioenergy crops, including grasses, shrubs, and trees, outlining their key characteristics and associated challenges.

5.3.3 Sustainable Practices for Carbon Sequestration in Plantations

Plantation farming strategies for CS integrate sustainable land management practices aimed at minimizing soil disturbance, promoting AFS and silvopasture, applying organic fertilizers, and implementing CA methods. These approaches not only balance commercial demands with environmental protection but also enhance livelihoods through improved soil health and productivity. For instance, organic practices in oil palm plantations in the Southeast Amazon have demonstrated significant improvements in soil OM (SOM) dynamics and increased sequestration of C and N compared to conventional methods (Mardegan et al. 2022). CA, coupled with cover crops, has proven effective in olive plantations, reducing runoff by 36.7% and erosion by 85.5% relative to conventional tillage practices, thereby minimizing SOC loss associated with sediment (Márquez-García et al. 2024). Similarly, organic management with cover crops over a 21-year period in Mediterranean citrus plantations has stabilized SOC contents and sequestered 61 Mg CO_2 ha^{-1}, with rates ranging from −1.10 to 1.89 Mg C ha^{-1} $yr.^{-1}$ (Novara et al. 2019). The age of plantations also significantly impacts soil characteristics. Studies on citrus plantations reveal that older plantations exhibit larger aggregates (>2 mm), which contribute significantly to soil aggregate stability and erosion resistance (Cao et al. 2021). In cocoa plantations, increasing SOC stocks with plantation age are attributed to enhanced leaf litter accumulation over the years, underscoring their potential for long-term CS (Arthur et al. 2022).

Intercropping strategies, such as using siratro (*Macroptilium atropurpureum*) as a cover crop in banana plantations, promote SOM and foster stable soil bacterial communities, thereby mitigating soil degradation compared to conventional tillage with bare soil or natural weed cover (Xu et al. 2024). Similarly, intercropping fruit trees like loquats (*Eriobotrya japonica*) and citrus in tea plantations enhances soil nutrient availability, aggregate stability, and influences tea quality components such as polyphenols and caffeine content, highlighting their role in sustainable

Table 5.3 Key characteristics and challenges of diverse bioenergy crops

Category	Bioenergy crop	Characteristics	Challenges	References
Grasses	Switchgrass (*Panicum virgatum*)	High yield, drought-tolerant, can grow on marginal lands, low input requirements, improve soil health	Requires significant land area, initial establishment costs	Soldatos (2015); Ahlswede et al. (2022)
	Miscanthus (*Miscanthus* × *giganteus*)	Tolerant to marginal and sandy soils, high growth rates, well-suited to various agro-climatic conditions, high biomass productivity, high lignin content for durability and calorific value, improves soil structure, reduces erosion, and requires minimal inputs.	Potential invasiveness, high initial costs	Mann et al. (2013); Soldatos (2015); Babich et al. (2019)
	Giant Reed (*Arundo donax*)	High lignocellulosic biomass production, soil adaptability (clay to sand), salt tolerance, hardiness in marginal lands, potential for phytoremediation.	Can be invasive, requires substantial water, high establishment costs	Ceotto and Di Candilo (2010); Zhang et al. (2021)
Shrubs	Jatropha (*Jatropha curcas*)	High biomass for biodiesel, grows on degraded land, remediates pollutants, C-neutral, controls soil erosion, suitable for marginal land use.	Inconsistent yield, susceptibility to pests and diseases, land and water requirements for optimal growth, limited market stability for biodiesel.	Pandey et al. (2012)
	Willow (*Salix spp.*)	Fast-growing, coppices well, can be cultivated on marginal lands, carbon sequestration, suitable for phytoremediation, enhances species diversity, reduces soil-water nitrate, low labor cost	Requires pest management, potential competition with food crops, water demand during growth periods.	Kakuk et al. (2021)
	Sea Buckthorn (*Hippophae rhamnoides*)	Nutrient-rich, high in vitamins, minerals, and antioxidants, improves soil fertility, nitrogen fixation, and minimizes soil erosion	Labor-intensive harvesting, limited market awareness, low harvest rates, land use conflicts.	Tamchos and Dorjey (2024)

(continued)

Table 5.3 (continued)

Category	Bioenergy crop	Characteristics	Challenges	References
Trees	Eucalyptus (*Eucalyptus spp.*)	Fast-growing, environmental stress tolerance, adaptability to different climates, potential for genetic improvement, valuable for timber production	Invasive potential in non-native regions, high water consumption	Al-Ahmad (2018)
	Poplar (*Populus spp.*)	Fast-growing, high productivity, low annual inputs, rich in cellulose and hemicellulose, ecologically beneficial, helps restore disturbed environments	Biomass yield variability due to genetic, environmental, and climatic factors, high biomass recalcitrance, expensive and energy-intensive pretreatment	An et al. (2021)
	Pine (*Pinus spp.*)	Resilient to nutrient-poor sites and extreme conditions, suitable for paper and pulp production, biofuel potential, adaptable to various climates and soils, and effective in intercropping systems.	Susceptible to pests and diseases, requires effective forest management for optimal performance.	Peter (2008); Minick et al. (2017)

agricultural practices (Duan et al. 2024). In sugarcane plantations, soil tillage practices significantly impact CO_2 emissions, with conventional tillage emitting 350.1 g m^{-2} compared to reduced tillage (51.7 g m^{-2}) and minimum tillage (5.5 g m^{-2}) methods over 44 days in southern Brazil, suggesting that adopting CA could enhance the sustainability of Brazilian ethanol production by reducing GHG emissions (Silva-Olaya et al. 2013). Moreover, long-term weed control and cover crop practices under coffee plantations contribute to increased SOC content by up to 26% at 0–10 cm depth, particularly in highly clayey soils, with the use of dwarf mucuna (*Mucuna deeringiana*) cover crop reducing weed density by 90.8% (Martins et al. 2015). In tropical plantations, soil microbial communities and their necro mass accumulation are critical for SOC storage, influenced by vegetation type and parent material variations (Li et al. 2024). Rubber plantations, particularly in clayey basalt-derived soils, exhibit higher microbial biomass and necro mass, significantly enhancing SOC accumulation. Conversely, in sandier soils from granite and marine sediments, fungal communities dominate, influencing SOC dynamics differently (Li et al. 2024). Introducing avocado (*Persea americana*) trees into coffee AFS in Mexico alters soil respiration dynamics, potentially enhancing SOC storage. However, this change is accompanied by a reduction in soil macrofauna due to decreased leaf litter, which in turn affects soil C and N dynamics (Ayala-Montejo et al. 2022). Overall, these diverse agricultural practices underscore the importance of tailored management approaches in enhancing soil health, CS, and sustainability across various plantation types, contributing to agricultural resilience and environmental stewardship.

5.4 Livestock Farming

5.4.1 Introduction

Livestock farming is crucial for enhancing agricultural productivity, sustaining rural incomes, and supporting the growing population dependent on the sector. It provides protein food, such as milk, meat, and eggs, and contributes to the production of valuable materials like wool, hides, and leather. Livestock waste is used as nutrient-rich manure (farmyard manure) and fuel (e.g. dung cakes, biogas) and animals serve as capital for many landless agricultural laborers. This agricultural practice proves indispensable for millions of small-scale farmers in mixed crop-livestock systems, particularly in regions where agricultural productivity is limited (Herrero et al. 2013). The integration of livestock into farming systems not only diversifies income sources but also enhances resilience against economic uncertainties and climatic variations (Sati and Singh 2010). However, the benefits of livestock farming come with significant environmental costs. Livestock are major contributors to GHG emissions, primarily through methane (CH_4) and nitrous oxide (N_2O) production, both potent contributors to global warming, accounting for nearly 14% of human-induced emissions (Gerber et al. 2013). The primary sources include feed production and processing, which account for 45%, and enteric fermentation from

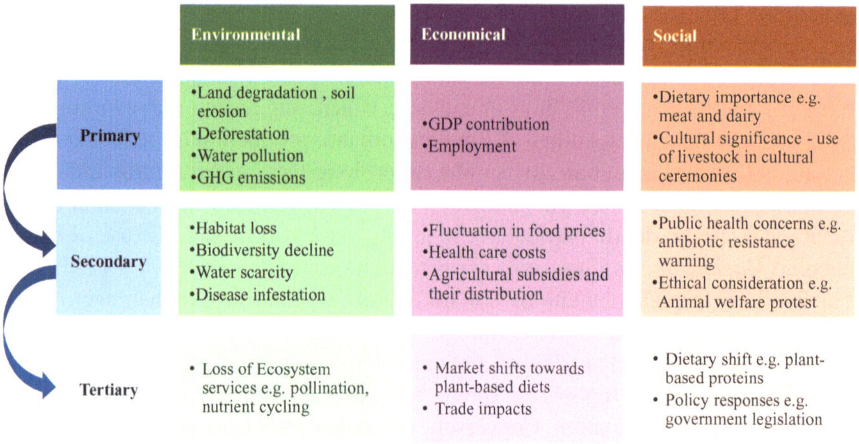

Fig. 5.2 Livestock farming impacts

ruminants, contributing 39% of sector emissions. Manure storage and processing represent 10% of these emissions, with the remainder attributed to the processing and transportation of animal products (Gerber et al. 2013). Moreover, the expansion of livestock operations can lead to soil degradation, deforestation, and water pollution (Cesoniene et al. 2019), exacerbating environmental challenges. Tobar-López et al. (2019) study underscores a significant transformation in La Vía Láctea, Nicaragua, where 93% of forested areas were converted to pasture for livestock, driven by socio-economic and political factors, highlighting livestock as the main cause of deforestation in the region.

Livestock farming impact can be categorized into primary, secondary, and tertiary effects, each influencing various aspects of the environment, economy, and society (Fig. 5.2). Despite these concerns, well-managed livestock systems can mitigate environmental impacts and even contribute positively to Sustainable Development Goals (SDGs). For instance, livestock manure, when properly managed, serves as a valuable organic fertilizer that enhances soil health and crop productivity (Rayne and Aula 2020). However, its effectiveness depends on factors such as manure properties, application rate and timing, soil type, and local climate conditions. In traditional smallholder settings, where integrated crop-livestock systems prevail, livestock play a crucial role in nutrient cycling, reducing the need for synthetic fertilizers and improving overall soil structure (Thorne and Tanner 2002). Improved livestock management aligns with the One Health concept endorsed by the WHO, which seeks to sustainably balance and optimize the health of people, animals, and ecosystems (WHO 2024).

5.4.2 Sustainable Livestock Farming

The concept of sustainable livestock farming, as outlined by Sati and Singh (2010), emphasizes balancing productivity with environmental stewardship. Strategies include enhancing livestock management practices to improve productivity and reduce GHG emissions per unit of product, enhancing soil CS through better grazing management, and adopting energy-efficient equipment with renewable energy integration to reduce and replace fossil fuel use (World Bank 2022). The Sustainable Agriculture Network (SAN) promotes sustainable livestock farming through a multi-level feed resource system and management plan aimed at improving productivity and resource efficiency (SAN 2024). SAN's approach includes minimizing environmental impacts by prohibiting ecosystem destruction, optimizing water and land use, and responsibly managing residual waters. It focuses on reducing GHG emissions and enhancing C sink capacity. Animal welfare is a top priority, ensuring animals are healthy, comfortable, well-fed, safe, and able to exhibit natural behaviors. Furthermore, SAN's protocols for food safety include rigorous sanitary practices, health monitoring, and traceability of animal origin and breeding processes throughout the supply chain. Effective disease management and improved husbandry practices can reduce GHG emissions by up to 30% (Man and Yee 2023). Additionally, optimizing animal nutrition through dietary adjustments and innovative products like CH_4-reducing supplements and vaccines further enhances sustainability of livestock farming.

Sustainable intensification practices, such as livestock based AFS and rotational grazing systems, aim to enhance productivity without expanding the environmental footprint, thereby preserving biodiversity and soil health (Scholten et al. 2013; Baronti et al. 2022). Grazing management plans globally employ rotational grazing to optimize livestock stocking rates in accordance with grassland carrying capacity, promoting sustainable land use and preserving ecosystem integrity (Pittarello et al. 2019). In semi-arid rangelands, rotational grazing strategies manage *Escherichia coli* (*E. coli*) concentrations in streams by adjusting grazing duration and timing (Hulvey et al. 2021). Shorter grazing periods and avoiding peak *E. coli* seasons, such as summer, mitigate water quality concerns without needing to fence off riparian areas or remove cattle entirely. AFS integrates multipurpose leguminous trees with crops and livestock, providing benefits such as shade, fodder, and CS (Devendra 1999; Lorenz and Lal 2014). Livestock-tree interactions in smallholding agriculture address fodder shortages (Alam and Jahiruddin 2007) and enhance soil fertility and crop growth through animal draught power, dung and urine recycling, and efficient use of crop residues (Devendra 1999). Integrating livestock with trees improves animal performance and effectively reduces weed control costs (Domiciano et al. 2020).

5.4.3 Precision Livestock Farming

Precision Livestock Farming (PLF) revolutionizes sustainable livestock management by integrating cutting-edge technologies to enhance efficiency and minimize environmental impact across farm operations (Aquilani et al. 2022). Advanced electronic devices such as sensors, biosensors, pedometers, computers, 2D and 3D surveillance cameras, thermal cameras, microphones, laser detectors, and automatic scales are collectively utilized in PLF systems for comprehensive animal monitoring on large farms (Domaćinović et al. 2023). Non-contact methods for acquiring livestock phenotypic data, including artificial Intelligence (AI), 3D reconstruction (via RGB cameras, laser scanning, and 3D cameras), and body dimension assessment (using RGB images and 3D point clouds), enhance operational efficiency while minimizing stress on animals (Ma et al. 2024b). These technologies optimize feed utilization, reduce energy consumption, and improve production efficiency, contributing significantly to environmental sustainability.

Understanding grazing behavior (influenced by herbage type, climate, and herd dynamics) and rumination efficiency (varied by herbage quality) is crucial for enhancing animal welfare, pasture management, and overall production efficiency in grazing-based systems. Global positioning system (GPS) and accelerometer technologies enable real-time monitoring of livestock locations, grazing patterns, and rumination behaviors, thereby optimizing grazing strategies, conserving natural resources, and supporting sustainable practices globally (Iqbal et al. 2021). GPS tracking helps prevent overgrazing in sensitive areas like riparian zones, while accelerometers detect health issues early, improving animal welfare and reducing veterinary costs (Bailey et al. 2021). Technologies like automatic feeders and sensors for feed intake monitoring reduce feed wastage and ensure optimal nutrition, promoting better growth rates and lowering the environmental footprint per unit of livestock production (Remus et al. 2021; Siberski-Cooper and Koltes 2021). Although PLF systems have reduced labor by 20% on extensive farms (Morgan-Davies et al. 2021), there is a trade-off with increased CO_2 emissions per kilogram of meat produced, necessitating careful management decisions to balance efficiency gains with environmental impacts. Advanced precision livestock tools facilitate the analysis of extensive datasets on livestock performance, health, and behavior, empowering informed decisions for sustainable livestock practices (Man and Yee 2023). The adoption and effectiveness of these tools rely on their affordability, user-friendliness, durability, and feasibility for widespread implementation on farms, which significantly influences their practical utility and adoption rates across the industry (Siberski-Cooper and Koltes 2021).

5.4.4 Role of Livestock Farming in Carbon Farming

Sustainable livestock farming plays a crucial role in C farming by employing practices that mitigate GHG emissions and enhance CS. Key practices include improving feed quality, managing livestock health, grazing management, and utilizing renewable energy sources. Improving feed quality is essential for enhancing livestock health and productivity, as optimizing diets with high-quality forages reduces CH_4 emissions from enteric fermentation. Research indicates that higher-quality grass silages and increased concentrate ratios significantly lower enteric CH_4 production in dairy cows while also reducing emissions per kilogram of milk (Chagunda et al. 2010). Supplementing yak with oats (*Avena sativa*) hay during the cold season significantly increases SOC stock by enhancing organic detritus from dung and litter (Yang et al. 2020). Additionally, feeding biochar can enhance animal health by improving nutrient absorption and productivity while converting manure into a valuable fertilizer that favors soil fertility and structure (Schmidt et al. 2019a). However, the type, amount of biochar used in feed, and associated costs are crucial factors to consider.

In pasturelands, greater forage yields promote biomass accumulation, which improves decomposition, soil moisture reserves, and microbial diversity (Yang et al. 2020). Studies have shown that pasturelands can sequester nearly as much C as forested areas, with average sequestration rates reaching 0.84 ± 0.11 Mg C ha^{-1} $yr.^{-1}$ when compared to cropland (Franzluebbers 2010). Perennial pastures enhance SOM (Franzluebbers et al. 2014), while organic amendments like compost (Cooper and DeMarco 2023) and biochar (Garbuz et al. 2021) improve soil structure, water-holding capacity, plant biomass, and root growth, all contributing to increased SOC. Organic amendments result in greater SOC sequestration, achieving 1.14 Mg C ha^{-1} $yr.^{-1}$ compared to 0.78 Mg C ha^{-1} $yr.^{-1}$ without amendments within the 0–0.3 m depth (Badgery et al. 2021). Additionally, integrating forage legumes can further enhance CS; for example, intercropping *Aeschynomena histrix* with maize (*Zea mays*) led to 64% increase in SOC in the Northern Guinea Savanna within 1 year, while *Stylosanthes hamata* with maize achieved a 30% increase (Odunze et al. 2004). Legume species such as Horse gram (*Macrotyloma uniflorum*) and Centurion (*Centrosema pascuorum*) also improved soil structure by reducing bulk density, further facilitating C storage (Odunze et al. 2004).

The concept of "C ranching" has emerged as a strategy to enhance soil CS through managed grazing practices. This approach aims to increase SOC content by optimizing grazing intensity and animal stocking rates. However, challenges remain in validating C ranching practices, as current research often relies on unrealistic scenarios, which limit the ability to demonstrate significant SOC increases (Reinhart et al. 2022a; Reinhart et al. 2022b). Key factors influencing the positive or negative effects of grazing management include climate, soil texture, soil depth, grass type, and grazing intensity (McSherry and Ritchie 2013). Most SOC is stored below ground, primarily in the soil rather than in roots (Jin et al. 2022). Heavy grazing reduces standing dead material and living root C, leading to a decline in aboveground SOC with higher stocking rates. However, intense grazing can increase SOC compared to light grazing, especially in C4-dominated grasslands, while potentially decreasing it in C3-dominated systems (McSherry and Ritchie 2013). Management-intensive rotational grazing has shown significant potential in enhancing SOC, with studies reporting average annual increases of 1.0 to 1.06 ± 0.24–0.69 Mg C ha^{-1} $yr.^{-1}$, in contrast to C losses observed under continuous grazing, haymaking, or land set aside (Oates and Jackson 2014). This method not only mitigates C losses but also achieves SOC levels comparable to ungrazed areas, indicating substantial climate change mitigation potential (Byrnes et al. 2018).

Sustainable manure management through converting livestock waste into bioenergy can significantly reduce GHG emissions and recycle nutrients effectively (Cantrell et al. 2008). For example, biogas production from manure has the potential to decrease Turkey's C footprint by 30%, underscoring the importance of such practices in mitigating GHG emissions (Dağlıoğlu et al. 2024). Additionally, integrating hybrid renewable energy systems—combining solar, biogas, diesel, and battery setups—can reduce CO_2 emissions by 778 tons annually from large industrial livestock farms (Jahangir et al. 2022). These efforts aim to decrease reliance on fossil fuels while addressing challenges like managing photovoltaic degradation and fluctuating diesel prices, which are crucial considerations for sustainable energy planning in the livestock sector.

5.5 Horticulture and Specialty Cropping

5.5.1 Introduction

Horticulture is the science and art of cultivating fruits, vegetables, flowers, medicinal and ornamental plants, and specialty crops. Unlike traditional agriculture, which often focuses on large-scale production of staple crops like wheat, maize, or rice (*Oryza sativa*), horticulture emphasizes the diverse cultivation of multiple crops, usually involving intensive care and management of high-value plants (Sillitoe 2001). Specialty crops, a specific subset within horticulture as defined by law, include fruits (e.g. apple, avocado, blueberry), vegetables (e.g. asparagus, cabbage, celery, radish, squash), tree nuts (e.g. chestnut, walnut), dried fruits (e.g. almond, apricot, cashew, raisin), and floriculture and nursery crops (e.g. begonia, carnation, magnolia, foliage plants) (USDA 2024). These crops hold significant economic, cultural, and nutritional value globally. This specialized category is distinguished for its economic and regulatory importance, fostering targeted agricultural policies and funding initiatives aimed at supporting the production and market competitiveness of these crops (USDA 2024). The practice of horticulture thus integrates scientific knowledge with practical techniques to optimize the growth and yield of diverse plant species, ensuring both aesthetic appeal and nutritional value in cultivated environments (Sillitoe 2001). It involves a deep understanding of plant biology, micro environmental variations, irrigation techniques, soil management, pest and disease control, and harvesting methods tailored to each crop's specific needs. Specialty crops contribute to dietary diversity, provide essential nutrients, enhance culinary experiences, and support local economies through niche markets and value-added products (Kim 2016). Since the 2014 Farm Bill, specialty crops and certified organic sectors have gained prominence due to expanded support through enhanced marketing, data collection, pest control, food safety measures, and integration across farm bill titles (Johnson 2014). Furthermore, increased USDA nutrition program support for fruit and vegetable purchases has significantly benefited the specialty crop industry (ERS USDA 2024). In addition to their nutritional and economic contributions, specialty crops have the potential to reduce GHG emissions through adopting organic farming methods, IPM, water-efficient irrigation systems, and CA practices (Gryze et al. 2010; Zhu-Barker et al. 2019). Life cycle assessment (LCA) and water footprint (WFP) analysis of specialty crops allow producers to assess and mitigate the environmental impacts of their practices (Majsztrik et al. 2019). Economic incentives, consumer preferences for sustainable products, and regulatory pressures further drive the adoption of environmentally friendly technologies and practices (Majsztrik et al. 2019). This integrated approach ensures that specialty crop production contributes positively to both economic viability and environmental protection and restoration.

5.5.2 Challenges and Innovation in Horticulture and Specialty Cropping

Horticulture and specialty cropping face several modern challenges despite their benefits. Climate change is a primary concern, leading to unpredictable weather, increased pest pressures, and water scarcity, which affect crop yields, quality, and production costs (Ahmed and Stepp 2016; Han et al. 2022). A study on specialty crops in California indicates significant potential declines in suitable areas for crops like almonds, citrus, and walnuts over the next 20 years, highlighting the need for collaboration among policymakers, researchers, and farmers to develop proactive adaptation strategies (Granco et al. 2023). Globalization of agricultural markets presents challenges as well, resulting in price volatility and competition from imports, requiring specialty crop growers to navigate complex supply chains, market fluctuations, and regulatory requirements to maintain profitability (Sandhu et al. 2020; Birthwright and Mighty 2023). Market planning is crucial for assessing the feasibility of producing and marketing specialty crops, considering factors such as sales options, market resources, target markets, and pricing objectives (German and Toensmeyer 1996). Value-added technologies can enhance profitability and market access, with grower networks and crop diversification driving their adoption (Rodriguez Izaba et al. 2023). Labeling new herbicides for specialty crops is challenging due to low profitability and limited industry investment, resulting in fewer available options (Fennimore and Doohan 2008). This situation highlights the need to explore alternative weed control methods and secure funding for ongoing research. Additionally, the spread of vector-borne diseases in U.S. specialty crops poses a significant threat to agriculture, necessitating integrated technologies for early detection and management, particularly for diseases like huanglong-bing (HLB) (Dandekar et al. 2010). Labor shortages further complicate the situation, as specialty crops require intensive labor for tasks such as planting, pruning, harvesting, and processing (Cutulle and Maja 2021). To address these labor constraints, innovations in mechanization and automation are becoming increasingly important for enhancing efficiency and reducing production costs.

Advancements in technology are significantly transforming horticulture and specialty cropping, providing solutions to several challenges faced by growers. Precision agriculture tools, such as drones, satellite imaging, and sensor-based monitoring systems, allow farmers to optimize inputs, monitor crop health in real-time, and make data-driven decisions to enhance productivity and resource efficiency (Lee et al. 2010; Jafarbiglu and Pourreza 2022). Additionally, biotechnological innovations like genetic engineering and marker-assisted breeding promise to develop crops with improved nutritional content, disease resistance, and environmental adaptability, which are essential for ensuring food security and addressing climate change challenges (Ramirez-Torres et al. 2021; Bate et al. 2021). However, despite the widespread application of biotechnology in agriculture, the adoption of bioengineered specialty crops remains low due to concerns about health, environmental, and socioeconomic impacts. A survey of Minnesota specialty crop growers indicated that their willingness to pay for bioengineered potatoes and petunias was significantly lower than for conventional options, with 25% of respondents never considering bioengineered crops (Abbey et al. 2024). The study also found that the method of bioengineering had less influence on adoption than price or end use, with bioengineered ornamental crops enjoying higher acceptance than food crops.

5.5.3 *Role of Horticulture and Specialty Cropping in Carbon Farming*

Horticulture and specialty cropping are crucial components of C farming, leveraging tree growth and biomass accumulation to sequester CO_2 effectively. Studies provide useful insights into their C assimilation and sequestration potential, as exemplified by apple and cherry orchards in Michigan where specific models estimate tree biomass and C footprints (Tozzini et al. 2015). This data is pivotal for assessing their eligibility for C credits on exchanges or payments for ESs, highlighting their role in sustainable agriculture and climate mitigation efforts. Cashew cultivation in Ghana's forest-savannah transitional zone expanded by 0.03 M ha between 2000 and 2020, resulting in an increase in above ground C from approximately 0.6 Tg to over 1 Tg, emphasizing its significant role in C assimilation (Ashiagbor et al. 2022). Ornamental plant production also contributes significantly to C storage, influencing urban forest C balances across the United States, where these forests sequester about 6.5 Mt. of C annually, factoring in losses from mortality and leaf drop to determine the net sequestration rate (Rowntree and Nowak

1991). Fruit orchards globally exhibit substantial potential for CS, with studies revealing significant C accumulation in biomass. For example, apple orchards in China sequestered between 14 and 32 Tg C from 1990 to 2010, contributing approximately 4.5% to China's terrestrial C sink during that period (Wu et al. 2012). Fruit tree plantations in the southern Philippines exhibit significant C storage capacities, with specific examples including a 32-year-old santol (*Sandoricum koetjape*) plantation storing the highest total C at 203.62 Mg ha^{-1}, followed by a 15-year-old mango (*Mangifera indica*) plantation at 122.34 Mg ha^{-1}, and a 12-year-old rambutan (*Nephelium lappaceum*) plantation at 112.18 Mg ha^{-1} (Janiola and Marin 2016). These findings highlight the significant role of soil C in these plantations, with santol and mango allocating 55% of their total C storage to soil C, while rambutan allocates 86% (Janiola and Marin 2016). Furthermore, the application of practices such as N supplementation and organic manures enhances C capture in orchard ecosystems (Sharma et al. 2018a; Khalsa et al. 2020). Biofertilizers in fruit orchards influence fruit quality attributes while enhancing soil N dynamics, microbial activity, and leaf nutrient content, thereby supporting sustainable fruit production (Perazzoli et al. 2020). Similarly, the use of soil amendments like biochar and chitosan in specialty crops such as lettuce (*Lactuca sativa*) enhances crop growth by improving leaf photosynthesis efficiency and electron transport rates (Xu and Mou 2018; Yoo et al. 2021).

Sustainable agronomic practices such as incorporating crop residues, cover cropping, and effective crop rotations enhance SOC content, improving the resilience of horticultural systems against climate change impacts. For instance, research on globe artichoke (*Cynara cardunculus*) management in Mediterranean regions compared conventional monoculture with dry residue incorporation, alternative monoculture with cover crops, and biannual rotation with cauliflower. Conventional and alternative monoculture achieved CS rates of 0.76 and 0.72 Mg C ha^{-1} $yr.^{-1}$ respectively, while biannual rotation showed a slight negative average rate of −0.2 Mg C ha^{-1} $yr.^{-1}$ despite higher C inputs (Deligios et al. 2021). These findings underscore the importance of sustainable residue management for maintaining soil fertility and enhancing resilience to climate change. In a 15-year-old apple orchard in Kashmir, a grass-legume bi-culture cover crop increased SOC by 20.7% compared to the control at 0–30 cm depth, accumulating 3.04 Mg C ha^{-1} over the study period (Dad et al. 2021). This highlights the significant role of cover crops in enhancing CS potential while promoting improved tree growth and fruit yield. Dessureault-Rompré et al. (2020) further emphasize the benefits of rye mulch in enhancing C pools and promoting sustainable soil management practices in vegetable crop production, calling for more research on N dynamics and long-term effects to optimize its application. By adopting these regenerative agricultural practices that enhance C storage, soil health, and ecosystem resilience, horticultural producers can contribute to global efforts to combat climate change while also promoting sustainable food production and rural development. However, addressing the challenges associated with intensive cultivation practices and water management will be essential for realizing the full potential of horticulture and specialty crops as C farming strategies.

5.6 Wetland Agriculture

5.6.1 Introduction

Wetlands are unique ecosystems that combine terrestrial and aquatic characteristics, characterized by frequent saturation or inundation creating anaerobic conditions (Mitsch et al. 2023). They are identified by indicators such as hydric soils and hydrophytic vegetation (NRC 1995). Wetlands provide essential ESs including biodiversity conservation, flood control, water pollution mitigation, CS, fish and biomass production, and recreation (Zedler and Kercher 2005; Delle Grazie and Gill 2022). However, wetlands are increasingly subjected to agricultural expansion through reclamation, involving drainage and alteration of water flow patterns like ditching, diking, or dredging to reduce water levels and promote drying. This transformation can significantly impact the environment, leading to habitat loss for species adapted to wetland conditions, compromising water quality, altering hydrological cycles, and potentially increasing flood risks by reducing natural water filtration (Day 2003; Loria et al. 2015). Recent studies have investigated the long-term impacts of wetland reclamation. In Hangzho Bay's coastal wetlands, reclamation over millennia has shown decreasing pH and electrical conductivity with time, alongside increased surface SOC due to agricultural practices content (Wang et al. 2023). However, SOC tends to decrease with depth, underscoring the need for careful management during early reclamation stages to preserve C levels. Similarly, conversion of natural wetlands near Xingkai Lake into soybean or paddy fields has been studied, revealing reduced microbial biomass and activity across various soil layers post-reclamation (Huo et al. 2015). Changes in plant species, litter accumulation, and waterlogging significantly influence microbial processes (Huo et al. 2015). These findings highlight the complex interplay between agricultural expansion and ecosystem health in wetlands, emphasizing the importance of sustainable land management practices to balance agricultural productivity with the preservation of wetland ecosystems. Figure 5.3 illustrates paddy transplantation following wetland conversion.

Global wetlands represent a substantial reservoir of C, storing between 520 and 710 Pg C above and below ground (Poulter et al. 2021). Including permafrost, this estimate expands to 1792–1882 Pg C. Adopting sustainable wetland agriculture practices enables farmers to harness wetlands' CS potential while maintaining agricultural productivity and ecosystem integrity. This approach balances food security with environmental conservation through advanced technologies like flood control in smaller catchment areas, high-yield farmland development, and resilient crop varieties for waterlogging and drought. Eco-friendly aquaculture methods, appropriate fertilization, environmentally friendly pesticides, controlled drainage systems, and innovative irrigation techniques (such as intermittent and dry-wet alternation methods) safeguard wetland ecosystems. Implementing zero-discharge aquaculture and recycling farmland drainage further reinforces sustainable approaches (Zhu et al. 2013). Drained agricultural peatlands are major CO_2 emitters due to deep drainage, and restoring them can reduce emissions, though balancing with agricultural needs is critical.

Fig. 5.3 Paddy transplantation in a wetland field (Photo taken by the author, Dr. Nancy Loria)

Paludiculture offers emission reductions but faces economic challenges (Tan et al. 2021). In Northeast Germany, implementing paludiculture through peatland rewetting could potentially mitigate up to 3 Tg CO_2eq annually, compared to current agricultural activities generating 4.5 Tg CO_2eq from 57% of the peatland area (Tanneberger et al. 2020). Rewetting disturbed peatlands alters biophysical and biogeochemical processes, affecting net radiation, surface water, turbulent fluxes, nighttime temperatures, and GHG emissions, necessitating integrated restoration strategies (Lee et al. 2021). Transitioning to less intense drainage systems and implementing agroforestry on partially drained peatlands are short-term solutions for restoration (Miller 2022; Freeman et al. 2022). Initially using pioneer or non-native species facilitates ecosystem recovery before transitioning to native species for sustainable management (Tan et al. 2021). Implementing wetland agriculture requires overcoming agronomic, economic, and water management hurdles, necessitating collaborative solutions involving stakeholders. Land managers face the primary risks associated with adopting new policies or technologies, despite the crucial roles of policymakers, funders, and researchers in the process.

5.6.2 Carbon Dynamics and Management Strategies in Wetlands

Wetlands are critical for CS due to their waterlogged conditions that slow OM decomposition. While wetland cultivation typically reduces SOC, specific management practices can balance SOC inputs and outputs, potentially increasing SOC (Xu et al. 2019). These areas are often utilized for growing rice and other crops essential for global food security, simultaneously serving as significant C sinks. Preserving paddy and other anthropogenic wetlands is crucial for maintaining and enhancing SOC storage, with paddy soils capable of sequestering significantly more C compared to adjacent upland soils. Studies from eastern China indicate that converting these wetlands to upland use could potentially release 504 Tg of C as CO_2 alone (Chen et al. 2021), highlighting the substantial C storage potential of wetland ecosystems. Moreover, integrating wetland species with high CS potential, such as reeds and sedges, into agricultural practices can further enhance CS while diversifying crops and improving agricultural resilience to environmental challenges (Wang et al. 2024). Beyond soil CS, wetland agriculture mitigates climate change by reducing GHG emissions. Flooded conditions promote anaerobic decomposition of OM while inhibiting CH_4 release through the activity of CH_4-consuming bacteria (Gilbert and Frenzel 1995). Effective water management techniques, such as intermittent flooding and drainage, can further minimize CH_4 emissions from these systems (Jiao et al. 2006; Ma et al. 2013). Studies in central Ohio's temperate riverine wetlands showed moderate CH_4 emissions (0.51–1.03 Mg C ha^{-1} $yr.^{-1}$) and higher CO_2 emissions (1.26–1.51 Mg C ha^{-1} $yr.^{-1}$), contrasting with negligible CH_4 emissions (0.01 Mg C ha^{-1} $yr.^{-1}$) but higher CO_2 emissions (3.24 Mg C ha^{-1} $yr.^{-1}$) in adjacent upland sites. Despite CH_4 emissions, these wetlands stored significant C (85–125 Mg C ha^{-1} up to 0.3 m depth) and exhibited a net CS rate of 5.55 Mg C ha^{-1} $yr.^{-1}$ (Nag et al. 2017), highlighting their role as substantial C sinks despite the unique environmental conditions that lead to CH_4 emissions.

Wetland management strategies significantly influence C and N dynamics in various ecosystems. For instance, drying-rewetting cycles in alpine wetlands, particularly with litter addition, accelerate C and N mineralization, which varies with soil moisture levels, underscoring implications for soil C dynamics under changing climatic conditions (Gao et al. 2016). In seasonal wetlands, a study over a 9-year period investigated the combined effects of pasture management intensity, livestock grazing, and prescribed fire on soil nutrients. It found delayed responses to fire but notable increases in total phosphorus (P) in managed, grazed, and burned pastures, likely reflecting historical fertilizer and cattle impacts. While N storage varied, total C stocks remained stable, emphasizing wetlands' critical role in nutrient processing within ranching landscapes (Ho et al. 2018).

Moreover, wetland restoration and land-use changes play a crucial role in recovering lost C. Research in southern Ontario highlighted the CS potential of restoring small, drained wetlands. After 40 years of restoration efforts, SOC stocks reached 35.60 Mg C ha^{-1}, with sequestration rates of 0.89 Mg C ha^{-1} $yr.^{-1}$, comparable to intact wetlands. Economic analysis suggests that enhancing CS rates or reducing restoration costs is pivotal for financial viability, emphasizing the role of C markets in promoting wetland conservation and restoration as climate solutions (Creed et al. 2022). Conversely, in the Sanjiang Plain of northeast China, converting native wetlands to drylands resulted in a 79% reduction in SOC at depths of 0–50 cm. Transitioning drylands to rice paddy fields increased SOC by 24%, although paddy fields exhibited 74% lower SOC compared to original wetlands. This study underscored the sensitivity of labile C fractions to land-use changes, highlighting the complex impacts on C storage in wetland ecosystems (Chen et al. 2023).

5.6.3 Variability in Carbon Sequestration Across Wetlands

Wetlands exhibit significant variability in CS rates influenced by diverse factors such as water inundation duration, plant community composition, and external C inputs. For instance, in the Colorado Rocky Mountains, depressional wetlands with seasonal water table fluctuations and high clay content show higher SOC content characterized by aliphatic C forms under anaerobic conditions, contrasting with slope wetlands receiving perennial groundwater inputs and dominated by aromatic C forms. However, depressional wetlands are increasingly vulnerable to drying and oxic conditions with drainage and climate change (Daugherty et al. 2019). Climate also plays a crucial role, with tropical wetlands exhibiting the highest CS rates, followed by temperate and boreal wetlands (Villa and Bernal 2018). Extreme climate events, such as droughts and floods, can reduce the CS capacity of coastal wetlands due to increased salinity (Chu et al. 2019). Influenced by temperature and precipitation variations, the Zoige alpine wetland has maintained its role as a significant C sink, with SOC storage increasing from 2.03 Pg in 1985 to 2.21 Pg in 2020, despite shrinking by 0.04 M ha over 35 years (Zhang et al. 2023).

Different wetland types and geomorphological factors also influence C stocks significantly. Mangroves and peat swamps generally sequester more C than marshes (Adame et al. 2015). Studies in Ohio highlight that depressional wetlands sequester more C annually (3.17 ± 0.93 Mg ha^{-1} yr. − 1) than riverine wetlands (1.40 ± 0.16 Mg ha^{-1} $yr.^{-1}$), emphasizing the role of wetland type and vegetation communities in CS (Bernal and Mitsch 2012). Research in Huascaran National Park, Peru, found that peatlands store significantly more C (~ 1092 Mg ha^{-1}) than wet meadows (~ 30 Mg ha^{-1}), with peatlands containing 97% of the park's total wetland C (Chimner et al. 2023). River deltas in Moreton Bay, Australia, harbor significantly higher SOC stocks (175–504 Mg ha^{-1}) compared to tide-dominated sand islands (44–271 Mg ha^{-1}), reflecting the influence of geomorphological factors and proximity to riverine outflows (Hayes et al. 2017). A study in southern Canada revealed that swamps and marshes accumulate SOC stocks averaging 60 to 136 Mg ha^{-1}, with swamps containing approximately 23% more SOC than marshes. Flooded wetlands showed 60% higher SOC stocks than seasonally wet systems, with an average accumulation rate of 1.12 ± 0.87 Mg ha^{-1} $yr.^{-1}$, underscoring the potential of wetland restoration through re-wetting to mitigate C loss from historical agricultural drainage (Pendea et al. 2023).

Geographical location strongly influences C stocks in wetlands globally. In the contiguous United States, wetlands store 11.52 Pg C, predominantly in soils deeper than 30 cm, with inland freshwater wetlands containing ten times more C than tidal saltwater sites (Nahlik and Fennessy 2016). Similarly, China's diverse wetlands sequester 16.87 Pg C, equivalent to 315.76 Mg C ha^{-1}, contributing 3.8% to global wetland C stocks (Xiao et al. 2019). In New Zealand, organic soils in freshwater wetlands hold approximately 11 Tg of C at 0.3 m depth, increasing to 144 Tg considering full peat depth, while mineral soils store around 23 Tg at the same depth (Ausseil et al. 2015). Victoria's wetlands store 68 Tg of organic C in the top meter of soil and sequester 3 Tg of CO_2eq annually, offsetting about 3% of the state's annual emissions (Carnell et al. 2018). In Arctic Norway, coastal wetlands vary widely in sequestration rates (0.19–6.03 Mg C ha^{-1} $yr.^{-1}$) and soil C stocks (3.67–13.79 Mg C ha^{-1}), influenced by historical sediment accretion outpacing sea level rise (Ward 2020). Ecuadorian Andean wetlands exhibit high C storage capacities (126–454 Mg ha^{-1} at 0–30 cm and 148–350 Mg ha^{-1} at 30–60 cm), emphasizing their importance for CO_2 emission mitigation (Beltrán-Dávalos et al. 2022).

In conclusion, the variability in CS across wetlands highlights the complex interplay of environmental, climatic, and geographical factors influencing these critical ecosystems' capacity to store C and mitigate climate change. Understanding these dynamics is crucial for developing effective wetland conservation and climate change mitigation strategies globally.

5.7 Knowledge Gaps and Research Priorities

Addressing C farming challenges across various agricultural systems requires targeted research to fill existing knowledge gaps and prioritize actionable strategies. In row cropping, there is a need to investigate the long-term impacts of crop rotations and cover crops on SOC sequestration, including how different crop species, rotation schedules, and cover crop types affect soil C dynamics and overall soil health. Additionally, research should focus on the integration of CA practices and their efficacy across diverse soil types and climates to enhance C storage.

In plantation farming, research must assess the CS potential of large-scale monocultures, such as oil palm and coffee plantations, compared to natural forests. This includes evaluating the impact of various management practices, such as agroforestry and reduced chemical inputs, on C stocks. It is crucial to understand the trade-offs between maintaining high productivity and achieving significant CS, particularly in tropical regions where plantation farming is prevalent.

For livestock farming, research should explore the effectiveness of different feed additives and dietary adjustments in reducing enteric CH_4 emissions, a significant contributor to GHGs. Additionally, there is a need to study the effects of rotational grazing, manure management, and agroecological practices on soil CS and overall soil health. The potential of integrating trees into pastureland (silvopasture) for enhancing CS also warrants further investigation.

In wetland agriculture, research should focus on the effects of water management practices, such as drainage and irrigation, on GHG emissions and CS. Understanding how to manage wetland soils to maximize C storage while maintaining agricultural productivity is essential. This includes developing practices that balance effective wetland management with CS goals.

Finally, in horticulture and specialty cropping, research is needed to evaluate the impact of various soil management techniques, including the use of organic amendments and mulch, on SOC levels. It is also important to compare the CS potential of intensive horticulture systems with more extensive agricultural practices. Exploring methods to integrate C farming practices into high-value crop production without compromising yield or quality is a key research priority.

5.8 Conclusion

C farming represents a sustainable approach that emphasizes storing C in soils over the long term. This concept is highly adaptable across diverse agricultural sectors, each offering unique opportunities for CS at scale. In row cropping systems, practices such as conservation tillage, cover cropping, and crop rotation improve soil health and OM content, thereby enhancing C storage. Plantation farming, including orchards and tree plantations, contributes significantly to CS through the accumulation of biomass and the promotion of deep root systems that stabilize soils.

Livestock farming, traditionally associated with significant GHG emissions, can adopt strategies like rotational grazing and improved pasture management to enhance C storage in soils and vegetation. Horticultural and specialty cropping systems, with their diverse crop rotations and intensive management practices, can optimize CS potential through soil amendments and enhanced OM accumulation. Wetland agriculture, including rice paddies and flooded fields, presents unique opportunities for CS through practices like water management and organic soil amendments, which promote anaerobic conditions conducive to C storage.

By integrating C farming practices across these diverse agricultural sectors, stakeholders can not only mitigate GHG emissions but also enhance agricultural sustainability and resilience to climate change. Supporting research, policy incentives, and farmer education are crucial for scaling up these practices and maximizing their potential impact on global C budgets.

References

Abbey M, Smith AG, Yue C et al (2024) Measuring specialty crop grower willingness to pay for genetic modification and genetic editing. Agribusiness. https://doi.org/10.1002/agr.21911

Adame MF, Santini NS, Tovilla C et al (2015) Carbon stocks and soil sequestration rates of tropical riverine wetlands. Biogeosciences 12:3805–3818. https://doi.org/10.5194/bg-12-3805-2015

Ahlswede BJ, O'Halloran TL, Forsythe JD, Thomas RQ (2022) A minimally managed switchgrass ecosystem in a humid subtropical climate is a source of carbon to the atmosphere. GCB Bioenergy 14:24–36. https://doi.org/10.1111/gcbb.12897

Ahmed S, Stepp JR (2016) Beyond yields: climate change effects on specialty crop quality and agroecological management. Elementa Sci Anthrop 4:10.12952/journal.elementa.000092
Al-Ahmad H (2018) Biotechnology for bioenergy dedicated trees: meeting future energy demands. Zeitschrift für Naturforschung C 73:15–32. https://doi.org/10.1515/znc-2016-0185
Alam M, Jahiruddin MD (2007) Agroforestry for sustainable forage and livestock production in a small-holding farming system. J Anim Feed Sci 16:76–81. https://doi.org/10.22358/jafs/74459/2007
Amsili JP, Kaye JP (2021) Root traits of cover crops and carbon inputs in an organic grain rotation. Renew Agr Food Syst 36:182–191. https://doi.org/10.1017/S1742170520000216
An Y, Liu Y, Liu Y et al (2021) Opportunities and barriers for biofuel and bioenergy production from poplar. GCB Bioenergy 13:905–913. https://doi.org/10.1111/gcbb.12829
Aquilani C, Confessore A, Bozzi R et al (2022) Review: precision livestock farming technologies in pasture-based livestock systems. Animal 16:100429. https://doi.org/10.1016/j.animal.2021.100429
Arthur A, Acquaye S, Cheng W et al (2022) Soil carbon stocks and main nutrients under cocoa plantations of different ages. Soil Sci Plant Nutr 68:99–103. https://doi.org/10.1080/00380768.2022.2029219
Ashiagbor G, Asare-Ansah AO, Laari PB, Asante WA (2022) Cashew expansion holds potential for carbon stocks enhancement in the forest-savannah transitional zone of Ghana. Land Use Policy 121:106318. https://doi.org/10.1016/j.landusepol.2022.106318
Ausseil A-GE, Jamali H, Clarkson BR, Golubiewski NE (2015) Soil carbon stocks in wetlands of New Zealand and impact of land conversion since European settlement. Wetl Ecol Manag 23:947–961. https://doi.org/10.1007/s11273-015-9432-4
Ayala-Montejo D, Valdés-Velarde E, Benedicto-Valdés GS et al (2022) Soil biological activity, carbon and nitrogen dynamics in modified coffee agroforestry systems in Mexico. Agronomy 12:1794. https://doi.org/10.3390/agronomy12081794
Babich O, Krieger O, Chupakhin E, Kozlova O (2019) Miscanthus plants processing in fuel, energy, chemical and microbiological industries. Foods Raw Mater:403–411. https://doi.org/10.21603/2308-4057-2019-2-403-411
Badgery W, Murphy B, Cowie A et al (2021) Soil carbon market-based instrument pilot—the sequestration of soil organic carbon for the purpose of obtaining carbon credits. Soil Res 59:12. https://doi.org/10.1071/SR19331
Bai L, Cui J, Jie W, Cai B (2015) Analysis of the community compositions of rhizosphere fungi in soybeans continuous cropping fields. Microbiol Res 180:49–56. https://doi.org/10.1016/j.micres.2015.07.007
Bailey DW, Trotter MG, Tobin C, Thomas MG (2021) Opportunities to apply precision livestock management on rangelands. Front Sustain Food Syst 5. https://doi.org/10.3389/fsufs.2021.611915
Baker JM, Albrecht KA, Feyereisen GW, Gamble JD (2022) A perennial living mulch substantially increases infiltration in row crop systems. J Soil Water Conserv 77:212–220. https://doi.org/10.2489/jswc.2022.00080
Bareja B (2022) What is row planting and how crops are arranged. https://www.cropsreview.com/row-planting/. Accessed 7 May 2024
Baronti S, Ungaro F, Maienza A et al (2022) Rotational pasture management to increase the sustainability of mountain livestock farms in the Alpine region. Reg Environ Chang 22:50. https://doi.org/10.1007/s10113-022-01896-1
Bate NJ, Dardick CD, de Maagd RA, Williams RW (2021) Opportunities and challenges applying gene editing to specialty crops. In Vitro Cell Dev Biol - Plant 57:709–719. https://doi.org/10.1007/s11627-021-10208-x
Baumber A (2017) Enhancing ecosystem services through targeted bioenergy support policies. Ecosyst Serv 26:98–110. https://doi.org/10.1016/j.ecoser.2017.06.012
Bautista EG, Regalado MJC (2023) Assessment of conventional and highly mechanized Rice production systems using life cycle assessment (LCA) approach. Philipp J Crop Sci 48:54–62

Beltrán-Dávalos AA, Ayala Izurieta JE, Echeverria Guadalupe MM et al (2022) Evaluation of soil organic carbon storage of Atillo in the Ecuadorian Andean wetlands. Soil Syst 6:92. https://doi.org/10.3390/soilsystems6040092
Bernal B, Mitsch WJ (2012) Comparing carbon sequestration in temperate freshwater wetland communities. Glob Chang Biol 18:1636–1647. https://doi.org/10.1111/j.1365-2486.2011.02619.x
Bichel A, Oelbermann M, Voroney P, Echarte L (2016) Sequestration of native soil organic carbon and residue carbon in complex agroecosystems. Carbon Manag 7:261–270. https://doi.org/10.1080/17583004.2016.1230441
Birthwright A-T, Mighty M (2023) Risky business: modeling the future of Jamaica's coffee production in a changing climate. Climate 11:122. https://doi.org/10.3390/cli11060122
Bissonnette J-F, De Koninck R (2017) Social and environmental implications of plantation agriculture in Malaysia and Indonesia. In: Oxford research encyclopedia of environmental science. Oxford University Press
Blösch S, Albrecht M, Jenny M et al (2023) Rows make the field: winter wheat fields with manipulated crop architecture show potential for ecological intensification based on higher natural pest and weed seed control. Agric Ecosyst Environ 348:108404. https://doi.org/10.1016/j.agee.2023.108404
Bonsch M, Humpenöder F, Popp A et al (2016) Trade-offs between land and water requirements for large-scale bioenergy production. GCB Bioenergy 8:11–24. https://doi.org/10.1111/gcbb.12226
Borger CPD, Hashem A, Pathan S (2010) Manipulating crop row orientation to suppress weeds and increase crop yield. Weed Sci 58:174–178. https://doi.org/10.1614/WS-09-094.1
Byrnes RC, Eastburn DJ, Tate KW, Roche LM (2018) A global meta-analysis of grazing impacts on soil health indicators. J Environ Qual 47:758–765. https://doi.org/10.2134/jeq2017.08.0313
Cantrell KB, Ducey T, Ro KS, Hunt PG (2008) Livestock waste-to-bioenergy generation opportunities. Bioresour Technol 99:7941–7953. https://doi.org/10.1016/j.biortech.2008.02.061
Cao S, Zhou Y, Zhou Y et al (2021) Soil organic carbon and soil aggregate stability associated with aggregate fractions in a chronosequence of citrus orchards plantations. J Environ Manag 293:112847. https://doi.org/10.1016/j.jenvman.2021.112847
Capa D, Pérez-Esteban J, Masaguer A (2015) Unsustainability of recommended fertilization rates for coffee monoculture due to high N2O emissions. Agron Sustain Dev 35:1551–1559. https://doi.org/10.1007/s13593-015-0316-z
Cardinael R, Chevallier T, Barthès BG et al (2015) Impact of alley cropping agroforestry on stocks, forms and spatial distribution of soil organic carbon — a case study in a Mediterranean context. Geoderma 259–260:288–299. https://doi.org/10.1016/j.geoderma.2015.06.015
Carnell PE, Windecker SM, Brenker M et al (2018) Carbon stocks, sequestration, and emissions of wetlands in south eastern Australia. Glob Chang Biol 24:4173–4184. https://doi.org/10.1111/gcb.14319
Ceotto E, Di Candilo M (2010) Shoot cuttings propagation of giant reed (Arundo donax L.) in water and moist soil: the path forward? Biomass Bioenergy 34:1614–1623. https://doi.org/10.1016/j.biombioe.2010.06.002
Cesoniene L, Dapkiene M, Sileikiene D (2019) The impact of livestock farming activity on the quality of surface water. Environ Sci Pollut Res 26:32678–32686. https://doi.org/10.1007/s11356-018-3694-3
Chagunda MGG, Flockhart JF, Roberts DJ (2010) The effect of forage quality on predicted enteric methane production from dairy cows. Int J Agric Sustain 8:250–256. https://doi.org/10.3763/ijas.2010.0490
Chen H, Shi F, Wang X et al (2023) Conversion of drylands to paddy fields on former wetlands restores soil organic carbon by accumulating labile carbon fractions in the Sanjiang Plain, Northeast China. J Sci Food Agric 103:599–605. https://doi.org/10.1002/jsfa.12171
Chen X, Hu Y, Xia Y et al (2021) Contrasting pathways of carbon sequestration in paddy and upland soils. Glob Chang Biol 27:2478–2490. https://doi.org/10.1111/gcb.15595

Chimner RA, Resh SC, Hribljan JA et al (2023) Mountain wetland soil carbon stocks of Huascarán National Park, Peru. Front Plant Sci 14. https://doi.org/10.3389/fpls.2023.1048609
Chu X, Han G, Xing Q et al (2019) Changes in plant biomass induced by soil moisture variability drive interannual variation in the net ecosystem CO2 exchange over a reclaimed coastal wetland. Agric For Meteorol 264:138–148. https://doi.org/10.1016/j.agrformet.2018.09.013
Clément R, Tuan D, Cuong V et al (2023) Transitioning from monoculture to mixed cropping systems: the case of coffee, pepper, and fruit trees in Vietnam. Ecol Econ 214:107980. https://doi.org/10.1016/j.ecolecon.2023.107980
Cooney D, Kim H, Quinn L et al (2017) Switchgrass as a bioenergy crop in the loess plateau, China: potential lignocellulosic feedstock production and environmental conservation. J Integr Agric 16:1211–1226. https://doi.org/10.1016/S2095-3119(16)61587-3
Cooper A, DeMarco J (2023) Composted biosolids amendments for enhanced soil organic carbon and water storage in perennial pastures in Colorado. Agric Ecosyst Environ 347:108401. https://doi.org/10.1016/j.agee.2023.108401
Creed IF, Badiou P, Enanga E et al (2022) Can restoration of freshwater mineral soil wetlands deliver nature-based climate solutions to agricultural landscapes? Front Ecol Evol 10. https://doi.org/10.3389/fevo.2022.932415
Cutulle MA, Maja JM (2021) Determining the utility of an unmanned ground vehicle for weed control in specialty crop systems. Ital J Agron 16. https://doi.org/10.4081/ija.2021.1865
Dad JM, Dand SA, Pala NA (2021) The effect of bi-culture cover crops on soil quality, carbon sequestration, and growth characteristics in apple orchards of North Western Himalayas. Agrofor Syst 95:1745–1758. https://doi.org/10.1007/s10457-021-00687-7
Dağlıoğlu ST, Taşkın R, Özteke Nİ et al (2024) Reducing strategies for carbon footprint of livestock in Izmir/Turkiye. Water Air Soil Pollut 235:243. https://doi.org/10.1007/s11270-024-07038-z
Dandekar AM, Martinelli F, Davis CE et al (2010) Analysis of early host responses for asymptomatic disease detection and management of specialty crops. Crit Rev Immunol 30:277–289. https://doi.org/10.1615/CritRevImmunol.v30.i3.50
Daugherty EE, McKee GA, Bergstrom R et al (2019) Hydrogeomorphic controls on soil carbon composition in two classes of subalpine wetlands. Biogeochemistry 145:161–175. https://doi.org/10.1007/s10533-019-00597-y
Day J (2003) Using Ecotechnology to address water quality and wetland habitat loss problems in the Mississippi basin: a hierarchical approach. Biotechnol Adv 22:135–159. https://doi.org/10.1016/j.biotechadv.2003.08.012
Deligios PA, Farina R, Tiloca MT et al (2021) C-sequestration and resilience to climate change of globe artichoke cropping systems depend on crop residues management. Agron Sustain Dev 41:20. https://doi.org/10.1007/s13593-021-00680-5
Delle Grazie FM, Gill LW (2022) Review of the ecosystem services of temperate wetlands and their valuation tools. Water (Basel) 14:1345. https://doi.org/10.3390/w14091345
Dessureault-Rompré J, Gloutney A, Caron J (2020) Nutrient availability under lettuce grown in Rye mulch in Histosols. Nitrogen 1:137–150. https://doi.org/10.3390/nitrogen1020012
Devendra C (1999) The relevance and implications of livestock-tree interactions in agroforestry systems in developing countries. Ann Arid Zone 38:399–414
Dincă LC, Grenni P, Onet C, Onet A (2022) Fertilization and soil microbial community: a review. Appl Sci 12:1198. https://doi.org/10.3390/app12031198
Domaćinović M, Mijić P, Novoselec J et al (2023) Advantages and threats of precise monitoring and management technology application on dairy farms. Poljoprivreda 29:70–77. https://doi.org/10.18047/poljo.29.2.9
Domiciano LF, Pedreira BC, da Silva NMF et al (2020) Agroforestry systems: an alternative to intensify forage-based livestock in the Brazilian Amazon. Agrofor Syst 94:1839–1849. https://doi.org/10.1007/s10457-020-00499-1
Dove MR (1993) Smallholder rubber and swidden agriculture in Borneo: a sustainable adaptation to the ecology and economy of the tropical forest. Econ Bot 47:136–147. https://doi.org/10.1007/BF02862016

Duan Y, Wang G, Liang L et al (2024) Intercropping fruit trees in tea plantation improves soil properties and the formation of tea quality components. Plant Physiol Biochem 210:108574. https://doi.org/10.1016/j.plaphy.2024.108574

Elouattassi Y, Ferioun M, El Ghachtouli N et al (2023) Agroecological concepts and alternatives to the problems of contemporary agriculture: monoculture and chemical fertilization in the context of climate change. J Agric Environ Int Dev 117:41–98. https://doi.org/10.36253/jaeid-14672

ERS USDA (2024) Specialty crops. https://www.ers.usda.gov/topics/farm-bill/2014-farm-bill/specialty-crops/. Accessed 14 Jul 2024

Fageria NK, Baligar VC, Bailey BA (2005) Role of cover crops in improving soil and row crop productivity. Commun Soil Sci Plant Anal 36:2733–2757. https://doi.org/10.1080/00103620500303939

Fennimore SA, Doohan DJ (2008) The challenges of specialty crop weed control, future directions. Weed Technol 22:364–372. https://doi.org/10.1614/WT-07-102.1

Ferchaud F, Peyrard C, Léonard J et al (2020) Large variations in N2O fluxes from bioenergy crops according to management practices and crop type. Atmosphere (Basel) 11:675. https://doi.org/10.3390/atmos11060675

Flaspohler DJ, Webster CR (2011) Plantations for bioenergy: principles for maintaining biodiversity in intensively managed forests. For Sci 57:516–524. https://doi.org/10.1093/forestscience/57.6.516

Franzluebbers AJ (2010) Achieving soil organic carbon sequestration with conservation agricultural systems in the Southeastern United States. Soil Sci Soc Am J 74:347–357. https://doi.org/10.2136/sssaj2009.0079

Franzluebbers AJ, Sawchik J, Taboada MA (2014) Agronomic and environmental impacts of pasture–crop rotations in temperate North and South America. Agric Ecosyst Environ 190:18–26. https://doi.org/10.1016/j.agee.2013.09.017

Freeman BWJ, Evans CD, Musarika S et al (2022) Responsible agriculture must adapt to the wetland character of mid-latitude peatlands. Glob Chang Biol 28:3795–3811. https://doi.org/10.1111/gcb.16152

Gao J, Feng J, Zhang X et al (2016) Drying-rewetting cycles alter carbon and nitrogen mineralization in litter-amended alpine wetland soil. Catena (Amst) 145:285–290. https://doi.org/10.1016/j.catena.2016.06.026

Gao L, Liu X, Du Y et al (2019) Effects of tobacco–peanut relay intercropping on soil bacteria community structure. Ann Microbiol 69:1531–1536. https://doi.org/10.1007/s13213-019-01537-9

Garbuz S, Mackay A, Camps-Arbestain M et al (2021) Biochar amendment improves soil physicochemical properties and alters root biomass and the soil food web in grazed pastures. Agric Ecosyst Environ 319:107517. https://doi.org/10.1016/j.agee.2021.107517

Gerbens-Leenes W, Hoekstra AY, van der Meer TH (2009) The water footprint of bioenergy. Proc Natl Acad Sci 106:10219–10223. https://doi.org/10.1073/pnas.0812619106

Gerber PJ, Steinfeld H, Henderson B et al (2013) Tackling climate change through livestock: a global assessment of emissions and mitigation opportunities. In: Food and Agriculture Organization of the United Nations (FAO), Rome

German CL, Toensmeyer UC (1996) Market planning for specialty crops: an applied economic approach. Acta Hortic:513–518. https://doi.org/10.17660/ActaHortic.1996.429.66

Gilbert B, Frenzel P (1995) Methanotrophic bacteria in the rhizosphere of rice microcosms and their effect on porewater methane concentration and methane emission. Biol Fertil Soils 20:93–100. https://doi.org/10.1007/BF00336586

Glaze-Corcoran S, Hashemi M, Sadeghpour A et al (2020) Understanding intercropping to improve agricultural resiliency and environmental sustainability. In: Advances in agronomy, pp 199–256

Gopalakrishnan G, Cristina Negri M, Salas W (2012) Modeling biogeochemical impacts of bioenergy buffers with perennial grasses for a row-crop field in Illinois. GCB Bioenergy 4:739–750. https://doi.org/10.1111/j.1757-1707.2011.01145.x

Goyal S, Chander K, Mundra MC, Kapoor KK (1999) Influence of inorganic fertilizers and organic amendments on soil organic matter and soil microbial properties under tropical conditions. Biol Fertil Soils 29:196–200. https://doi.org/10.1007/s003740050544

Granco G, He H, Lentz B et al (2023) Mid- and end-of-the-century estimation of agricultural suitability of California's specialty crops. Land (Basel) 12:1907. https://doi.org/10.3390/land12101907

Gross A, Glaser B (2021) Meta-analysis on how manure application changes soil organic carbon storage. Sci Rep 11:5516. https://doi.org/10.1038/s41598-021-82739-7

Gryze SD, Wolf A, Kaffka SR et al (2010) Simulating greenhouse gas budgets of four California cropping systems under conventional and alternative management. Ecol Appl 20:1805–1819. https://doi.org/10.1890/09-0772.1

Guillen S, López G, Ormaza P et al (2023) Soil health and dragon fruit cultivation: assessing the impact on soil organic carbon. Sci Agropecu 14:519–528. https://doi.org/10.17268/sci.agropecu.2023.043

Han G, Schoolman ED, Arbuckle JG, Morton LW (2022) Weather, values, capacity and concern: toward a social-cognitive model of specialty crop farmers' perceptions of climate change risk. Environ Behav 54:327–362. https://doi.org/10.1177/00139165211026607

Hartemink AE (2005) Plantation agriculture in the tropics. Outlook Agric 34:11–21. https://doi.org/10.5367/0000000053295150

Hartemink AE (2008) Sugarcane for bioethanol. In: Sparks DL (ed) Advances in agronomy. Academic Press INC Elsevier Science, San Diego, CA, pp 125–182

Hayami Y (2010) Plantations agriculture. In: Pingali PL, Evenson RE (eds) Handbook of agricultural economics. Elsevier, Amsterdam, Netherlands, pp 3305–3322

Hayes MA, Jesse A, Hawke B et al (2017) Dynamics of sediment carbon stocks across intertidal wetland habitats of Moreton Bay, Australia. Glob Chang Biol 23:4222–4234. https://doi.org/10.1111/gcb.13722

Herrero M, Grace D, Njuki J et al (2013) The roles of livestock in developing countries. Animal 7:3–18. https://doi.org/10.1017/S1751731112001954

Ho J, Boughton EH, Jenkins DG et al (2018) Ranching practices interactively affect soil nutrients in subtropical wetlands. Agric Ecosyst Environ 254:130–137. https://doi.org/10.1016/j.agee.2017.11.031

Hombegowda HC, Adhikary PP, Jakhar P et al (2020) Hedge row intercropping impact on runoff, soil erosion, carbon sequestration and millet yield. Nutr Cycl Agroecosyst 116:103–116. https://doi.org/10.1007/s10705-019-10031-2

Horel Á, Zsigmond T (2023) Plant growth and soil water content changes under different inter-row soil management methods in a sloping vineyard. Plants 12:1549. https://doi.org/10.3390/plants12071549

Hulvey KB, Mellon CD, Kleinhesselink AR (2021) Rotational grazing can mitigate ecosystem service trade-offs between livestock production and water quality in semi-arid rangelands. J Appl Ecol 58:2113–2123. https://doi.org/10.1111/1365-2664.13954

Huo L, Guo J, Zou Y, Lu X (2015) Effect of reclamation on microbial biomass and activity in peat mire soil around Xingkai Lake in Northeast China. Fresenius Environ Bull 24:1098–1107

Iqbal MW, Draganova I, Morel PCH, Morris ST (2021) Validation of an accelerometer sensor-based collar for monitoring grazing and rumination behaviours in grazing dairy cows. Animals 11:2724. https://doi.org/10.3390/ani11092724

Jafarbiglu H, Pourreza A (2022) A comprehensive review of remote sensing platforms, sensors, and applications in nut crops. Comput Electron Agric 197:106844. https://doi.org/10.1016/j.compag.2022.106844

Jahangir MH, Montazeri M, Mousavi SA, Kargarzadeh A (2022) Reducing carbon emissions of industrial large livestock farms using hybrid renewable energy systems. Renew Energy 189:52–65. https://doi.org/10.1016/j.renene.2022.02.022

Janiola MDC, Marin A (2016) Carbon sequestration potential of fruit tree plantations in southern Philippines. J Biodivers Environ Sci 8:164–174

Jaskulska I, Jaskulski D, Gałęzewski L (2022) Peas and barley grown in the strip-till one pass technology as row intercropping components in sustainable crop production. Agriculture 12:229. https://doi.org/10.3390/agriculture12020229

Javad Faraji (2011) Wheat cultivar blends: a step forward to sustainable agriculture. Afr J Agric Res 6. https://doi.org/10.5897/AJARX11.047

Jiang M, Dong C, Bian W et al (2024) Effects of different fertilization practices on maize yield, soil nutrients, soil moisture, and water use efficiency in northern China based on a meta-analysis. Sci Rep 14:6480. https://doi.org/10.1038/s41598-024-57031-z

Jiao Z, Hou A, Shi Y et al (2006) Water management influencing methane and nitrous oxide emissions from rice field in relation to soil redox and microbial community. Commun Soil Sci Plant Anal 37:1889–1903. https://doi.org/10.1080/00103620600767124

Jin D, Yan R, Li L et al (2022) Stocking rate changed the magnitude of carbon sequestration and flow within the plant-soil system of a meadow steppe ecosystem. Plant Soil 473:33–47. https://doi.org/10.1007/s11104-021-05213-3

Johnson R (2014) Specialty crop provisions in the 2014 farm bill (P.L. 113-79). Washington, DC

Kaiyong W, Hua F, Ranab T et al (2011) Changes in soil carbon and nitrogen under long-term cotton plantations in China. J Agric Sci 149:497–505. https://doi.org/10.1017/S0021859611000049

Kakuk B, Bagi Z, Rákhely G et al (2021) Methane production from green and woody biomass using short rotation willow genotypes for bioenergy generation. Bioresour Technol 333:125223. https://doi.org/10.1016/j.biortech.2021.125223

Kantola IB, Masters MD, DeLucia EH (2017) Soil particulate organic matter increases under perennial bioenergy crop agriculture. Soil Biol Biochem 113:184–191. https://doi.org/10.1016/j.soilbio.2017.05.023

Khalsa SDS, Smart DR, Muhammad S et al (2020) Intensive fertilizer use increases orchard N cycling and lowers net global warming potential. Sci Total Environ 722:137889. https://doi.org/10.1016/j.scitotenv.2020.137889

Khamzina A, Lamers JPA, Vlek PLG (2012) Conversion of degraded cropland to tree plantations for ecosystem and livelihood benefits. In: Martius C, Rudenko I, Lamers J, Vlek P (eds) Cotton, water, salts and soums: economic and ecological restructuring in Khorezm, Uzbekistan. Springer, Netherlands, Dordrecht, pp 235–248

Kim H-J (2016) Opportunities and challenges of alternative specialty crops: the global picture. HortScience 51:1316–1319. https://doi.org/10.21273/HORTSCI10659-16

Lee S, Black TA, Nyberg M et al (2021) Biophysical impacts of historical disturbances, restoration strategies, and vegetation types in a peatland ecosystem. J Geophys Res Biogeosci 126. https://doi.org/10.1029/2021JG006532

Lee WS, Alchanatis V, Yang C et al (2010) Sensing technologies for precision specialty crop production. Comput Electron Agric 74:2–33. https://doi.org/10.1016/j.compag.2010.08.005

Lehrsch GA, Whisler FD, Buehring NW (1994) Cropping system influences on extractable water for mono- and double-cropped soybean. Agric Water Manag 26:13–25. https://doi.org/10.1016/0378-3774(94)90021-3

Li C, Aluko OO, Yuan G et al (2022) The responses of soil organic carbon and total nitrogen to chemical nitrogen fertilizers reduction base on a meta-analysis. Sci Rep 12:16326. https://doi.org/10.1038/s41598-022-18684-w

Li X, Mu Y, Cheng Y et al (2013) Effects of intercropping sugarcane and soybean on growth, rhizosphere soil microbes, nitrogen and phosphorus availability. Acta Physiol Plant 35:1113–1119. https://doi.org/10.1007/s11738-012-1148-y

Li Y-Z, Bao X-L, Tang S-X et al (2024) Toward soil carbon storage: the influence of parent material and vegetation on profile-scale microbial community structure and necromass accumulation. Soil Biol Biochem 193:109399. https://doi.org/10.1016/j.soilbio.2024.109399

Liang L, Ridoutt BG, Wang L et al (2021) China's tea industry: net greenhouse gas emissions and mitigation potential. Agriculture 11:363. https://doi.org/10.3390/agriculture11040363

Lithourgidis AS, Dordas CA, Damalas CA, Vlachostergios DN (2011) Annual intercrops: an alternative pathway for sustainable agriculture. Aust J Crop Sci 5:396–410

Lorenz K, Lal R (2014) Soil organic carbon sequestration in agroforestry systems. A review. Agron Sustain Dev 34:443–454. https://doi.org/10.1007/s13593-014-0212-y

Loria N, Verma KS, Bhardwaj KS, Brahmi MK (2015) Impact of catchment land use on water quality of pong wetland of Himachal Pradesh. Indian J Ecol 42:21–26

Ma K, Conrad R, Lu Y (2013) Dry/wet cycles change the activity and population dynamics of methanotrophs in rice field soil. Appl Environ Microbiol 79:4932–4939. https://doi.org/10.1128/AEM.00850-13

Ma Q, Wu Y, Liu Y et al (2024a) Interspecific interaction and productivity in a dryland wheat/alfalfa strip intercropping. Field Crop Res 309:109335. https://doi.org/10.1016/j.fcr.2024.109335

Ma W, Qi X, Sun Y et al (2024b) Computer vision-based measurement techniques for livestock body dimension and weight: a review. Agriculture 14:306. https://doi.org/10.3390/agriculture14020306

Majsztrik JC, Behe B, Hall CR et al (2019) Social and economic aspects of water use in specialty crop production in the USA: a review. Water (Basel) 11:2337. https://doi.org/10.3390/w11112337

Man N, Yee SS (2023) Sustainable livestock practices, pp 25–35

Mann JJ, Kyser GB, Barney JN, DiTomaso JM (2013) Assessment of aboveground and belowground vegetative fragments as propagules in the bioenergy crops Arundo donax and Miscanthus × giganteus. Bioenergy Res 6:688–698. https://doi.org/10.1007/s12155-012-9286-z

Mardegan SF, de Castro AF, Chaves SSNF et al (2022) Organic farming enhances soil carbon and nitrogen dynamics in oil palm crops from Southeast Amazon. Soil Sci Plant Nutr 68:104–113. https://doi.org/10.1080/00380768.2022.2031285

Márquez-García F, Hayas A, Peña A et al (2024) Influence of cover crops and tillage on organic carbon loss in Mediterranean olive orchards. Soil Tillage Res 235:105905. https://doi.org/10.1016/j.still.2023.105905

Martin T, Culman S, Sprunger CD (2022) Quality or quantity? Determining the impact of fine root traits on soil health in row crop agriculture. J Soil Sci Plant Nutr 22:2322–2333. https://doi.org/10.1007/s42729-022-00811-1

Martins BH, Araujo-Junior CF, Miyazawa M et al (2015) Soil organic matter quality and weed diversity in coffee plantation area submitted to weed control and cover crops management. Soil Tillage Res 153:169–174. https://doi.org/10.1016/j.still.2015.06.005

McSherry ME, Ritchie ME (2013) Effects of grazing on grassland soil carbon: a global review. Glob Chang Biol 19:1347–1357. https://doi.org/10.1111/gcb.12144

Miller MA (2022) Market-based commons: social agroforestry, fire mitigation strategies, and green supply chains in Indonesia's peatlands. Trans Inst Br Geogr 47:77–91. https://doi.org/10.1111/tran.12472

Minick KJ, Leggett ZH, Sucre EB et al (2017) Soil and aggregate-associated carbon in a young loblolly pine plantation. Soil Sci 182:233–240. https://doi.org/10.1097/SS.0000000000000215

Mitsch WJ, Gosselink JG, Anderson CJ, Fennessy MS (2023) Wetlands, Sixth. John Wiley & Sons, Inc., Hoboken, NJ

Montalvo M, Pajares G, Guerrero JM et al (2012) Automatic detection of crop rows in maize fields with high weeds pressure. Expert Syst Appl 39:11889–11897. https://doi.org/10.1016/j.eswa.2012.02.117

Morgan-Davies C, Kyle J, Boman IA et al (2021) A comparison of farm labour, profitability, and carbon footprint of different management strategies in Northern European grassland sheep systems. Agric Syst 191:103155. https://doi.org/10.1016/j.agsy.2021.103155

Nadaraja D, Lu C, Islam MM (2021) The sustainability assessment of plantation agriculture – a systematic review of sustainability indicators. Sustain Prod Consum 26:892–910. https://doi.org/10.1016/j.spc.2020.12.042

Nag SK, Liu R, Lal R (2017) Emission of greenhouse gases and soil carbon sequestration in a riparian marsh wetland in Central Ohio. Environ Monit Assess 189:580. https://doi.org/10.1007/s10661-017-6276-9

Nahlik AM, Fennessy MS (2016) Carbon storage in US wetlands. Nat Commun 7:13835. https://doi.org/10.1038/ncomms13835
Novara A, Pulido M, Rodrigo-Comino J et al (2019) Long-term organic farming on a citrus plantation results in soil organic carbon recovery. Cuadernos de Investigación Geográfica 45:271–286. https://doi.org/10.18172/cig.3794
NRC (1995) Wetlands: characteristics and boundaries. National Academies Press, Washington, D.C.
Oates LG, Jackson RD (2014) Livestock management strategy affects net ecosystem carbon balance of subhumid pasture. Rangel Ecol Manag 67:19–29. https://doi.org/10.2111/REM-D-12-00151.1
Odunze AC, Tarawali SA, de Haan NC et al (2004) Forage legumes for soil productivity enhancement and quality fodder production. Food Agr Environ 2:201–209
Oelbermann M, Echarte L, Marroquin L et al (2017) Estimating soil carbon dynamics in intercrop and sole crop agroecosystems using the century model. J Plant Nutr Soil Sci 180:241–251. https://doi.org/10.1002/jpln.201600578
Pandey VC, Singh K, Singh JS et al (2012) Jatropha curcas: a potential biofuel plant for sustainable environmental development. Renew Sust Energ Rev 16:2870–2883. https://doi.org/10.1016/j.rser.2012.02.004
Pendea IF, Kanavillil N, Kurissery S, Chmura GL (2023) Carbon stocks and recent rates of carbon sequestration in nutrient-rich freshwater wetlands from Lake Simcoe watershed (southern Canada). J Geophys Res Biogeosci 128. https://doi.org/10.1029/2023JG007561
Perazzoli BE, Pauletti V, Quartieri M et al (2020) Changes in leaf nutrient content and quality of pear fruits by biofertilizer application in northeastern Italy. Rev Bras Frutic 42. https://doi.org/10.1590/0100-29452020530
Peter GF (2008) Southern pines: a resource for bioenergy. In: Genetic improvement of bioenergy crops. Springer, New York, NY, pp 421–449
Phoofolo MW, Giles KL, Elliott NC (2010) Effects of relay-intercropping sorghum with winter wheat, alfalfa, and cotton on lady beetle (Coleoptera: Coccinellidae) abundance and species composition. Environ Entomol 39:763–774. https://doi.org/10.1603/EN09122
Pittarello M, Probo M, Perotti E et al (2019) Grazing management plans improve pasture selection by cattle and forage quality in sub-alpine and alpine grasslands. J Mt Sci 16:2126–2135. https://doi.org/10.1007/s11629-019-5522-8
Poulter B, Fluet-Chouinard E, Hugelius G et al (2021) A review of global wetland carbon stocks and management challenges. In: Krauss KW, Zhu Z, Stagg CL (eds) Wetland carbon and environmental management, pp 1–20
Rahman MM, Mostafiz SB, Wahid MA (2017) Widespread production of bioenergy: land and water availability factors. Biofuels 8:623–632. https://doi.org/10.1080/17597269.2016.1242693
Ramirez-Torres F, Ghogare R, Stowe E et al (2021) Genome editing in fruit, ornamental, and industrial crops. Transgenic Res 30:499–528. https://doi.org/10.1007/s11248-021-00240-3
Rayne N, Aula L (2020) Livestock manure and the impacts on soil health: a review. Soil Syst 4:64. https://doi.org/10.3390/soilsystems4040064
Regassa TH, Wortmann CS (2014) Sweet sorghum as a bioenergy crop: literature review. Biomass Bioenergy 64:348–355. https://doi.org/10.1016/j.biombioe.2014.03.052
Reinhart KO, Rinella MJ, Sanni Worogo HS et al (2022a) Lessons from a next generation carbon ranching experiment. Geoderma 425:116061. https://doi.org/10.1016/j.geoderma.2022.116061
Reinhart KO, Sanni Worogo HS, Rinella MJ (2022b) Ruminating on the science of carbon ranching. J Appl Ecol 59:642–648. https://doi.org/10.1111/1365-2664.14100
Remus A, Hauschild L, Létourneau-Montminy M-P, Pomar C (2021) Estimating amino acid requirements in real-time for precision-fed pigs: the challenge of variability among individuals. Animals 11:3354. https://doi.org/10.3390/ani11123354
Robertson AD, Zhang Y, Sherrod LA et al (2018) Climate change impacts on yields and soil carbon in row crop dryland agriculture. J Environ Qual 47:684–694. https://doi.org/10.2134/jeq2017.08.0309

Rodrigues GS, Martins CR, de Barros I (2018) Sustainability assessment of ecological intensification practices in coconut production. Agric Syst 165:71–84. https://doi.org/10.1016/j.agsy.2018.06.001
Rodriguez Izaba OF, Torres AP, Marshall MI, Thompson AW (2023) Market access and value-added strategies in the specialty crops industry. HortScience 58:32–39. https://doi.org/10.21273/HORTSCI16909-22
Ronchetti G, Mayer A, Facchi A et al (2020) Crop row detection through UAV surveys to optimize on-farm irrigation management. Remote Sens 12:1967. https://doi.org/10.3390/rs12121967
Rowntree R, Nowak D (1991) Quantifying the role of urban forests in removing atmospheric carbon dioxide. J Arboricul 17:269–275. https://doi.org/10.48044/jauf.1991.061
SAN (2024) Sustainable livestock production. Sustainable Agriculture Network (SAN). https://www.sustainableagriculture.eco/impact-areas/sustainable-livestock-production. Accessed 9 Jul 2024
Sandhu RK, Boyd NS, Sharpe S et al (2020) Management of relay-cropped Strawberry and Eggplant to maximize yield and economic return. HortScience 55:1083–1089. https://doi.org/10.21273/HORTSCI14932-20
Sartori F, Lal R, Ebinger MH, Parrish DJ (2006) Potential soil carbon sequestration and CO_2 offset by dedicated energy crops in the USA. CRC Crit Rev Plant Sci 25:441–472. https://doi.org/10.1080/07352680600961021
Sati VP, Singh RB (2010) Prospects of sustainable livestock farming in the Uttarakhand Himalaya, India. J Livest Sci 1:9–16
Schmidt H-P, Hagemann N, Draper K, Kammann C (2019a) The use of biochar in animal feeding. PeerJ 7:e7373. https://doi.org/10.7717/peerj.7373
Schmidt R, Mitchell J, Scow K (2019b) Cover cropping and no-till increase diversity and symbiotroph:saprotroph ratios of soil fungal communities. Soil Biol Biochem 129:99–109. https://doi.org/10.1016/j.soilbio.2018.11.010
Scholten MCT, de Boer IJM, Gremmen B, Lokhorst C (2013) Livestock farming with care: towards sustainable production of animal-source food. NJAS Wagening J Life Sci 66:3–5. https://doi.org/10.1016/j.njas.2013.05.009
Sharma S, Rana VS, Kumari M, Mishra P (2018a) Biofertilizers: boon for fruit production. J Pharmacogn Phytochem 7:3244–3247
Sharma V, Irmak S, Padhi J (2018b) Effects of cover crops on soil quality: part I. Soil chemical properties—organic carbon, total nitrogen, pH, electrical conductivity, organic matter content, nitrate-nitrogen, and phosphorus. J Soil Water Conserv 73:637–651. https://doi.org/10.2489/jswc.73.6.637
Siberski-Cooper CJ, Koltes JE (2021) Opportunities to harness high-throughput and novel sensing phenotypes to improve feed efficiency in dairy cattle. Animals 12:15. https://doi.org/10.3390/ani12010015
Sillitoe P (2001) Horticultural societies. In: International encyclopedia of the social & behavioral sciences. Elsevier, pp 6909–6915
Silva-Olaya AM, Cerri CEP, La Scala JN et al (2013) Carbon dioxide emissions under different soil tillage systems in mechanically harvested sugarcane. Environ Res Lett 8:015014. https://doi.org/10.1088/1748-9326/8/1/015014
Singh M, Joshi NL (1994) Performance of pearl millet-based intercropping systems under drought conditions. Arid Soil Res Rehabil 8:277–283. https://doi.org/10.1080/15324989409381402
Singh S, Jaiswal DK, Krishna R et al (2020) Restoration of degraded lands through bioenergy plantations. Restor Ecol 28:263–266. https://doi.org/10.1111/rec.13095
Smith DR, Livingston SJ, Zuercher BW et al (2008) Nutrient losses from row crop agriculture in Indiana. J Soil Water Conserv 63:396–409. https://doi.org/10.2489/jswc.63.6.396
Smith MA, Carter PR (1998) Strip intercropping corn and alfalfa. J Prod Agric 11:345–353. https://doi.org/10.2134/jpa1998.0345

Snapp SS, Gentry LE, Harwood R (2010) Management intensity—not biodiversity—the driver of ecosystem services in a long-term row crop experiment. Agric Ecosyst Environ 138:242–248. https://doi.org/10.1016/j.agee.2010.05.005

Soldatos P (2015) Economic aspects of bioenergy production from perennial grasses in marginal lands of South Europe. Bioenergy Res 8:1562–1573. https://doi.org/10.1007/s12155-015-9678-y

Sprunger CD, Culman SW, Peralta AL et al (2019) Perennial grain crop roots and nitrogen management shape soil food webs and soil carbon dynamics. Soil Biol Biochem 137:107573. https://doi.org/10.1016/j.soilbio.2019.107573

Strom N, Hu W, Haarith D et al (2020) Interactions between soil properties, fungal communities, the soybean cyst nematode, and crop yield under continuous corn and soybean monoculture. Appl Soil Ecol 147:103388. https://doi.org/10.1016/j.apsoil.2019.103388

Tamchos S, Dorjey K (2024) Hippophae rhamnoides subsp. turkestanica: a potential species for sustainable propagation in the Cold Desert of Ladakh. National Academy Science Letters. https://doi.org/10.1007/s40009-024-01431-2

Tan Y, Hu F, Chai Q et al (2020) Expanding row ratio with lowered nitrogen fertilization improves system productivity of maize/pea strip intercropping. Eur J Agron 113:125986. https://doi.org/10.1016/j.eja.2019.125986

Tan ZD, Lupascu M, Wijedasa LS (2021) Paludiculture as a sustainable land use alternative for tropical peatlands: a review. Sci Total Environ 753:142111. https://doi.org/10.1016/j.scitotenv.2020.142111

Tanneberger F, Schröder C, Hohlbein M et al (2020) Climate change mitigation through land use on rewetted peatlands—cross-sectoral spatial planning for Paludiculture in Northeast Germany. Wetlands 40:2309–2320. https://doi.org/10.1007/s13157-020-01310-8

Testa G, Reyneri A, Blandino M (2016) Maize grain yield enhancement through high plant density cultivation with different inter-row and intra-row spacings. Eur J Agron 72:28–37. https://doi.org/10.1016/j.eja.2015.09.006

Thorne PJ, Tanner JC (2002) Livestock and nutrient cycling in crop–animal systems in Asia. Agric Syst 71:111–126. https://doi.org/10.1016/S0308-521X(01)00039-7

Tobar-López D, Bonin M, Andrade HJ et al (2019) Deforestation processes in the livestock territory of La Vía Láctea, Matagalpa, Nicaragua. J Land Use Sci 14:225–241. https://doi.org/10.1080/1747423X.2019.1671907

Tozzini L, Lakso AN, Flore JA (2015) Estimating the carbon footprint of Michigan apple and cherry trees – lifetime dry matter accumulation. Acta Hortic:85–90. https://doi.org/10.17660/ActaHortic.2015.1068.9

USDA (2024) USDA definition of specialty crop. https://www.nifa.usda.gov/sites/default/files/resources/definition_of_specialty_crops.pdf. Accessed 14 Jul 2024

van der Meulen A, Chauhan BS (2017) A review of weed management in wheat using crop competition. Crop Prot 95:38–44. https://doi.org/10.1016/j.cropro.2016.08.004

van Eijck J, Smeets E, Faaij A (2012) Jatropha: a promising crop for Africa's biofuel production? In: Janssen R, Rutz D (eds) Bioenergy for sustainable development in Africa. Springer, Netherlands, Dordrecht, pp 27–40

Velmourougane K, Bhat R (2017) Sustainability challenges in the coffee plantation sector. In: Bhat R (ed) Sustainability challenges in the Agrofood sector. Wiley, West Sussex, pp 616–642

Villa JA, Bernal B (2018) Carbon sequestration in wetlands, from science to practice: an overview of the biogeochemical process, measurement methods, and policy framework. Ecol Eng 114:115–128. https://doi.org/10.1016/j.ecoleng.2017.06.037

Wang F, Wang T, Gustave W et al (2023) Spatial-temporal patterns of organic carbon sequestration capacity after long-term coastal wetland reclamation. Agric Ecosyst Environ 341:108209. https://doi.org/10.1016/j.agee.2022.108209

Wang J, Fu Z, Qiao H et al (2024) Linking rhizospheric microbiota and metabolite interactions with harvested aboveground carbon and soil carbon of lakeshore reed wetlands in a subtropical region. Plant Soil 499:439–455. https://doi.org/10.1007/s11104-023-06471-z

Ward RD (2020) Carbon sequestration and storage in Norwegian Arctic coastal wetlands: impacts of climate change. Sci Total Environ 748:141343. https://doi.org/10.1016/j.scitotenv.2020.141343

Whitaker J, Field JL, Bernacchi CJ et al (2018) Consensus, uncertainties and challenges for perennial bioenergy crops and land use. GCB Bioenergy 10:150–164. https://doi.org/10.1111/gcbb.12488

WHO (2024) One health. WHO. https://www.who.int/health-topics/one-health#tab=tab_1. Accessed 8 Jul 2024

Willers JL, Riggins JJ (2010) Geographical approaches for integrated pest management of arthropods in forestry and row crops. In: Precision crop protection – the challenge and use of heterogeneity. Springer, Netherlands, Dordrecht, pp 183–202

Williams A, Davis AS, Jilling A et al (2017) Reconciling opposing soil processes in row-crop agroecosystems via soil functional zone management. Agric Ecosyst Environ 236:99–107. https://doi.org/10.1016/j.agee.2016.11.012

World Bank (2022) Moving towards sustainability: the livestock sector and the World Bank. World Bank. https://www.worldbank.org/en/topic/agriculture/brief/moving-towards-sustainability-the-livestock-sector-and-the-world-bank. Accessed 8 Jul 2024

Wu T, Wang Y, Yu C et al (2012) Carbon sequestration by fruit trees – Chinese apple orchards as an example. PLoS One 7:e38883. https://doi.org/10.1371/journal.pone.0038883

Xiao D, Deng L, Kim D et al (2019) Carbon budgets of wetland ecosystems in China. Glob Chang Biol 25:2061–2076. https://doi.org/10.1111/gcb.14621

Xiong W, Li Z, Liu H et al (2015) The effect of long-term continuous cropping of black pepper on soil bacterial communities as determined by 454 pyrosequencing. PLoS One 10:e0136946. https://doi.org/10.1371/journal.pone.0136946

Xu C, Mou B (2018) Chitosan as soil amendment affects lettuce growth, photochemical efficiency, and gas exchange. HortTechnology 28:476–480. https://doi.org/10.21273/HORTTECH04032-18

Xu S, Liu X, Li X, Tian C (2019) Soil organic carbon changes following wetland cultivation: a global meta-analysis. Geoderma 347:49–58. https://doi.org/10.1016/j.geoderma.2019.03.036

Xu S, Zhang W, Goodwin PH et al (2024) Effect of cover crop on soil fertility and bacterial diversity in a banana plantation in southwestern China. Soil Tillage Res 240:106092. https://doi.org/10.1016/j.still.2024.106092

Yan L, Zhang Z-D, Zhang J-J et al (2016) Effects of improving nitrogen management on nitrogen utilization, nitrogen balance, and reactive nitrogen losses in a Mollisol with maize monoculture in Northeast China. Environ Sci Pollut Res 23:4576–4584. https://doi.org/10.1007/s11356-015-5684-z

Yang C, Hou F, Sun Y et al (2020) Oats hay supplementation to yak grazing alpine meadow improves carbon return to the soil of grassland ecosystem on the Qinghai-Tibet Plateau. China Glob Ecol Conserv 23:e01158. https://doi.org/10.1016/j.gecco.2020.e01158

Yang F, Liao D, Wu X et al (2017) Effect of aboveground and belowground interactions on the intercrop yields in maize-soybean relay intercropping systems. Field Crop Res 203:16–23. https://doi.org/10.1016/j.fcr.2016.12.007

Yoo J-H, Luyima D, Lee J-H et al (2021) Effects of brewer's spent grain biochar on the growth and quality of leaf lettuce (Lactuca sativa L. var. crispa.). Appl Biol Chem 64:10. https://doi.org/10.1186/s13765-020-00577-z

Yuan X, Yang B, Liu W et al (2024) Intercropping with cash crops promotes sustainability of rubber agroforestry: insights from litterfall production and associated carbon and nutrient fluxes. Eur J Agron 154:127071. https://doi.org/10.1016/j.eja.2023.127071

Zaidelman FR, Belichenko MV, Bibin AS (2013) Degradation and restoration of soils in the Moscow River floodplain for the last fifty years. Eurasian Soil Sci 46:1097–1106. https://doi.org/10.1134/S1064229313110070

Zalles V, Hansen MC, Potapov PV et al (2019) Near doubling of Brazil's intensive row crop area since 2000. Proc Natl Acad Sci 116:428–435. https://doi.org/10.1073/pnas.1810301115

Zedler JB, Kercher S (2005) Wetland resources: status, trends, ecosystem services, and restorability. Annu Rev Environ Resour 30:39–74. https://doi.org/10.1146/annurev.energy.30.050504.144248
Zhang D, Jiang Q, Liang D et al (2021) The potential application of giant reed (Arundo donax) in ecological remediation. Front Environ Sci 9. https://doi.org/10.3389/fenvs.2021.652367
Zhang H, Wang J, Zhang Y et al (2023) Soil organic carbon dynamics and influencing factors in the Zoige alpine wetland from the 1980s to 2020 based on a random Forest model. Land (Basel) 12:1923. https://doi.org/10.3390/land12101923
Zhang S, Sun L, Wang Y et al (2020) Cow manure application effectively regulates the soil bacterial community in tea plantation. BMC Microbiol 20:190. https://doi.org/10.1186/s12866-020-01871-y
Zhao Q, Xiong W, Xing Y et al (2018) Long-term coffee monoculture alters soil chemical properties and microbial communities. Sci Rep 8:6116. https://doi.org/10.1038/s41598-018-24537-2
Zhu J-Q, Yu H, He Q-Y (2013) Development of wetland agriculture and its prospect. In: 2013 third international conference on intelligent system design and engineering applications. IEEE, pp 596–601
Zhu-Barker X, Easter M, Swan A et al (2019) Soil management practices to mitigate nitrous oxide emissions and inform emission factors in arid irrigated specialty crop systems. Soil Syst 3:76. https://doi.org/10.3390/soilsystems3040076

Chapter 6
Measuring and Monitoring Soil Carbon Sequestration

Abstract This chapter provides a comprehensive overview of the methods used for measuring and monitoring soil carbon (C), with a primary focus on soil organic C (SOC). It begins by introducing key concepts of C sequestration and its role in improving soil health and agricultural productivity. The chapter then details various SOC measurement techniques, including wet oxidation, dry combustion, spectroscopic and colorimetric methods, as well as the growing use of remote sensing technologies. In addition, it explores approaches for monitoring SOC, including soil C modeling and addresses the challenges involved in long-term monitoring. Recent advancements, such as the integration of remote sensing, machine learning, and other technological innovations, are also discussed to highlight how these tools can enhance the accuracy, efficiency, and scalability of soil C assessments.

6.1 Introduction

Measuring and monitoring soil carbon (C) sequestration are essential components of C farming, a practice aimed at enhancing the soil's capacity to capture and store atmospheric carbon-dioxide (CO_2) through sustainable land management (SLM) practices. Soil C sequestration (CS) includes both soil organic C (SOC) and soil inorganic C (SIC), with SOC being the primary component influenced by SLM practices. Techniques such as conservation agriculture (CA), cover cropping, agroforestry, and rotational grazing help increase SOC levels and improve soil health (Lal 2016; Pezzuolo et al. 2017; Naorem et al. 2023). By quantifying the C stored in the soil, farmers can demonstrate the environmental benefits of their practices, supporting climate change mitigation and enhancing the appeal of C farming initiatives for potential C credit markets (Nevalainen et al. 2022; Hayo and Hasegawa 2024). Accurate data on soil C is crucial for verifying and validating C credits, which can be sold to companies or organizations looking to offset their CO_2 emissions. This creates a financial incentive for farmers to adopt and maintain sustainable agriculture practices (SAPs), aligning economic benefits with environmental goals.

N. Loria, R. Lal, *Carbon Farming*, https://doi.org/10.1007/978-3-032-00842-8_6

Monitoring soil CS is vital for understanding the impact of these practices on soil health and productivity. SAPs that increase CS often lead to improved soil structure, water retention, and nutrient availability (Meena et al. 2020). By tracking changes in SOC levels, farmers can assess how these improvements contribute to enhanced crop yields, reduced need for chemical inputs, and overall better farm productivity. Soil CS is a gradual process, and the benefits of C farming practices may take years or even decades to fully materialize. Regular monitoring allows farmers to track progress over time, adjust practices as needed, and ensure that CS goals are being met (Nevalainen et al. 2022). It also helps in identifying any potential issues or challenges that may arise, enabling timely interventions to maintain or improve soil C stocks. Furthermore, robust data on soil CS supports broader policy and research efforts (Lorenz and Lal 2018a). Governments and organizations developing climate policies or research programs rely on accurate soil C data to design effective strategies and interventions. By providing clear evidence of the benefits of C farming, monitoring contributes to the formulation of policies that support sustainable agriculture and climate resilience. The objectives of this chapter are to discuss in detail the various methods used for measuring and monitoring soil C, with a particular focus on SOC. It will also cover foundational concepts related to CS, emphasizing the significance of soil CS in enhancing soil health and agricultural productivity. Furthermore, the chapter will highlight recent advancements in soil C measurement techniques and innovative monitoring approaches, as well as the integration of technological innovations such as remote sensing (RS) and machine learning to enhance the accuracy and efficiency of soil C assessments.

6.2 Carbon Sequestration

CS is the process of capturing and securely storing atmospheric CO_2 to mitigate risks of its re-emission into the atmosphere. This process can be achieved through both natural and engineered methods (Fig. 6.1), with C being stored in various reservoirs, including biotic, oceanic, pedologic, and atmospheric pools (Lal 2008). Biotic reservoirs encompass C storage in vegetation through photosynthesis. Practices such as afforestation and reforestation contribute to this pool, which holds an estimated 560 PgC (Lal 2008). Pedologic reservoirs, primarily consisting of soils, store approximately 2500 PgC, forming part of a terrestrial pool with a total capacity of about 3060 PgC (Lal 2008). Peatlands, which are C-rich wetlands covering roughly four million km^2, store over 750 Pg of C globally, highlighting their critical role in the global C cycle and climate regulation (Lorenz and Lal 2018b). Oceans, the largest C storage reservoir with 38,000 Pg of C, including 670 Pg in the surface layer (Lal 2008), involves several mechanisms: the biological pump, which sequesters CO_2 through phytoplankton photosynthesis and the sinking of organic matter (OM); the microbial C pump, where microbes decompose and transform organic C, exporting it to the deep ocean for long-term storage; and coastal and marine vegetation, such as seagrasses, mangroves, and salt marshes, which capture

Carbon sequestration	Natural	Terrestrial	Soil Carbon sequestration
			Afforestation/ reforestation
			Peatlands
		Oceanic	Marine and coastal vegetation
			Biological pump
			Microbial Carbon pump
	Engineered	Bioenergy with carbon capture and storage (BECCS)	Growing biomass for energy, capturing the resulting CO_2 emissions, and storing it in geological formations
		Direct air capture	Physisorption
			Chemisorption
		Ocean based engineered methods	Ocean fertilization; Artificial upwelling
		Mineral Carbonation	Reacting CO_2 with naturally occurring alkaline minerals to form stable carbonates

Fig. 6.1 Overview of carbon sequestration methods and reservoirs

and store C (Zhang et al. 2022). The ocean's biological pump exports about 10 Pg C annually from the surface and sequesters around 1300 Pg C as dissolved inorganic C (DIC), with average sequestration times of 140 years for the gravitational pump, 150 years for the migrant pump, and 50 years for the mixing pump (Nowicki et al. 2022; Siegel et al. 2023).

Engineered methods for CS include bioenergy with C capture and storage (BECCS), direct air capture (DAC), mineral carbonation, and ocean-based sequestration (Fig. 6.1). BECCS involves planting bioenergy crops that, after being used for fuel, have their residual C stored in geological formations (Jeswani et al. 2022). DAC captures CO_2 directly from the atmosphere using physisorption and chemisorption techniques (Sun et al. 2023). Physisorption techniques involve adsorption on C-based materials, zeolites, and metal-organic frameworks (MOFs), while chemisorption includes adsorption on alkali carbonate-based adsorbents (e.g., NaOH, CaO, or $Ca(OH)_2$), solid amine-based adsorbents, silica-based materials, and amine-mixed metal oxide composites. Mineral carbonation involves reacting CO_2 with naturally occurring alkaline or alkaline earth oxides, such as MgO or CaO, to form carbonates ($MgCO_3$ or $CaCO_3$) (Snæbjörnsdóttir et al. 2020). Ocean-based engineered methods include artificial ocean fertilization, which enhances CS by increasing phytoplankton growth through the supply of limiting nutrients like iron, and artificial upwelling, which involves bringing nutrient-rich deep ocean water to the surface to boost phytoplankton growth and increase CS (Pan et al. 2015).

Soil CS is vital for mitigating climate change by capturing and storing atmospheric CO_2 within the soil matrix. Globally, soils hold nearly three times the C stored in the atmosphere (~760 Pg) and five times that in vegetation (~560 Pg). Of this, SOC makes up approximately 1550 Pg C, while SIC accounts for about 950 Pg C to 1 m depth (Lal 2008). SOC is more sensitive to changes in agricultural practices and can be significantly enhanced through sustainable methods such as no-till farming, cover cropping, and crop rotation (Lal 2004). Consequently, research has predominantly focused on SOC, with SIC receiving only about 4% of soil C research attention (Raza et al. 2024). SIC is typically categorized into three types: geogenic (derived from parent materials like limestone), biogenic (formed in living organisms), and pedogenic (authigenically formed in soils through weathering and precipitation processes) (Sharififar et al. 2023). A comprehensive approach to soil CS requires integrating knowledge of both SOC and SIC. SOC's role in C storage is influenced by OM decomposition and stabilization, while SIC contributes through mineral formation and weathering processes (Manning et al. 2024). Effective sequestration strategies must consider both components to enhance soil C stocks, improve soil health, and support climate change mitigation goals.

6.3 Measuring Soil Organic Carbon

To effectively understand and manage soil CS, it is essential to accurately measure and monitor soil C stocks including SOC and SIC. This process begins with sampling, where soil samples are collected from different depths and locations to ensure a representative measure of the C content. Following collection, the samples are analyzed in a laboratory to determine their C concentration, typically using methods such as dry combustion or wet chemistry (Bisutti et al. 2004). The next step involves modeling, where mathematical models are employed to estimate soil C stocks across different areas based on the sample data. These models are crucial for predicting changes over time and under diverse management practices (Zhang et al. 2024).

Accurate measurement of SOC is essential for understanding soil fertility, assessing C storage, and implementing effective soil management practices. Several methods are used to measure SOC, including direct sampling and analysis, indirect methods, and emerging technologies (Lal et al. 2001). SOC estimation methods range from qualitative and semi-quantitative approaches to precise quantitative techniques, with each having its own set of advantages and limitations (Table 6.1). Quantitative methods offer precise, accurate measurements of SOC but require more sophisticated equipment and procedures (Sleutel et al. 2007). Semi-quantitative methods provide approximate estimates that are easier to perform and analyze but are less precise (Ussiri et al. 2014). The choice of method depends on the specific goals of the study, available resources, cost, and the need for accuracy versus practicality (Zhang and Zhang 2014). Measuring SOC accurately requires a clear separation of SOC from SIC. Carbonates in soil are commonly detected using a fizz test: applying drops of 10% hydrochloric acid (HCl) to the soil core will produce strong effervescence if carbonates are present (Morgan et al. 2020). In soils without carbonates, SOC equals total C (TC) since there is no inorganic C to account for. However, in carbonate-containing soils, it is crucial to remove inorganic C before measuring TC to accurately estimate SOC (Shamrikova et al. 2024). This can be achieved by pretreating the sample with a strong acid, such as HCl and H_3PO_4, to

Table 6.1 Overview of soil organic carbon estimation methods

Method type	Method	Description	Precision	Destructive/non-destructive	Advantages	Limitations	References
Qualitative	Visual assessment	Estimation based on soil color and texture	Low	Non-destructive	Simple and inexpensive	Subjective and imprecise; Ignores moisture variations	Pretorius et al. (2017); Rubinić et al. (2021); Herts et al. (2022)
Semi-quantitative	Loss-on-ignition (LoI)	Measures weight loss after heating to burn off OM	Moderate	Destructive	Inexpensive and convenient; no use of toxic chemicals	Affected by carbonates; less precise; high energy use; limited for diverse soils; requires standard SOC calibration	Konen et al. (2002); Westman et al. (2006); Zhang and Zhang (2014); Burgos Hernández et al. (2023)
	Wet oxidation (Walkley-Black)	Uses chemical oxidation to estimate SOC by measuring dichromate reduction	Moderate	Destructive	Commonly used and relatively quick.	May not measure all SOC fractions; requires correction factor for incomplete digestion; involve harmful chemicals	Ulmer et al. (1992); Chatterjee et al. (2009); Loria et al. (2024)
	Infrared spectroscopy	Analyzes soil using near-infrared or mid-infrared light to estimate SOC based on spectral data	Moderate to high	Non-destructive	Rapid and accurate; cost-effective; less labor-intensive	Requires calibration and spectral interpretation	McCarty et al. (2010); Li et al. (2022); Loria et al. (2024)
	Remote sensing	Uses satellite imagery and airborne sensors to estimate SOC over large areas	Moderate to high	Non-destructive	Provides broad spatial coverage; minimal soil disturbance	Less precise at local scale; dependent on calibration	Chen et al. (2005); Ladoni et al. (2010); Croft et al. (2012); Angelopoulou et al. (2019); Loria et al. (2024)

(continued)

Table 6.1 (continued)

Method type	Method	Description	Precision	Destructive/non-destructive	Advantages	Limitations	References
Quantitative	Dry combustion (elemental analysis)	Burns soil samples at high temperatures to convert SOC to CO_2, which is then quantified	High	Destructive	Highly accurate; considered the gold standard	Requires sophisticated equipment; costly.	Ulmer et al. (1992); Santi et al. (2006); Vitti et al. (2016); Loria et al. (2024)
	Chemical extractions	Extracts specific SOC fractions using solvents or reagents for detailed analysis	High	Destructive	Provides detailed insights into SOC fractions (resistant and active C).	Requires multiple steps and can be labor-intensive; requires hazardous chemicals (e.g., sodium polytungstate)	Tan et al. (2007); Silveira et al. (2008); Poeplau et al. 2018

remove inorganic C (Bisutti et al. 2004; Dhillon et al. 2015) or by indirectly measuring SIC using the calcimeter method and subtracting it from TC (Sherrod et al. 2002; Apesteguia et al. 2018). The volumetric calcimetry is a reliable method for measuring SIC, effectively handling diverse conditions such as solid-to-liquid ratios, unweathered silicate phases, and high OM content. This method involves treating the samples with HCl to release CO_2, which is then captured in a graduated burette. The volume of displaced liquid in the burette is used to determine the CO_2 mass in the soil sample (Silva et al. 2024). Measuring methods for SOC and SIC are influenced by soil chemistry, accuracy needs, cost, calibration, and environmental factors, and discrepancies can arise from differences in sampling, preparation, measurement techniques, and storage conditions, complicating long-term monitoring and data consistency (Shamrikova et al. 2024). Accurate C assessment requires careful consideration of these factors to ensure reliable and comparable results.

6.3.1 Wet Oxidation

The Walkley-Black chromic acid wet oxidation method, introduced in 1934, offers a rapid and less precise estimation of SOC. It relies on the oxidation of SOC by using potassium dichromate ($K_2Cr_2O_7$) in concentrated sulfuric acid (H_2SO_4). During digestion, SOC is oxidized to CO_2, while potassium dichromate is reduced to chromium (III) ions. The remaining unreacted dichromate is then quantified by titration with a ferrous sulfate ($FeSO_4$) solution, using diphenylamine as an indicator. The amount of dichromate reduced is proportional to the amount of SOC, allowing for its estimation (Nelson and Sommers 1996; Ravindranath and Ostwald 2008b). This method, on average, recovers about 76% of the SOC, with a correction factor of 1.29 applied to account for the unrecovered portion (Walkley and Black 1934). Variability in recovery rates, which can range from 60–80%, is influenced by soil type, management practices, and soil depth, necessitating the development of a corrected Walkley-Black recovery factor (WBRFc) for more accurate SOC measurements across different bioclimatic regions (Bhattacharyya et al. 2015). Díaz-Zorita (1999) proposed correction factors of 1.69 for perennial pasture and 1.35 for agricultural soils in Typic Hapludolls in Argentina, whereas Mikhailova et al. (2003) suggested a uniform factor of 1.63 for a Pachic Hapludoll in Russia across different management regimes, emphasizing the need for soil-specific correction factors. Additionally, SOC recovery is influenced by the size and decomposition rates of organic particles, with smaller, more complex particles having longer residence times and larger, undecomposed fragments degrading more quickly (Buyanovsky et al. 1994). This method may be inaccurate for high-C soils due to incomplete oxidation and is unsuitable for soils with significant carbonized materials; using smaller sample weights can help improve accuracy in such cases (FAO 2019).

The Tyurin titrimetric technique, used primarily in Russia and parts of Central, Eastern Europe, and Asia, measures SOC by oxidizing soil OM (SOM) with potassium dichromate and sulfuric acid, followed by titration of the excess dichromate

(Jankauskas et al. 2006). SOC is calculated using a 1.724 factor, which assumes OM contains 58% C, though this may not accurately reflect variations in C content across different soils (Allison 2016). Studies have shown that the conventional factor of 1.724 for converting OM to C is too low, with a factor of 2 being more accurate (Gortner 1916; Pribyl 2010). Variations in soil composition and methods suggest no single conversion factor applies universally across soils (Pribyl 2010). The key difference between the Tyurin and Walkley-Black methods is that Walkley-Black uses a more concentrated dichromate and sulfuric acid mixture, which is heated, while Tyurin involves cooling the mixture and boiling it for 60 minutes (Shamrikova et al. 2023). Tyurin's higher reaction temperatures and longer oxidation times result in greater dispersion of soil particles, potentially leading to higher SOC values. Although largely phased out in some laboratories, the Tyurin method remains in use for historical data (Shamrikova et al. 2022, 2023). The Tyurin method is advantageous due to its simplicity, low cost, and lack of need for specialized equipment, but it suffers from incomplete oxidation of OM, particularly in peaty, decayed, or forest soils, resulting in SOC values about 90% of those obtained by dry combustion (Jankauskas et al. 2006). A conversion factor of 1.15 is mostly applied to account for unoxidized organic C in the Tyurin method (Shamrikova et al. 2022). Despite its broad application, the Tyurin method's accuracy depends on consistent procedures and careful calibration. Both the Tyurin and Walkley-Black methods involve the use of hazardous reagents, raising concerns about human and environmental health (Shamrikova et al. 2023). Alternatively, the modified Tyurin spectrophotometric method, endorsed by the Global Soil Laboratory Network (GLOSOLAN) of the Global Soil Partnership (FAO 2021), provides a safer alternative by measuring SOC through the absorbance of chromium (III) complexes at 590 nm, where Cr^{3+} levels correlate directly with C content in soil samples. Offers a safer approach. Additionally, a modified Walkley-Black method utilizing spectrophotometric determination has been proposed for its enhanced reproducibility, ease of use, and cost-effectiveness (Bahadori and Tofighi 2016; FAO 2019).

6.3.2 Dry Combustion

6.3.2.1 Loss-on-Ignition

Loss-on-Ignition (LoI) is a simple and cost-effective method for estimating SOM by measuring the weight loss of a sample after heating in a muffle furnace, where the lost weight reflects the OM content and the remaining weight represents the ash content (Konen et al. 2002; Schulte and Hopkins 2015). Ignition temperature and duration are crucial factors, with various studies employing different conditions: 400 °C for 16 hours (Nelson and Sommers 1996), 360 °C for 2 hours (Schulte and Hopkins 2015), 375 °C for 17 hours (Wang et al. 1996), 400 °C for 8 hours (Ben-Dor and Banin 1989), and 430 °C for 24 hours (Davies 1974). The goal is to choose temperatures that effectively combust SOC while minimizing the loss of mineral components (Salehi et al. 2011). Wang et al. (2013) validated the LoI method as reliable for estimating SOC and SIC in arid soils using a two-step approach with combustion at 375 °C for 17 hours and 800 °C for 12 hours, showing strong correlations and minimal errors. Zhang et al. (2005) found that LoI, when performed at 105 °C for drying and 500 °C for 12 hours of ignition, was most consistent with the Walkley and Black method for measuring OM in composts and mulches. Optimal heating conditions, typically below 500 °C, help avoid errors from carbonate CO_2 loss, structural water loss from clays, making LoI a reliable method when carefully controlled (Schulte and Hopkins 2015). However, LoI is less reliable for soils with SOC < 15 mg C/g compared to other methods but can be accurate if clay content is factored into the calibration (Soon and Abboud 1991). Without direct calibration, LoI results are indicative and temperature-based, and may include losses from non-organic components like water and mineral oxides (Konen et al. 2002; Sun et al. 2009; Johns et al. 2015). Caution is needed when samples contain interfering substances such as gypsum or clays that affect mass loss at relevant temperatures (Smith 2003). Clay content must be considered when converting LoI results to OM percentages in heavily textured soils, as the traditional conversion factor of 1.724 is invalid, indicating the need for region-specific factors (Sleutel et al. 2007). For accurate SOC measurement, LoI should ideally be calibrated against a standard method such as dry combustion.

To determine SOM, the soil sample is first weighed after being oven-dried at 105 °C for 2 hours to establish a base weight. The sample is then heated in muffle furnace at high temperatures, ranging from 300 °C to 550 °C, and the weight loss is measured (Salehi et al. 2011). The percentage of SOM is calculated using the following equation (Nelson and Sommers 1996):

$$SOM(\%) = \frac{\begin{array}{c} Weight\ of\ soil\ sample\ after\ combustion - weight\ of\ oven \\ -dried\ soil\ sample \end{array}}{Weight\ of\ oven-dried\ soil\ sample} \times 100$$

To calculate SOC from weight differences, individual conversion factors or regression equations must be tailored to specific soil types, land management practices, and regional conditions (Bisutti et al. 2004). A single regression equation is not accurate across different soils, requiring detailed soil knowledge and correct correction values for precise analysis (Konen et al. 2002). The suitability of LoI as a substitute for direct SOC measurement remains unresolved; logratio-based regression analysis indicated that LoI is effective at high SOC levels but unreliable at lower levels due to the loss of structural water, making direct measurement preferable despite its higher cost (Szava-Kovats 2009). Accurately calculating SOC from LOI requires using multiple predictive equations instead of a single factor (Baker et al. 2023). Various studies have developed different regression equations for various land uses to enhance SOC predictions from LoI analysis, with these equations reflecting the relationship between SOC concentrations determined by standard methods (primarily dry combustion) and estimates derived from LoI-based OM content (Table 6.2).

Table 6.2 Accuracy of soil organic carbon predictions using LoI equations across different regions

Region	Land use	Ignition temperature (°C)	Duration (hours)	Regression equation	R^2	Reference
West Africa	Agriculture	350	–	SOC = 0.36 LoI + 0.03	0.89	Konare et al. (2010)
North Belgium	Forest	550	3	SOC = 0.59 LoI - 0.1 clay	0.98	De Vos et al. (2005)
Lake Michigan, USA	Wetland	550	4	SOC = 0.53 LoI + 0.81	0.75	Braun et al. (2020)
Central Spain	Wetland	550	4	SOC = 0.63 LoI - 1.83	0.97	Santisteban et al. (2004)
South Florida	Wetland	550	4	SOC = 1.01 LoI - 18.10	0.98	Wright et al. (2008)
Mexico, USA	Forest	550	4	SOC = 0.55 LoI - 2.90	0.96	Campos C. (2010)
Florida, USA	Wetland	500	12	SOC = 1.05 LoI + 2.99	0.99	Wang et al. (2011)
Northwestern Washington (USA) and British Columbia (Canada)	Agriculture, pasture, and woodland	600	6	SOC = 0.39 LoI—0.58	0.98	Goldin (1987)
Flander, Belgium	Agriculture, pasture	800	4	–	0.99	Sleutel et al. (2007)

6.3.2.2 Automated Dry Combustion

In 1831, Justus von Liebig introduced a groundbreaking method for determining C content in organic compounds. By burning samples with copper oxide (Cuo) and measuring the resulting water vapor and CO_2 using a tube of calcium chloride and a specialized five-bulb apparatus with caustic potash, Liebig significantly improved both precision and efficiency. This approach reduced the processing time for seven samples from a week to one day (Brock 2013). Despite its significant advancements, the method's complexity and need for specialized equipment limited its practicality and led to the development of simpler, more efficient alternatives over time (Johns et al. 2015). Subsequent modifications to Liebig's method by researchers like Dennsted, Gustavson, Walkley, Robinson, Kataoka, and Kosaka aimed to simplify the process. These improvements, which included innovations such as electric furnaces and alternative reagents, addressed some practical issues but still struggled with limitations related to time and equipment size (Kosaka et al. 1959). Modern dry combustion operates by heating soil samples in a furnace at high temperatures, typically above 1000 °C, in the presence of pure oxygen. The high temperature facilitates the complete oxidation of SOC to CO_2, which is then measured to calculate SOC as a percentage of the soil's total weight (Bisutti et al. 2004; Johns et al. 2015). At lower temperatures, complete oxidation requires catalysts: Cr_2O_3,

CoO, and CuO are effective at 950 °C, while transition metal oxides like Pt, Cu, Ir, and Ni are suitable for 680 °C (Bisutti et al. 2004; O′ Rourke and Holden 2011). These methods measure TC but often include carbonates in the results unless pre-treatment is used to remove them. Soil samples should be dried before analysis, and precision is enhanced by using finely ground, representative samples to minimize measurement errors (Wang and Anderson 1998; Bisutti et al. 2004).

Commercial analyzers for the dry combustion offer advanced and efficient methods for analyzing SOC content. Automated dry combustion (ADC) systems often incorporate advanced analyzers, such as Non-Dispersive Infrared (NDIR) gas analyzers, to detect the released CO_2 accurately (Wang and Anderson 1998; Vitti et al. 2016). NDIR analyzers operate on the principle of measuring the absorption of infrared light by CO_2 molecules, enabling precise quantification of C content in soil samples (Yasuda et al. 2008). This technology allows for real-time monitoring, facilitating immediate feedback on SOC levels. Some systems also utilize thermal conductivity detectors (TCDs) to measure gas concentrations more effectively (Santi et al. 2006). TCDs analyze the thermal conductivity of gases, providing an additional layer of accuracy and reliability in the measurement process (Kowalenko 2001). In addition to CO_2 analyzers, modern ADC setups may include automated sample handling and processing systems that streamline workflows and reduce the potential for human error (Hatté et al. 2003; Dhaliwal et al. 2011). This automation not only increases efficiency but also ensures consistent results across multiple samples. Modern instruments are highly automated and can measure additional elements like nitrogen (N) and sulfur (S), improving the utility of the analysis (Matejovic 1993; Kowalenko 2001). They operate at temperatures between 950 °C and 1400 °C, sometimes using tin containers to facilitate combustion. Although they offer rapid and accurate results, they are expensive, require careful handling, and need regular maintenance, including catalyst replacement, to maintain precision (Tam and Yao 1998). Zhou et al. (2023) present a dry-combustion CO_2 detection system utilizing electrical heating for measuring SOM that analyzes samples in about 1.5 minutes, much faster than the 10 minutes required by TC analyzers. While the method involves simple pretreatment steps that take approximately one minute each and requires no expensive consumables, it currently lacks full automation due to the absence of an integrated rapid soil sampling device. Future improvements could reduce the overall testing time to around 15 seconds. However, challenges remain with varying soil textures, leading to higher detection error rates compared to TC methods. Table 6.3 summarizes the advanced instruments used in ADC for measuring SOC, emphasizing their shared principles and key features.

Modern dry combustion using an automated analyzer is the most used and reliable method for estimating SOC due to its high accuracy and precision compared to other traditional methods (Kerven et al. 2000; Santi et al. 2006; Silva-Olaya et al. 2013). It is also environmentally safe (Da Silva Dias et al. 2013) and serves as the standard for calibrating and evaluating alternative methods (Shamrikova et al. 2022). Many studies have compared various methods to ADC for measuring SOC, yielding both significant and non-significant correlations. For instance, Vitti et al. (2016) found a strong consistency and a high correlation of approximately 0.96

Table 6.3 Instruments for automated dry combustion of soil organic carbon

Manufacturer	Model	Key features	Detector	Reference
Laboratory equipment corporation (LECO), St. Joseph, MI, USA www.leco.com	LECO TruMac series C/N analyser	Convenient sample handling Fully automated with autoloading for 50 preweighed samples Easy access for combustion gas handling maintenance Supports macro sample analysis	Infrared detection	Strydom et al. (2019)
	Leco CNS-2000	Adjustable parameters (oven temperature, gas flow rates) optimize measurements for various sample types.	Non-dispersive infrared (NDIR)	Matejovic (1997); Kowalenko (2001)
	Leco CHN-600 and CHN-800 analyzers	The sample is combusted in a resistance furnace with oxygen, and the CO_2 produced is measured.	Infrared	Matejovic (1993)
Costech analytical technologies, Valencia, CA, USA www.costechanalytical.com	ECS 4010 - CHNS-O Elemental Combustion System	Fully automated analysis system User friendly detector that does not require reference gas High accuracy and precision	Thermal conductivity detector (TCD)	Orr et al. (2017)
PerkinElmer instruments, Shelton, CT, USA www.perkinelmer.com	Perkin Elmer 2400 elemental analyser	Optimized combustion flexibility for enhanced performance Simpler, more reliable, and accurate determination of CO_2	TCD	Fadeeva et al. (2008)
Thermo fisher scientific, Waltham, MA www.thermofisher.com	Carlo Erba CHN EA 1108	Simple design and usability Improved CHNS/O workflow analysis Higher sample throughput	TCD	Burgos Hernández et al. (2023)

Elementar Analysensysteme GmbH, Germany www.elementar.com	Vario MACRO resistance furnace CNS elemental analyzer	Simultaneous CHNS determination with highest precision and accuracy Minimum sample preparation Highest sample throughput Low maintenance design Advanced purge and trap (APT) technology measures samples with challenging C elemental ratios up to 7000:1.	TCD	Da Silva Dias et al. (2013)
	Elementar Variomax CNS-analyzer	Largest weighing range and sample volume Simplified sample preparation Maximum independence in carrier gas	TCD	Sleutel et al. (2007)
	Elemental analyzer Vario EL III	Easy operation and maintenance No requirement for costly special catalysts Affordable consumable expenses	TCD	Dhaliwal et al. (2011)
Skalar analytical, Breda, the Netherlands www.skalar.com	Skalar Primacs^{SLC} TOC analyser	Can process large volumes of samples daily in one batch.	NDIR	O′ Rourke and Holden (2011)
HekaTech, Wegberg, Germany www.hekatech.com	Euro EA - CHNSO elemental Analyser	Spontaneous combustion followed by chromatographic separation	TCD	Fadeeva et al. (2008); Hutengs et al. (2018)

between the Springer-Klee wet digestion and ADC methods in Southern Italy, with slight variations due to different soil types. Similarly, Ciric et al. (2021) found considerable variability in SOC concentrations and recoveries based on sample origin, highlighting the uncertainty and inconsistency across diverse soil types and conditions. Additionally, Sato et al. (2014) emphasized how mineralogy affects the accuracy of SOC determination in Brazilian savannah soils, leading to inconsistencies in C stock comparisons.

6.3.3 Spectroscopic Methods

Spectroscopic methods represent a significant advancement in SOC measurement, offering rapid, accurate, and non-destructive analysis compared to traditional combustion methods (Chatterjee et al. 2009; Acharya et al. 2022; Li et al. 2022; Loria et al. 2024). These methods, which include visible near-infrared (VNIR) spectroscopy, mid-infrared (MIR) spectroscopy, Laser-Induced Breakdown Spectroscopy (LIBS) and Inelastic-neutron scattering (INS), leverage the interaction of light with soil components, yielding valuable insights into SOC content and its implications for soil health and C cycling. Table 6.4 offers a comparative overview of these methods, highlighting their unique underlying principles, advantages, and limitations.

VNIR spectroscopy operates within the wavelength range of 0.8 to 2.5 μm, measuring reflectance that is sensitive to moisture and particle size (Nocita et al. 2013; Seidel et al. 2022). This method is particularly advantageous for in-situ studies, providing rapid assessments with minimal sample preparation (Johns et al. 2015). However, VNIR spectroscopy produces more diffuse spectra, which may lack specificity compared to other techniques. Its cost-effectiveness improves significantly with larger sample size, making it a popular choice for broader studies (Li et al. 2022). MIR ranges from 2.5 to 25 μm and focuses on the molecular vibrations of soil components. This technique delivers more specific spectra with clearer absorption bands, enhancing its accuracy in quantifying soil properties when samples are dried and ground (Bellon-Maurel and McBratney 2011; Johns et al. 2015). However, MIR typically requires moderate sample preparation, making it less suitable for field applications. A study by McCarty et al. (2002) found that MIR outperformed VNIR in measuring TC, SOC, and SIC, particularly when SIC was removed through acid treatment. Diffuse reflectance infrared spectroscopy, whether in the near or MIR spectrum, demonstrated potential for accurately predicting SOC and texture in low-C soils of the Sahel region, making it a valuable tool for resource-limited laboratories, unlike traditional methods such as wet oxidation and loss-on-ignition (McCarty et al. 2010).

LIBS utilizes high-energy laser pulses to create a plasma on the soil surface, which emits light that can be analyzed to determine the elemental composition of the soil. Covering a wavelength range from 200 nm to 980 nm, LIBS offers real-time analysis and high spatial resolution, with minimal sample preparation required compared to traditional methods (Izaurralde et al. 2013; Senesi and Senesi 2016). It

Table 6.4 Overview of spectroscopic methods for soil organic carbon estimation

Feature	Near-Infrared (NIR)	Mid-Infrared (MIR)	Laser-Induced Breakdown Spectroscopy (LIBS)	Inelastic Neutron Scattering (INS)	Reference (s)
Wavelength range	0.8–2.5 μm	2.5–25 μm	200 nm - 980 m	Depends on neutron energy (~1 Å wavelength)	Kearton and Mattley (2008); Johns et al. (2015); Kashyap (2022)
Principle	Measures reflectance in the visible and near-infrared regions; sensitive to moisture and particle size	Analyzes mid-infrared reflectance; focuses on molecular vibrations of OM and soil properties	Excites atoms with plasma energy, measuring emitted radiation for elemental composition analysis	Uses fast neutrons to interact with C atoms, detecting emitted gamma rays for SOC quantification	Loria et al. (2024)
Depth of penetration	Deeper penetration into samples	Shallow penetration; more specific spectra	Affects a small volume (~0.1–0.5 mm wide, 0.01–0.2 mm deep), varying based on pulse number and instrumentation	Bulk samples can be analysed (~0.4 m deep)	Johns et al. (2015)
Sample preparation	Minimal sample preparation	Moderate; may require preparation	Require sample preparation	Minimal to no sample preparation; bulk samples preferred	Chatterjee et al. (2009); Wielopolski et al. (2011); Johns et al. (2015)
Signal specificity	Less specific, more diffuse spectra	More specific with clearer absorption bands	Weak signals due to external environmental effects and inhomogeneous sample surfaces	Ability to distinguish between different elements based on their unique radiation signatures	Johns et al. (2015); Zhang et al. (2021)

(continued)

Table 6.4 (continued)

Feature	Near-Infrared (NIR)	Mid-Infrared (MIR)	Laser-Induced Breakdown Spectroscopy (LIBS)	Inelastic Neutron Scattering (INS)	Reference (s)
Advantages	Fast, often non-destructive, allows for in-situ application	Detailed Chemical information	High spatial resolution, real-time analysis	Non-destructive in-situ analysis; requires only a few measurements to interpolate across an entire field; capable of measuring large soil volumes (~300 kg); can be maneuvered over various field types	Wielopolski et al. (2010); Johns et al. (2015); Yakubova et al. (2017)
Disadvantages	Limited to overtone and combination bands; requires calibration across diverse soil types	Requires drying and grinding for consistency; requires reference methods; calibration can be complex	Matrix effects from other soil elements can complicate results; require signal enhancement	May lose lower C concentrations in soil; requires specialized equipment and expertise	Johns et al. (2015); Yakubova et al. (2017); Zhang et al. (2021); Loria et al. (2024)

can also perform simultaneous multi-element analysis. However, challenges remain, including control of plasma formation, self-absorption effects, and matrix influences from soil properties. Robust calibration models across diverse soil types are essential for effective application (Senesi and Senesi 2016). Studies such as Wangeci et al. (2024) demonstrate LIBS's superiority over VNIR spectroscopy in predicting SOC and other properties across diverse regions, while Villas-Boas et al. (2016) highlighted its efficacy in estimating soil texture in Brazilian samples. INS involves the interaction of fast neutrons with soil particles, enabling bulk sample analysis to a depth of approximately 0.4 m (Johns et al. 2015). This non-destructive method requires minimal sample preparation and can be utilized for in-situ analysis. However, INS has limitations, including lower sensitivity for detecting small variations in SOC contents and the necessity for specialized equipment (Yakubova et al. 2017). Research by Izaurralde et al. (2013) indicated that INS, alongside LIBS and other methods, can effectively measure SOC density, although improvements in calibration are still needed.

Recent advancements in soil science have significantly transformed the landscape of SOC analysis, particularly with the development of handheld spectroscopic devices. These portable instruments enable efficient real-time monitoring of SOC, offering researchers and practitioners the ability to conduct assessments directly in the field without the need for extensive laboratory setups (Loria et al. 2024). This accessibility not only saves time but also allows for immediate decision-making in C farming. Moreover, advancements in calibration strategies, such as spiking methods, have further enhanced the predictive capabilities of these handheld devices (Clairotte et al. 2016; Jiang et al. 2017; Nawar and Mouazen 2019; Hu et al. 2022; Babos et al. 2024). By incorporating these innovative calibration techniques, researchers can improve the accuracy and reliability of SOC measurements across diverse soil types and conditions. This adaptability is crucial, as soil characteristics can vary significantly within short distances, and effective monitoring requires a nuanced understanding of these variations. The integration of these technological advancements also underscores the importance of user-friendly designs, enabling non-specialists to effectively utilize these tools. Overall, each spectroscopic method has its own advantages and limitations, which affect their applications based on specific research needs and the unique environmental and geographical conditions of each site.

6.3.3.1 Reliability of Spectroscopic Methods

Spectroscopic methods hold significant promise for in situ studies, enabling real-time estimates of SOC for monitoring, reporting, and verification (MRV) programs. However, the accuracy of these methods is influenced by interrelated factors that affect the quality of results. Key considerations include sample preparation—such as drying, grinding, and achieving uniform particle sizes—which can introduce variability in spectral readings. For instance, Homhuan et al. (2016) reported that sieving samples through 2 mm mesh optimizes SOC estimation with VNIR spectroscopy, while Stumpe et al. (2011) demonstrated that lightly grinding soil for two minutes enhances predictions in MIR spectroscopy.

Regular calibration of instruments is crucial; inadequate calibration can lead to systematic errors (Morris and Langari 2012). Calibration methods, including pre-processing techniques and multivariate regression analysis, are essential for accurate SOC estimation using spectroscopy (Guo et al. 2021b; Wang et al. 2022). Common calibration methods in spectroscopic analysis include multiple linear regression (MLR), support vector machine (SVM), random forest (RF), and partial least squares regression (PLSR), with PLSR often preferred for its interpretability and efficiency (Loria et al. 2024). Effective validation is crucial, using independent samples and metrics like R^2 and RMSE to accurately assess prediction performance and reliability (Jia et al. 2017; Loria et al. 2024). Advanced methods, like fractional order derivatives combined with memory-based learning, have been shown to significantly improve SOC estimation (Hong et al. 2019). Furthermore, the number of samples used in developing calibration models is critical, as a larger sample size significantly reduces prediction errors (Kuang and Mouazen 2012).

Spectral interference from overlapping bands of soil components complicates result interpretation. Mirzaeitalarposhti et al. (2016) highlighted how carbonate interference affects MIR spectroscopy, proposing a spectral correction method to enhance reliability. Additionally, moisture content strongly influences spectral data, with variations misleading conclusions about soil composition (Jiang et al. 2016; Seidel et al. 2022). Seidel et al. (2022) investigated the effects of soil moisture on VNIR and MIR spectra, finding that MIR is more sensitive to moisture variability, while VNIR performs better under uniform moisture conditions. Integrating both VNIR and MIR data enhances predictive accuracy for SOC and clay content across different moisture levels. To mitigate moisture interference, algorithms like Direct Standardization (DS), External Parameter Orthogonalization (EPO), Orthogonal Signal Correction (OSC), Piecewise Direct Standardization (PDS), and Generalized Least Squares Weighting (GLSW) have been effective (Ji et al. 2015; Jiang et al. 2016; Veum et al. 2018; Wang et al. 2019; Liu et al. 2020). Yang et al. (2024) specifically noted that EPO is particularly effective at 25–48% moisture levels, while DS excels at lower moisture ranges. Other factors affecting spectral data include surface roughness (Römer et al. 2014; Piekarczyk et al. 2016), particle size (Ge et al. 2014; Wu et al. 2021), soil depth (Zhang et al. 2017; Shahrayini et al. 2022), and environmental conditions like ambient temperature (Li et al. 2023) and humidity.

A high signal-to-noise ratio is essential for accurate readings, as excessive noise can obscure critical spectral features (Stevens et al. 2008). Proper instrument maintenance and controlled measurement environments can help address this issue. Moreover, the quality and representativeness of the calibration dataset used to develop predictive models are critical; a dataset that inadequately captures the variability of the target population may result in biased predictions and poor model performance (Minasny et al. 2009; Kuang and Mouazen 2012; Dorantes et al. 2022). Finally, the resolution of the spectroscopic instrument plays a role in distinguishing closely spaced spectral features; higher resolution provides more detailed information, improving differentiation in complex mixtures (Stenberg and Rossel 2010). By addressing these factors through careful experimental design and robust methodology, the reliability and accuracy of spectroscopic measurements can be enhanced, leading to more confident interpretations in C farming.

6.3.4 *Colorimetric Methods*

These methods estimate SOC based on soil color. This technique involves correlating soil color parameters with SOC content, typically using statistical models to establish this relationship. This connection is based on the principle that darker soils generally contain higher levels of OM content, which affects their color (Konen et al. 2003). Common approaches include Munsell color system analysis and CIE-Lab color space analysis. The Munsell Color System, developed by A.H. Munsell in 1905, estimates SOC by analyzing soil color through the Munsell color chart. This system characterizes color based on three parameters: Hue (H), Value (V), and Chroma (C), which correspond to perceived hue, lightness, and color saturation, respectively (Luo 2015). Soil samples are compared to this chart to record their color parameters, which are then correlated with SOC content through regression models (Wills et al. 2007). These models predict SOC based on soil color measurements, offering a cost-effective and efficient alternative to direct chemical analysis. While this method is practical for field use and large-scale surveys, it has notable limitations (Johns et al. 2015). Effective color analysis demands meticulous calibration, which can be complex and vary by site, potentially impacting the consistency of results across different soil types. Traditional color charts, while useful, are constrained by their discrete nature, failing to capture the full spectrum of continuous color variations. Additionally, methods used may not adequately account for the diverse forms of organic C present in soils, which can affect the precision of SOC measurements. Furthermore, environmental factors such as soil moisture, texture, iron content, and lighting conditions can significantly influence color measurements, necessitating careful adjustments to ensure accurate predictions (Demattê et al. 2011; Fan et al. 2017; Swetha et al. 2022).

The CIE-Lab color space system, developed by the International Commission on Illumination in 1976, provides a precise method for quantifying soil color using Cartesian coordinates: Lightness (L*), Red-Green axis (a*), and Yellow-Blue axis (b*) (Vodyanitskii and Savichev 2017). This system provides a detailed numerical description of soil color, which is advantageous for accurately linking color to SOC content (Luo 2015). Using a colorimeter or spectrophotometer, soil color parameters are measured and correlated with SOC levels through regression analysis (Gozukara et al. 2023). This approach enhances the precision and reliability of SOC estimation by creating predictive models that account for color variations (Chen et al. 2018). The CIE-Lab system's standardization and uniformity facilitate consistent comparisons across studies and regions (Fan et al. 2017; Yang et al. 2021), making it a valuable tool for soil health monitoring, land management, and agricultural planning. However, its effectiveness can be limited by factors such as moisture content and light conditions, which can affect color measurements and model accuracy (Kirillova et al. 2021).

Several studies have investigated using digital devices to estimate SOC from images, emphasizing the role of soil color analysis. Accurate SOC estimation is influenced by factors such as soil moisture, texture, lighting, device characteristics, and iron mineralogy, making proper calibration essential for reliable results (Stiglitz et al. 2017; Sahwan et al. 2021; Yang et al. 2021; Gozukara et al. 2023). For instance, the Nix Pro color sensor, combined with soil depth, reliably predicts SOC in both dry and moist soils (Stiglitz et al. 2017). Research by Swetha & Chakraborty (2021) found that integrating Nix color sensor data with soil texture information improved SOC prediction accuracy (R^2 = 0.81) compared to using soil texture alone (R^2 = 0.70). Herts et al. (2022) demonstrated that combining visual soil color assessment with NixPro and Our Sci Reflectometer data significantly enhanced SOC estimation, with luminosity (L*) and the 500–632 nm spectral range improving prediction accuracy by 8–10%. Dai et al. (2024) showed that using the ASD FieldSpec 3 spectrometer and iron-removal learning machine (ILM) method substantially enhanced SOC estimation accuracy and real-time predictions by incorporating iron indices. Heil et al. (2022) highlighted that high-resolution digital images effectively map SOC and iron content at a fine scale, offering a cost-effective method for detailed soil property mapping.

6.3.5 Remote Sensing

Remote sensing (RS) has emerged as a powerful tool for assessing and monitoring SOC, providing spatial data on ecosystem C dynamics at local, regional, and global scales. This capability is crucial for tracking SOC stocks and fluxes, informing climate change mitigation strategies and resource management (César Izaurralde et al. 2001; Kumar et al. 2016; Guo et al. 2021a). The primary types of RS—optical, thermal, and microwave—each serve distinct functions in SOC assessment (Silvero et al. 2023). Optical RS captures sunlight reflected in the visible to short-wave infrared spectrum (400–2500 nm), making it effective for large-scale assessments despite potential atmospheric interference (Escribano et al. 2017). It frequently employs spectral indices like Normalized Difference Vegetation Index (NDVI) and Soil Adjusted Vegetation Index (SAVI) to indirectly estimate SOC levels by assessing vegetation cover and health (Kumar et al. 2016). Thermal RS measures emitted thermal radiation (3–14 μm) to monitor land surface temperature and infer soil characteristics, often requiring integration with other data for enhanced accuracy. Wang et al. (2023a) highlighted thermal RS role in tracking permafrost retreat and improving SOC estimates in vulnerable cold regions affected by global warming. In contrast to optical and thermal RS, microwave RS operates in the long wavelength range (0.5–100 cm) and is less susceptible to atmospheric interference, making it particularly effective for estimating soil moisture content due to the stark difference in dry soil and water dielectric properties (Mulder et al. 2011). Shi et al. (2008)

introduces new microwave vegetation indices (MVIs) derived from Advanced Microwave Scanning Radiometer (AMSR-E) observations, which effectively model vegetated surfaces by minimizing background soil emission issues. MVIs offer complementary information to NDVI, enhancing monitoring of global vegetation and phenology, though further research is needed to address limitations related to polarization effects and atmospheric influences. Microwave RS also has the advantage of penetrating deeper into forest canopies, allowing for accurate assessment of vegetation height and estimation of above-ground biomass (Lu et al. 2016).

RS technologies are utilized across diverse platforms—including spaceborne, airborne, and unmanned aerial systems (UAS) for monitoring SOC through imaging spectroscopy (Angelopoulou et al. 2019; Mandal et al. 2022; Pouladi et al. 2023; Das and Ghimire 2024). These platforms each bring unique strengths and limitations (Table 6.5). Remote sensing sensors vary in resolution, which is largely determined by the number of spectral bands they encompass and can be categorized into panchromatic, multispectral, and hyperspectral types. Panchromatic sensors capture data in a single band, multispectral sensors analyze 3 to 10 bands for land cover

Table 6.5 Comparison of soil organic carbon measurement platforms using remote sensing

Platform type	Key features	Advantages	Limitations	Case studies
Spaceborne sensors	High spatial coverage Frequent revisit times Multispectral and hyperspectral capabilities	Suitable for regional and global scale mapping Freely available data (e.g., Landsat, Sentinel-2) Extensive historical data	Atmospheric corrections needed Limited adaptability to conditions Ground validation needed	Biney et al. (2021); Gholizadeh et al. (2018); Vågen et al. (2016)
Airborne sensors	High spatial (0.5 to 2 m) and spectral resolution (400 to 2500 nm) Diverse revisit times High payload capacity	High accuracy for local-scale mapping Capable of scanning large areas Flexible flight conditions	Higher operational costs Weather-dependent Requires specific flight permissions Limited to top soils	Stevens et al. (2010); Pascucci et al. (2014); Gholizadeh et al. (2018); Dvorakova et al. (2020); Mandal et al. (2022)
UAV	Lower cost Autonomous data acquisition Flexible navigation Low spatial coverage and high spatial resolution	Suitable for complex and inaccessible areas Quick deployment Can capture fine-scale data	Limited payload Short flight duration Stability issues in wind	Angelopoulou et al. (2019); Pouladi et al. (2023); Tsouros et al. (2019)

variations, while hyperspectral sensors use 10 to hundreds of narrow bands for detailed feature identification (Das and Ghimire 2024). The choice of sensor depends on the specific requirements of the analysis. Soil sensor data can be leveraged both directly and indirectly to assess SOC levels (Silvero et al. 2023). The direct method provides insights into SOC composition by analyzing the interaction of electromagnetic energy with soil (Crucil et al. 2019), while the indirect approach uses laboratory data to calibrate models that estimate SOC from spectral data and environmental covariates (Campos et al. 2019). This process, known as Digital Soil Mapping (DSM), relates these data to SOC levels, facilitating improved characterization and monitoring of SOC across various landscapes. Bao et al. (2023) advances SOC mapping by integrating satellite and proximal sensing data, significantly enhancing prediction accuracy and spatial resolution across diverse geographical scales.

RS technologies provide valuable data related to SOC across spatial and temporal scales, enabling researchers to capture the dynamic nature of SOC in different environments. Key data derived from RS include vegetation indices like the NDVI and Enhanced Vegetation Index (EVI) which reflect plant health and biomass (Madugundu et al. 2022). These indices are useful for extracting phenological parameters that are closely linked to SOC levels (Yang et al. 2020). Additionally, RS technologies provide critical insights into soil properties, including land surface temperature and moisture content, which are vital for understanding SOC dynamics (Wang et al. 2020; Sodango et al. 2021). Climate variables, such as precipitation and temperature patterns captured over time, also play a significant role in influencing SOC accumulation and turnover rates (Mishra et al. 2010). Moreover, topographic features derived from Digital Elevation Models (DEMs) help assess how landscape characteristics, such as slope and aspect, impact soil erosion and deposition processes that affect SOC distribution (Delarmelina et al. 2022). Geological properties, including soil mineral composition and parent material, are important factors that shape SOC content and its stability within different ecosystems (Niwa et al. 2011).

Relying solely on RS data can introduce uncertainties due to spatial and temporal variability, which may lead to inaccurate estimates (Abdoli et al. 2023). To address this challenge, data fusion can be employed by integrating diverse data types—such as vegetation, soil properties, climate, topography, and geology—using advanced modeling techniques like machine learning and statistical analysis (Bou Kheir et al. 2010). For instance, Rengma et al. (2023) employed the SCORPAN model alongside over 100 environmental covariates to effectively map SOC and texture in the mid-Himalayas, achieving significant predictive accuracy with coefficient of determination (R^2) values of up to 0.84. Bou Kheir et al. (2010) analyzed SOC distribution across Denmark using RS, GIS, and decision-tree modeling, revealing high coincidence values and overall accuracies of approximately 68–70% compared to field observations. In another study, He et al. (2021) demonstrated that integrating phenological parameters and RS data from Sentinel-2 significantly improved SOC prediction accuracy in farmland, enhancing predictability by 171% and reducing RMSE by 14%. Additionally, Rasel et al. (2017) utilized LiDAR and WorldView 2 imagery to estimate SOC stocks in the subtropical forests of Chitwan, Nepal, achieving a R^2 of 0.69 based on variables like above-ground biomass, elevation, and forest type. Lamichhane et al. (2021) assessed the effectiveness of terrain variables—including elevation, maximum elevation deviation, and mid-slope position—alongside Sentinel-2 image-based indices (e.g., brightness index and SeLI) from three cropping seasons for SOC mapping. Their findings indicate that incorporating both terrain features and multi-season RS indices, enhances SOC prediction accuracy, tailored to local cropping patterns.

6.4 Monitoring Soil Organic Carbon

SOC is a dynamic component of soil ecosystems, constantly changing due to various factors such as management practices, climate fluctuations, and the migration of C from the topsoil to deeper layers over time, as well as the methods used for measurement (as noted in Sect. 6.3). This variability highlights the importance of regular monitoring to effectively maintain and enhance SOC levels. Regular SOC monitoring allows land managers and farmers to evaluate the impact of their agricultural practices and adjust strategies accordingly. For instance, the adoption of CA over eight years in Vertisols of Central India resulted in a significant increase in SOC levels, yielding long-term benefits for soil fertility, agricultural productivity, and environmental sustainability (Naorem et al. 2023). Conversely, SOC under continuous cultivation decreased due to heightened OM decomposition linked to practices such as tillage, irrigation, and nitrogen applications, along with lower clay and silt content (Stevenson et al. 2024). Additionally, the same study found that afforestation of abandoned cultivated lands contributed to increased SOC by enhancing OM accumulation. Furthermore, tracking SOC trends aids informed decision-making that enhances biodiversity and ecosystem services like water filtration, nutrient cycling, and habitat provision for various organisms (Blanco-Canqui 2024). In the context of C markets, monitoring SOC is crucial for earning C credits. As organizations and governments seek to reduce greenhouse gas (GHG) emissions, verified C credits have become a valuable commodity (Badgery et al. 2021; Okoli and Birkenberg 2024). Regular and accurate SOC assessments provide the necessary data to demonstrate CS efforts, enabling stakeholders to participate in C trading programs. This not only incentivizes SLM practices but also contributes to broader climate change mitigation goals. However, establishing standardized monitoring protocols is crucial to ensure reliable assessments and promote effective management strategies that enhance SOC levels and overall soil health (Jandl et al. 2014).

Several methods are available for monitoring SOC, each with distinct strengths and limitations that depend on the monitoring program's objectives, available resources, and spatial scale (Sect. 6.3). Traditional soil sampling and laboratory analysis yield accurate SOC data but can be labor-intensive and costly, often failing to capture spatial variability unless a well-designed sampling strategy is employed (Chatterjee et al. 2009). Advances in RS technologies enable SOC assessment over larger areas through satellite imagery and aerial surveys, providing valuable data on land cover and vegetation types that correlate with SOC levels; however, this method requires ground-truthing for accuracy (Sodango et al. 2021; Das and Ghimire 2024). Various soil C models, such as RothC, and Century, simulate SOC dynamics based on climate, land use, and management practices, offering insights for long-term monitoring, though their accuracy hinges on the quality of input data and assumptions (Ravindranath and Ostwald 2008a). Emerging technologies, including soil sensors and automated data collection systems, facilitate continuous monitoring of SOC and related properties, providing real-time data on factors like soil moisture and temperature, thus allowing for timely interventions (Pei et al. 2019; Conway et al. 2023). Additionally, engaging local communities and citizen scientists in SOC monitoring can expand data collection efforts, empowering stakeholders and enhancing environmental stewardship (Viscarra Rossel et al. 2008; McKinley et al. 2017), particularly in remote or underserved areas with limited monitoring resources. Together, these methods contribute to effective SOC tracking, supporting sustainable practices and climate mitigation through integrated monitoring networks (Arrouays et al. 2014). For instance, the International Soil C Network (ISCN) database includes data from over 257,000 soil layers across more than 41,000 profiles globally, primarily in North America, with downloadable soil stock data from 30,691 points available at http://iscn.fluxdata.org/ (Tifafi et al. 2018). Establishing a global monitoring network demands significant infrastructure, ongoing funding, and institutional collaboration to integrate new data and identify SOC change hotspots effectively (Jandl et al. 2014).

6.4.1 Soil Organic Carbon Modelling

SOC models bridge theoretical principles and empirical data by integrating biogeochemical understanding with real-world observations, enhancing predictive capabilities and guiding effective soil management practices (Le Noë et al. 2023). They are essential for estimating SOC stock trends and calculating GHG budgets, with performance influenced by land use and management practices (Riggers et al. 2019). By simulating C inputs such as plant residues, root biomass, and organic amendments alongside outputs like decomposition and erosion (Benbi and Nisar 2020), these models estimate SOC levels across different soil types and land use systems to establish baseline soil health and C reserves. Additionally, SOC models can predict changes over time in response to various management practices, climate scenarios, and land-use changes, evaluating the long-term impacts of techniques like reduced tillage and cover cropping (Álvaro-Fuentes et al. 2012), which are crucial for enhancing SOC sequestration. They also explore the biochemical processes governing SOC dynamics, such as decomposition, mineralization, and stabilization, helping to identify key factors influencing SOC levels. Two primary approaches in modeling various environmental and ecological systems, including SOC dynamics are empirical models and process-based models (Table 6.6).

Table 6.6 Comparison between empirical and process-based models (Stockmann et al. 2013; Cuddington et al. 2013; Paustian et al. 2019; Choquet et al. 2021; Ermolieva et al. 2024)

Approach	Empirical models	Process-based models
Definition	Based on statistical relationships from field observations	Based on biogeochemical processes and scientific understanding
Data requirements	Relies on historical data to establish correlations.	Requires extensive on ecological mechanisms and processes.
Modeling approach	Statistical, aggregated data	Comprehensive, dynamic simulation
Complexity	Simpler; uses scaling factors to adjust reference SOC stocks	More complex; employs differential equations to describe dynamics
Transparency	Assumptions are less clear; based on statistical correlations.	Assumptions grounded in ecological theory; more transparent.
Predictive capability	Limited to conditions represented in the observational data	Better suited for extrapolating beyond known data, focusing on causal mechanisms.
Applications	Primarily for national-scale inventory and reporting	Used for both research and GHG inventory purposes at multiple scales
Handling uncertainty	Often does not account for uncertainty well.	Allows for partitioning of uncertainty among ecological processes.
Flexibility	Less flexible; requires specific data for local applications	More flexible; integrates various environmental and management factors
Examples	Regression models (Wang et al. 2023b), Cool Farm Tool (Schipanski et al. 2024)	DayCent, DNDC, EPIC, COMET farm

Empirical models are based on observed data and statistical relationships. They do not explicitly simulate the underlying processes that govern the system but instead rely on correlations and patterns identified through historical data (Stockmann et al. 2013). Process-based models simulate the underlying physical, chemical, and biological processes governing a system. They aim to understand how different factors interact to influence outcomes (Paustian et al. 2019). Table 6.7 provides examples of different process-based models utilized in SOC studies, along with their specific characteristics (Stockmann et al. 2013; Mandal et al. 2022).

Recent studies highlight the effectiveness of various models in assessing SOC dynamics and the impacts of management practices on CS. A simple empirical model applied to Chernozemic soils on the Canadian prairies demonstrated significant SOC sequestration benefits from no-till and continuous cropping, sequestering 1.23 million Mg of C in 1996 and 1.72 million Mg in 2001, thereby serving as a valuable tool for guiding agricultural policy on GHG emissions (Liang et al. 2005). In Spain, the validation of the Century model for Mediterranean semiarid agroecosystems showed that reduced tillage and intensified cropping effectively enhance SOC sequestration (Álvaro-Fuentes et al. 2009). Similarly, the Century model in northern China showed that CA, along with sufficient residue return and manure application, is crucial for maintaining SOC balance (Wang et al. 2005). The RothC model effectively simulated SOC in stubble-burnt treatments but faced challenges

Table 6.7 Examples of process-based models used in soil organic carbon studies

Model	Expanded Form	Origin	Focus	Input requirements	Time step	Reference study
RothC	Rothamsted C Model	UK	Models SOC dynamics, focusing on decomposition processes and C cycling in soils, particularly under different land management practices	Organic C inputs including above-ground stubble and below-ground roots	Monthly	Liu et al. (2009); Morais et al. (2019); Teixeira et al. (2021)
CENTURY	–	USA	Simulates long-term dynamics of C and nitrogen (N), with capabilities for phosphorus (P) and sulfur (S)	Soil texture, climate variables, plant N and lignin content; atmospheric N inputs, and initial SOC content	Monthly	Pennock and Frick (2001)
DNDC	Denitrification–Decomposition	USA	Simulate the interactions between crops, soil, and climate in relation to C and N cycling	Soil (temperature, pH etc), weather (temperature, precipitation) and crop residue amount	Daily	Smith et al. (2010); Li et al. (2016)
SALUS	System Approach to Land Use Sustainability	USA, evolved from CERES (crop environment resource synthesis) model	Modeling framework that integrates biophysical processes and management practices to simulate crop growth, soil water dynamics, and nutrient cycling during growing seasons	Information on soil properties, climate data, genotype characteristics, and agronomic management	Daily	Pezzuolo et al. (2017); Cillis et al. (2018)
EPIC	Environmental Policy Integrated Climate	USA	Simulates agricultural production, soil erosion, SOC, and the impacts of various agricultural practices on the environment	Requires weather data, soil characteristics, and management inputs like crop residues and organic amendments for simulations	Daily	Causarano et al. (2011)
DAYCENT	Daily CENTURY	USA	A daily time-step model that focuses on soil C and N cycling, GHG emissions, and the impacts of land management practices on these dynamics	Weather data, a site-specific management schedule (e.g., fertilization, tillage), and crop-specific parameters	Daily	Lemma et al. (2021)

DAISY	–	Denmark	Soil-plant-atmosphere system model for agro-ecosystems	Weather data, soil water characteristics (bulk density, water content at field capacity and wilting point), OM characteristics (C-to-N ratios, turnover rates), crop management details (tillage, sowing, harvesting), and nitrogen dynamics (fertilizer application, N-fixation estimates)	Daily meteorological data	Jensen et al. (1997); Laub et al. (2021)
APSIM	Agricultural Production Systems sIMulator	Australia	Simulates SOC dynamics by integrating interconnected crop and soil modules that automatically account for biomass inputs, decomposition processes, and environmental factors influencing C cycling	Soil properties, management practices, and crop parameters (e.g. crop biomass)	Daily	O'Leary et al. (2016)
CANDY	CArbon and Nitrogen DYnamics	Germany	Simulates the effects of management practices, crop residues, and environmental factors on the overall C balance	Soil physical properties, climate data and land management information	Daily	Smith et al. (2007)
ECOSYS	Ecosystem System Model	Canada	Models complex ecosystem behavior by simulating the interactions of C,N,P, water, and energy cycles among plants, soil, and the atmosphere; simulates C transformations in different SOC pools	Soil properties, climate data, and crop information (e.g. cover crop type, planting time etc.)	Daily	Qin et al. (2023)

with stubble-retained practices, indicating significant loss of retained stubble prior to SOC sequestration (Liu et al. 2009). Additionally, pasture management studies using RothC demonstrated accurate SOC predictions, particularly for annual pastures, and suggested that linking total above-ground pasture production to SOC input could enhance CS outcomes (Liu et al. 2011). The AMG model was evaluated for predicting SOC changes from cereal straw export across various climates, indicating that systematic straw removal could reduce SOC by 2.5–10.9% over 50 years, depending on specific site conditions (Saffih-Hdadi and Mary 2008). The C-TOOL model assessed SOC dynamics in Danish farming from 1986 to 2012, revealing that practices like organic manure application and cover crops can enhance SOC, though potential increases remained limited (Taghizadeh-Toosi and Olesen 2016). Long-term SOC dynamics simulated with the DNDC model, compared to field observations across 11 plots in Europe and Australia, showed significant SOC loss at two Australian sites due to grassland conversion to cropland, with overall simulations closely aligning with field data and average percent differences under 0.07% for most plots (Li et al. 1997). An ensemble of crop models in Italy, including APSRVI-NWheat and DSSAT, indicated a decline in SOC stocks under future climate scenarios, although CA practices maintained higher SOC levels, with no-till showing potential for necessary annual increases (Iocola et al. 2017). Furthermore, a recent study in southern Jiangsu, China, developed integrated models combining process-based approaches (RothC and MIMICS) with machine learning techniques, revealing that the RF-RothC-MIMICS model outperformed others for spatially explicit SOC predictions (Zhang et al. 2023).

6.4.2 Challenges in Monitoring Soil Organic Carbon

Monitoring SOC is crucial for understanding its impact on ecosystem services and climate change mitigation; however, several challenges impede effective assessment. One major issue is spatial variability. SOC levels can differ significantly across various soil types, land uses, and management practices, making data interpretation complex (Bai and Zhou 2020). This spatial heterogeneity necessitates dense sampling, often requiring sampling intervals of less than 50 meters to capture variability adequately (Das and Ghimire 2024). Such demands complicate the logistics and increase the costs of SOC monitoring efforts. In addition to spatial variability, temporal factors also play a significant role. Environmental changes, such as seasonal variations and shifts in land management practices, require continuous monitoring to accurately capture SOC trends (Orton et al. 2023). The introduction of management practices like biochar application, while aimed at enhancing SOC levels, can also increase heterogeneity and complicate representative sampling (Lotz et al. 2024). The dynamic nature of SOC makes it essential to have long-term datasets; however, existing studies often rely on infrequent soil sampling, which can lead to gaps in understanding how SOC levels respond over time (Jandl et al. 2014).

This is particularly important considering the influence of climate change on SOC dynamics, which introduces further uncertainty (He et al. 2024). Another critical challenge is the lack of standardized protocols for SOC sampling and analysis, which hampers comparability across different studies and regions (Okoli and Birkenberg 2024). Variations in methodologies can lead to discrepancies in results, complicating efforts to establish reliable baselines or monitor changes over time. This is further exacerbated by the high costs associated with traditional soil sampling and laboratory analyses, which can limit the scope and frequency of monitoring programs (Viscarra Rossel and Brus 2018).

Innovative approaches like RS and continuous monitoring can help address some of the challenges associated with SOC assessment (Escribano et al. 2017). However, the site-specific nature of relationships between RS variables and SOC introduces additional complexity (Croft et al. 2012). Variability in sensor characteristics complicates the transfer of predictive models across different contexts, and data accuracy issues from airborne and satellite sensors can result in measurement uncertainties that surpass the actual spatial and temporal variations in SOC content. This disparity underscores the need for greater validation of RS techniques, particularly in scenarios where traditional methods have demonstrated significant accuracy. Additionally, while global interest in spectroscopy has led to the creation of diverse soil spectral libraries (SSLs) for modeling SOC changes, the challenge of generalizing these libraries to local contexts persists, limiting precise modeling at the field scale (Shen et al. 2022). To overcome this, deep transfer learning has been proposed as a method to leverage information from large-scale SSLs, facilitating more accurate, rapid, and cost-effective SOC monitoring. However, a significant barrier to the widespread adoption of satellite-based spectral measurements is their lack of approval from major C certification standards (Okoli and Birkenberg 2024). This absence of approval raises concerns about the reliability and validity of data derived from these measurements, as certification standards are essential for ensuring that C accounting methods meet rigorous scientific and regulatory requirements. Collectively, these challenges necessitate a multifaceted approach that combines technological advancements, institutional collaborations, and the development of standardized methodologies (Jandl et al. 2014). By addressing these issues, the scientific community can enhance the accuracy and reliability of SOC data, ultimately supporting better decision-making for soil health initiatives and C farming.

6.5 Conclusion

Estimated to 1 m depth, world soils store nearly three times the C found in the atmosphere and five times that in vegetation, with approximately 1550 Pg as SOC and 950 Pg as SIC. Both SOC and SIC are crucial for CS; however, SOC is more influenced by changing agricultural practices and sustainable land management, while SIC primarily originates from parent minerals and is less affected by these practices, except when acidifying fertilizers like ammonium sulfate or urea are used. Consequently, measuring SOC is essential for guiding farmers in adopting C farming techniques that not only aid in climate change mitigation but also enhance soil health and farm productivity, ultimately providing financial benefits through C credits. Regular monitoring of SOC allows farmers to track progress, adjust practices, and address challenges, ensuring that CS goals are met, and SOC stocks are maintained or improved.

The verification and validation of SOC measurements involve selecting appropriate methods based on available resources, cost, and the need for accuracy versus practicality in specific site conditions. Common methods for SOC measurement include wet and dry combustion methods. Wet combustion estimate SOC using traditional approaches like the Walkley-Black and Tyurin methods, each with specific recovery factors and limitations influenced by soil type and site conditions. Dry combustion methods for estimating SOC include LoI and ADC, with LoI being cost-effective but less reliable for low SOC levels and requiring calibration for accurate results. In contrast, ADC offers high precision and efficiency, serving as the standard for SOC measurement, although it can be costly and necessitates careful handling and regular maintenance. Spectroscopic methods, such as VNIR, MIR, LIBS, and INS, provide rapid in situ SOC analysis, each with unique advantages and challenges. Recent advancements in handheld devices have further enhanced real-time monitoring and decision-making in C farming. Colorimetric methods estimate SOC by correlating soil color with OM levels, using systems like Munsell and CIE-Lab for analysis. Although there are challenges related to calibration and environmental sensitivity, advancements in digital technology have improved the accuracy of SOC predictions. Finally, RS is particularly advantageous for monitoring SOC in inaccessible areas, employing optical, thermal, and microwave methods to gather data on SOC dynamics across various spatial scales. By integrating diverse data sources, such as vegetation indices and environmental covariates, RS enhances SOC mapping accuracy and informs climate change mitigation strategies. However, challenges related to spatial and temporal variability necessitate careful data fusion and modeling approaches for effective monitoring and reporting in C management.

SOC models like RothC, Century, and DNDC integrate theoretical principles and empirical data to estimate SOC trends and GHG budgets, guiding effective soil management. However, monitoring SOC faces challenges, including spatial variability, temporal changes, and a lack of standardized sampling protocols. Dense sampling increases costs, while discrepancies in methodologies hinder comparability. Innovative methods like RS show promise but face site-specific variability issues. A multifaceted approach, combining technology, collaboration, and standardized practices, is essential to enhance SOC assessment accuracy and support better soil health and C farming decisions.

References

Abdoli P, Khanmirzaei A, Hamzeh S et al (2023) Use of remote sensing data to predict soil organic carbon in some agricultural soils of Iran. Remote Sens Appl 30:100969. https://doi.org/10.1016/j.rsase.2023.100969

Acharya U, Lal R, Chandra R (2022) Data driven approach on in-situ soil carbon measurement. Carbon Manag 13:401–419. https://doi.org/10.1080/17583004.2022.2106310

Allison LE (2016) Organic carbon. In: Norman AG (ed) Methods of soil analysis: part 2 chemical and microbiological properties. pp 1367–1378

Álvaro-Fuentes J, López MV, Arrúe JL et al (2009) Tillage and cropping effects on soil organic carbon in Mediterranean semiarid agroecosystems: testing the Century model. Agric Ecosyst Environ 134:211–217. https://doi.org/10.1016/j.agee.2009.07.001

Álvaro-Fuentes J, Morell FJ, Plaza-Bonilla D et al (2012) Modelling tillage and nitrogen fertilization effects on soil organic carbon dynamics. Soil Tillage Res 120:32–39. https://doi.org/10.1016/j.still.2012.01.009

Angelopoulou T, Tziolas N, Balafoutis A et al (2019) Remote sensing techniques for soil organic carbon estimation: a review. Remote Sens 11:676. https://doi.org/10.3390/rs11060676

Apesteguia M, Plante AF, Virto I (2018) Methods assessment for organic and inorganic carbon quantification in calcareous soils of the Mediterranean region. Geoderma Reg 12:39–48. https://doi.org/10.1016/j.geodrs.2017.12.001

Arrouays D, Marchant BP, Saby NPA et al (2014) On soil carbon monitoring networks. In: Soil Carbon. Springer International Publishing, Cham, pp 59–68

Babos DV, Guedes WN, Freitas VS et al (2024) Laser-induced breakdown spectroscopy as an analytical tool for total carbon quantification in tropical and subtropical soils: evaluation of calibration algorithms. Frontiers in Soil Science 3. https://doi.org/10.3389/fsoil.2023.1242647

Badgery W, Murphy B, Cowie A et al (2021) Soil carbon market-based instrument pilot—the sequestration of soil organic carbon for the purpose of obtaining carbon credits. Soil Research 59:12. https://doi.org/10.1071/SR19331

Bahadori M, Tofighi H (2016) A modified Walkley-Black method based on spectrophotometric procedure. Commun Soil Sci Plant Anal 47:213–220. https://doi.org/10.1080/00103624.2015.1118118

Bai Y, Zhou Y (2020) The main factors controlling spatial variability of soil organic carbon in a small karst watershed, Guizhou Province, China. Geoderma 357:113938. https://doi.org/10.1016/j.geoderma.2019.113938

Baker TJ, Kaiser M, Turk JK (2023) Predicting soil organic carbon from loss-on-ignition across four regions of Nebraska. Commun Soil Sci Plant Anal 54:2301–2312. https://doi.org/10.1080/00103624.2023.2217841

Bao Y, Yao F, Meng X et al (2023) Predicting soil organic carbon in cultivated land across geographical and spatial scales: integrating sentinel-2A and laboratory Vis-NIR spectra. ISPRS J Photogramm Remote Sens 203:1–18. https://doi.org/10.1016/j.isprsjprs.2023.07.020

Bellon-Maurel V, McBratney A (2011) Near-infrared (NIR) and mid-infrared (MIR) spectroscopic techniques for assessing the amount of carbon stock in soils—critical review and research perspectives. Soil Biol Biochem 43:1398–1410. https://doi.org/10.1016/j.soilbio.2011.02.019

Benbi DK, Nisar S (2020) Developments in measurement and modelling of soil organic carbon. In: Carbon management in tropical and sub-tropical terrestrial systems. Springer Singapore, Singapore, pp 379–402

Ben-Dor E, Banin A (1989) Determination of organic matter content in arid-zone soils using a simple "loss-on-ignition" method. Commun Soil Sci Plant Anal 20:1675–1695. https://doi.org/10.1080/00103628909368175

Bhattacharyya T, Chandran P, Ray SK et al (2015) Walkley-Black recovery factor to reassess soil organic matter: indo-Gangetic Plains and Black soil region of India case studies. Commun Soil Sci Plant Anal 46:2628–2648. https://doi.org/10.1080/00103624.2015.1089265

Biney JKM, Saberioon M, Borůvka L et al (2021) Exploring the suitability of UAS-based multispectral images for estimating soil organic carbon: comparison with proximal soil sensing and Spaceborne imagery. Remote Sens 13:308. https://doi.org/10.3390/rs13020308

Bisutti I, Hilke I, Raessler M (2004) Determination of total organic carbon—an overview of current methods. TrAC Trends Anal Chem 23:716–726. https://doi.org/10.1016/j.trac.2004.09.003

Blanco-Canqui H (2024) Assessing the potential of nature-based solutions for restoring soil ecosystem services in croplands. Sci Total Environ 921:170854. https://doi.org/10.1016/j.scitotenv.2024.170854

Bou Kheir R, Greve MH, Bøcher PK et al (2010) Predictive mapping of soil organic carbon in wet cultivated lands using classification-tree based models: the case study of Denmark. J Environ Manag 91:1150–1160. https://doi.org/10.1016/j.jenvman.2010.01.001

Braun KN, Theuerkauf EJ, Hurtgen MT et al (2020) Loss-on-ignition estimates for soil organic carbon in Great Lakes freshwater coastal wetlands. Wetlands 40:1201–1206. https://doi.org/10.1007/s13157-020-01270-z

Brock WH (2013) Justus, baron von Liebig. Encyclopaedia Britanica

Burgos Hernández TD, Slater BK, Shaffer JM, Basta N (2023) Comparison of methods for determining organic carbon content of urban soils in Central Ohio. Geoderma Reg 34:e00680. https://doi.org/10.1016/j.geodrs.2023.e00680

Buyanovsky GA, Aslam M, Wagner GH (1994) Carbon turnover in soil physical fractions. Soil Sci Soc Am J 58:1167–1173. https://doi.org/10.2136/sssaj1994.03615995005800040023x

Campos CA (2010) Analyzing the relation between loss-on-ignition and other methods of soil organic carbon determination in a tropical cloud Forest (Mexico). Commun Soil Sci Plant Anal 41:1454–1462. https://doi.org/10.1080/00103624.2010.482168

Campos AR, Giasson E, Costa JJF et al (2019) Selection of environmental covariates for classifier training applied in digital soil mapping. Rev Bras Cienc Solo 42. https://doi.org/10.1590/18069657rbcs20170414

Causarano HJ, Doraiswamy PC, Muratova N et al (2011) Improved modeling of soil organic carbon in a semiarid region of Central East Kazakhstan using EPIC. Agron Sustain Dev 31:275–286. https://doi.org/10.1051/agro/2010028

César Izaurralde R, Rosenberg NJ, Lal R (2001) Mitigation of climatic change by soil carbon sequestration: issues of science, monitoring, and degraded lands. In: Advances in Agronomy, pp 1–75

Chatterjee A, Lal R, Wielopolski L et al (2009) Evaluation of different soil carbon determination methods. CRC Crit Rev Plant Sci 28:164–178. https://doi.org/10.1080/07352680902776556

Chen F, Kissel DE, West LT et al (2005) Mapping surface soil organic carbon for crop fields with remote sensing. J Soil Water Conserv 60:51–57

Chen Y, Zhang M, Fan D et al (2018) Linear regression between CIE-lab color parameters and organic matter in soils of tea plantations. Eurasian Soil Sci 51:199–203. https://doi.org/10.1134/S1064229318020011

Choquet P, Gabrielle B, Chalhoub M et al (2021) Comparison of empirical and process-based modelling to quantify soil-supported ecosystem services on the Saclay plateau (France). Ecosyst Serv 50:101332. https://doi.org/10.1016/j.ecoser.2021.101332

Cillis D, Maestrini B, Pezzuolo A et al (2018) Modeling soil organic carbon and carbon dioxide emissions in different tillage systems supported by precision agriculture technologies under current climatic conditions. Soil Tillage Res 183:51–59. https://doi.org/10.1016/j.still.2018.06.001

Ciric V, Manojlovic M, Svarc-Gajic J, Kumar Sitaula B (2021) Comparison, limitations and uncertainty of wet chemistry techniques, loss on ignition and dry combustion in soil organic carbon analysis. Agrochimica 65:295–314. https://doi.org/10.12871/00021857202136

Clairotte M, Grinand C, Kouakoua E et al (2016) National calibration of soil organic carbon concentration using diffuse infrared reflectance spectroscopy. Geoderma 276:41–52. https://doi.org/10.1016/j.geoderma.2016.04.021

Conway LS, Sudduth KA, Kitchen NR, Anderson SH (2023) Repeatability of commercially available visible and near infrared proximal soil sensors. Precis Agric 24:1014–1029. https://doi.org/10.1007/s11119-022-09985-1

Croft H, Kuhn NJ, Anderson K (2012) On the use of remote sensing techniques for monitoring spatio-temporal soil organic carbon dynamics in agricultural systems. Catena (Amst) 94:64–74. https://doi.org/10.1016/j.catena.2012.01.001

Crucil G, Castaldi F, Aldana-Jague E et al (2019) Assessing the performance of UAS-compatible multispectral and hyperspectral sensors for soil organic carbon prediction. Sustainability 11:1889. https://doi.org/10.3390/su11071889

Cuddington K, Fortin M-J, Gerber LR et al (2013) Process-based models are required to manage ecological systems in a changing world. Ecosphere 4:1–12. https://doi.org/10.1890/ES12-00178.1

Da Silva DR, De Abreu CA, De Abreu MF et al (2013) Comparison of methods to quantify organic carbon in soil samples from São Paulo State, Brazil. Commun Soil Sci Plant Anal 44:429–439. https://doi.org/10.1080/00103624.2013.742345

Dai Z, Liu X, Ding Y (2024) Iron-removal learning machine for multicolor determination of soil organic carbon. J Soils Sediments 24:2058–2067. https://doi.org/10.1007/s11368-024-03770-5

Das S, Ghimire D (2024) Soil organic carbon: measurement and monitoring using remote sensing data. In: Remote Sensing of Soils. Elsevier, pp 395–409

Davies BE (1974) Loss-on-ignition as an estimate of soil organic matter. Soil Sci Soc Am J 38:150–151. https://doi.org/10.2136/sssaj1974.03615995003800010046x

De Vos B, Vandecasteele B, Deckers J, Muys B (2005) Capability of loss-on-ignition as a predictor of Total organic carbon in non-calcareous Forest soils. Commun Soil Sci Plant Anal 36:2899–2921. https://doi.org/10.1080/00103620500306080

Delarmelina WM, Caldeira MVW, Gomes Junior D et al (2022) Soil attributes and spatial variability of soil organic carbon stock under the Atlantic Forest, Brazil. Ciência Florestal 32:1528–1551. https://doi.org/10.5902/1980509867028

Demattê JAM, Bortoletto MAM, Vasques GM, Rizzo R (2011) Quantification of soil organic matter using mathematical models based on colorimetry in the Munsell color system. Bragantia 70:590–597. https://doi.org/10.1590/S0006-87052011005000006

Dhaliwal GS, Gupta N, Kukal SS, Kaur M (2011) Standardization of automated Vario EL III CHNS analyzer for Total carbon and nitrogen determination in soils. Commun Soil Sci Plant Anal 42:971–979. https://doi.org/10.1080/00103624.2011.558965

Dhillon GS, Amichev BY, de Freitas R, Van Rees K (2015) Accurate and precise measurement of organic carbon content in carbonate-rich soils. Commun Soil Sci Plant Anal 46:2707–2720. https://doi.org/10.1080/00103624.2015.1089271

Díaz-Zorita M (1999) Soil organic carbon recovery by the Walkley-Black method in a typic hapludoll. Commun Soil Sci Plant Anal 30:739–745. https://doi.org/10.1080/00103629909370242

Dorantes MJ, Fuentes BA, Miller DM (2022) Calibration set optimization and library transfer for soil carbon estimation using soil spectroscopy—a review. Soil Sci Soc Am J 86:879–903. https://doi.org/10.1002/saj2.20435

Dvorakova K, Shi P, Limbourg Q, van Wesemael B (2020) Soil organic carbon mapping from remote sensing: the effect of crop residues. Remote Sens 12:1913. https://doi.org/10.3390/rs12121913

Ermolieva T, Havlik P, Lessa-Derci-Augustynczik A et al (2024) Tracking the dynamics and uncertainties of soil organic carbon in agricultural soils based on a novel robust meta-model framework using multisource data. Sustainability 16:6849. https://doi.org/10.3390/su16166849

Escribano P, Schmid T, Chabrillat S et al (2017) Optical remote sensing for soil mapping and monitoring. In: Soil Mapping and Process Modeling for Sustainable Land Use Management. Elsevier, pp 87–125

Fadeeva VP, Tikhova VD, Nikulicheva ON (2008) Elemental analysis of organic compounds with the use of automated CHNS analyzers. J Anal Chem 63:1094–1106. https://doi.org/10.1134/S1061934808110142

Fan Z, Herrick JE, Saltzman R et al (2017) Measurement of soil color: a comparison between smartphone camera and the Munsell color charts. Soil Sci Soc Am J 81:1139–1146. https://doi.org/10.2136/sssaj2017.01.0009

FAO (2019) Standard operating procedure for soil organic carbon: Walkley-Black method
FAO (2021) Standard operating procedure for soil organic carbon: Tyurin spectrophotometric method, Rome
Ge Y, Thomasson JA, Morgan CLS (2014) Mid-infrared attenuated total reflectance spectroscopy for soil carbon and particle size determination. Geoderma 213:57–63. https://doi.org/10.1016/j.geoderma.2013.07.017
Gholizadeh A, Žižala D, Saberioon M, Borůvka L (2018) Soil organic carbon and texture retrieving and mapping using proximal, airborne and Sentinel-2 spectral imaging. Remote Sens Environ 218:89–103. https://doi.org/10.1016/j.rse.2018.09.015
Goldin A (1987) Reassessing the use of loss-on-ignition for estimating organic matter content in noncalcareous soils. Commun Soil Sci Plant Anal 18:1111–1116. https://doi.org/10.1080/00103628709367886
Gortner RA (1916) The organic matter of the soil 1. Some data on humus, humus carbon and humus nitrogen. Soil Sci 2:395–442
Gozukara G, Hartemink AE, Zhang Y (2023) Illumination levels affect the prediction of soil organic carbon using smartphone-based digital images. Comput Electron Agric 204:107524. https://doi.org/10.1016/j.compag.2022.107524
Guo L, Fu P, Shi T et al (2021a) Exploring influence factors in mapping soil organic carbon on low-relief agricultural lands using time series of remote sensing data. Soil Tillage Res 210:104982. https://doi.org/10.1016/j.still.2021.104982
Guo P, Li T, Gao H et al (2021b) Evaluating calibration and spectral variable selection methods for predicting three soil nutrients using Vis-NIR spectroscopy. Remote Sens 13:4000. https://doi.org/10.3390/rs13194000
Hatté C, Poupeau J-J, Tannau J-F, Paterne M (2003) Development of an automated system for preparation of organic samples. Radiocarbon 45:421–430. https://doi.org/10.1017/S0033822200032781
Hayo L, Hasegawa H (2024) Enhancing emission reductions in South African agriculture: the crucial role of carbon credits in incentivizing climate-smart farming practices. Sustainable Futures 8:100260. https://doi.org/10.1016/j.sftr.2024.100260
He X, Yang L, Li A et al (2021) Soil organic carbon prediction using phenological parameters and remote sensing variables generated from Sentinel-2 images. Catena (Amst) 205:105442. https://doi.org/10.1016/j.catena.2021.105442
He X, Abramoff RZ, Abs E et al (2024) Model uncertainty obscures major driver of soil carbon. Nature 627:E1–E3. https://doi.org/10.1038/s41586-023-06999-1
Heil J, Jörges C, Stumpe B (2022) Evaluation of using digital photography as a cost-effective tool for the rapid assessment of soil organic carbon at a regional scale. Soil Security 6:100023. https://doi.org/10.1016/j.soisec.2021.100023
Herts AI, Khomenchuk VO, Kononchuk OB et al (2022) Use of visual-diagnostic color parameters of soils and optical reflectometry for determination of organic carbon content. J Geol Geogr Geoecol 31:260–272. https://doi.org/10.15421/112224
Homhuan S, Pansak W, Lawawirojwong S, Narongrit C (2016) Laboratory spectroscopy assessments of Rainfed Paddy soil samples on visible and near-infrared spectroscopy reflectance for estimating soil organic carbon. Air, Soil and Water Research 9:ASWR.S40173. https://doi.org/10.4137/ASWR.S40173
Hong Y, Chen S, Liu Y et al (2019) Combination of fractional order derivative and memory-based learning algorithm to improve the estimation accuracy of soil organic matter by visible and near-infrared spectroscopy. Catena (Amst) 174:104–116. https://doi.org/10.1016/j.catena.2018.10.051
Hu M, Ma F, Li Z et al (2022) Sensing of soil organic matter using laser-induced breakdown spectroscopy coupled with optimized self-adaptive calibration strategy. Sensors 22:1488. https://doi.org/10.3390/s22041488
Hutengs C, Ludwig B, Jung A et al (2018) Comparison of portable and bench-top spectrometers for mid-infrared diffuse reflectance measurements of soils. Sensors 18:993. https://doi.org/10.3390/s18040993

Iocola I, Bassu S, Farina R et al (2017) Can conservation tillage mitigate climate change impacts in Mediterranean cereal systems? A soil organic carbon assessment using long term experiments. Eur J Agron 90:96–107. https://doi.org/10.1016/j.eja.2017.07.011

Izaurralde RC, Rice CW, Wielopolski L et al (2013) Evaluation of three field-based methods for quantifying soil carbon. PLoS One 8:e55560. https://doi.org/10.1371/journal.pone.0055560

Jandl R, Rodeghiero M, Martinez C et al (2014) Current status, uncertainty and future needs in soil organic carbon monitoring. Sci Total Environ 468–469:376–383. https://doi.org/10.1016/j.scitotenv.2013.08.026

Jankauskas B, Slepetiene A, Jankauskiene G et al (2006) A comparative study of analytical methodologies to determine the soil organic matter content of Lithuanian Eutric Albeluvisols. Geoderma 136:763–773. https://doi.org/10.1016/j.geoderma.2006.05.015

Jensen LS, Mueller T, Nielsen NE et al (1997) Simulating trends in soil organic carbon in long-term experiments using the soil-plant-atmosphere model DAISY. Geoderma 81:5–28. https://doi.org/10.1016/S0016-7061(97)88181-5

Jeswani HK, Saharudin DM, Azapagic A (2022) Environmental sustainability of negative emissions technologies: a review. Sustain Prod Consum 33:608–635. https://doi.org/10.1016/j.spc.2022.06.028

Ji W, Viscarra Rossel RA, Shi Z (2015) Improved estimates of organic carbon using proximally sensed vis-NIR spectra corrected by piecewise direct standardization. Eur J Soil Sci 66:670–678. https://doi.org/10.1111/ejss.12271

Jia X, Chen S, Yang Y et al (2017) Organic carbon prediction in soil cores using VNIR and MIR techniques in an alpine landscape. Sci Rep 7:2144. https://doi.org/10.1038/s41598-017-02061-z

Jiang Q, Chen Y, Guo L et al (2016) Estimating soil organic carbon of cropland soil at different levels of soil moisture using VIS-NIR spectroscopy. Remote Sens 8:755. https://doi.org/10.3390/rs8090755

Jiang Q, Li Q, Wang X et al (2017) Estimation of soil organic carbon and total nitrogen in different soil layers using VNIR spectroscopy: effects of spiking on model applicability. Geoderma 293:54–63. https://doi.org/10.1016/j.geoderma.2017.01.030

Johns TJ, Angove MJ, Wilkens S (2015) Measuring soil organic carbon: which technique and where to from here? Soil Research 53:717. https://doi.org/10.1071/SR14339

Kashyap YS (2022) Introduction to neutron physics. In: Neutron imaging. Springer Singapore, Singapore, pp 1–31

Kearton B, Mattley Y (2008) Sparking new applications. Nat Photonics 2:537–540. https://doi.org/10.1038/nphoton.2008.173

Kerven GL, Menzies NW, Geyer MD (2000) Analytical methods and quality assurance. Commun Soil Sci Plant Anal 31:1935–1939. https://doi.org/10.1080/00103620009370551

Kirillova NP, Zhang Y, Hartemink AE et al (2021) Calibration methods for measuring the color of moist soils with digital cameras. Catena (Amst) 202:105274. https://doi.org/10.1016/j.catena.2021.105274

Konare H, Yost RS, Doumbia M et al (2010) Loss on ignition: measuring soil organic carbon in soils of the Sahel, West Africa. Afr J Agric Res 5:3088–3095

Konen ME, Jacobs PM, Burras CL et al (2002) Equations for predicting soil organic carbon using loss-on-ignition for North Central U.S. soils. Soil Sci Soc Am J 66:1878–1881. https://doi.org/10.2136/sssaj2002.1878

Konen ME, Burras CL, Sandor JA (2003) Organic carbon, texture, and quantitative color measurement relationships for cultivated soils in North Central Iowa. Soil Sci Soc Am J 67:1823–1830. https://doi.org/10.2136/sssaj2003.1823

Kosaka J, Honda C, Iseki A (1959) A new rapid and accurate method for the determination of carbon in soils. Soil Sci Plant Nutr 5:77–83. https://doi.org/10.1080/00380768.1959.10430898

Kowalenko CG (2001) Assessment of Leco CNS-2000 analyzer for simultaneously measuring total carbon, nitrogen, and Sulphur in soil. Commun Soil Sci Plant Anal 32:2065–2078. https://doi.org/10.1081/CSS-120000269

Kuang B, Mouazen AM (2012) Influence of the number of samples on prediction error of visible and near infrared spectroscopy of selected soil properties at the farm scale. Eur J Soil Sci 63:421–429. https://doi.org/10.1111/j.1365-2389.2012.01456.x

Kumar P, Pandey PC, Singh BK et al (2016) Estimation of accumulated soil organic carbon stock in tropical forest using geospatial strategy. Egypt J Remote Sens Space Sci 19:109–123. https://doi.org/10.1016/j.ejrs.2015.12.003

Ladoni M, Alavipanah SK, Bahrami HA, Noroozi AA (2010) Remote sensing of soil organic carbon in semi-arid region of Iran. Arid Land Res Manag 24:271–281. https://doi.org/10.1080/15324982.2010.502917

Lal R, Kimble JM, Follett RF, Stewart BA (eds) (2001) Assessment Methods for Soil Carbon. CRC Press, Boca Raton, FL

Lal R (2004) Soil carbon sequestration to mitigate climate change. Geoderma 123:1–22. https://doi.org/10.1016/j.geoderma.2004.01.032

Lal R (2008) Carbon sequestration. Philos Trans R Soc B Biol Sci 363:815–830. https://doi.org/10.1098/rstb.2007.2185

Lal R (2016) Soil health and carbon management. Food Energy Secur 5:212–222. https://doi.org/10.1002/fes3.96

Lamichhane S, Adhikari K, Kumar L (2021) Use of multi-seasonal satellite images to predict SOC from cultivated lands in a montane ecosystem. Remote Sens 13:4772. https://doi.org/10.3390/rs13234772

Laub M, Ali RS, Demyan MS et al (2021) Modeling temperature sensitivity of soil organic matter decomposition: splitting the pools. Soil Biol Biochem 153:108108. https://doi.org/10.1016/j.soilbio.2020.108108

Le Noë J, Manzoni S, Abramoff R et al (2023) Soil organic carbon models need independent time-series validation for reliable prediction. Commun Earth Environ 4:158. https://doi.org/10.1038/s43247-023-00830-5

Lemma B, Williams S, Paustian K (2021) Long term soil carbon sequestration potential of smallholder croplands in Southern Ethiopia with DAYCENT model. J Environ Manag 294:112893. https://doi.org/10.1016/j.jenvman.2021.112893

Li C, Frolking S, Crocker GJ et al (1997) Simulating trends in soil organic carbon in long-term experiments using the DNDC model. Geoderma 81:45–60. https://doi.org/10.1016/S0016-7061(97)00080-3

Li S, Li J, Li C et al (2016) Testing the RothC and DNDC models against long-term dynamics of soil organic carbon stock observed at cropping field soils in North China. Soil Tillage Res 163:290–297. https://doi.org/10.1016/j.still.2016.07.001

Li S, Viscarra Rossel RA, Webster R (2022) The cost-effectiveness of reflectance spectroscopy for estimating soil organic carbon. Eur J Soil Sci 73. https://doi.org/10.1111/ejss.13202

Li Q, Guo G, Singh BP et al (2023) Associations of soil Fe oxides and organic carbon vary in different aggregate fractions under warming. J Soils Sediments 23:2744–2755. https://doi.org/10.1007/s11368-023-03532-9

Liang BC, Campbell CA, McConkey BG et al (2005) An empirical model for estimating carbon sequestration on the Canadian prairies. Can J Soil Sci 85:549–556. https://doi.org/10.4141/S04-089

Liu DL, Chan KY, Conyers MK (2009) Simulation of soil organic carbon under different tillage and stubble management practices using the Rothamsted carbon model. Soil Tillage Res 104:65–73. https://doi.org/10.1016/j.still.2008.12.011

Liu DL, Chan KY, Conyers MK et al (2011) Simulation of soil organic carbon dynamics under different pasture managements using the RothC carbon model. Geoderma 165:69–77. https://doi.org/10.1016/j.geoderma.2011.07.005

Liu Y, Deng C, Lu Y et al (2020) Evaluating the characteristics of soil vis-NIR spectra after the removal of moisture effect using external parameter orthogonalization. Geoderma 376:114568. https://doi.org/10.1016/j.geoderma.2020.114568

Lorenz K, Lal R (2018a) Importance of soils of agroecosystems for climate change policy. In: Carbon sequestration in agricultural ecosystems. Springer International Publishing, Cham, pp 357–386

Lorenz K, Lal R (2018b) Carbon sequestration in wetland soils. In: Carbon sequestration in agricultural ecosystems. Springer International Publishing, Cham, pp 211–234

Loria N, Lal R, Chandra R (2024) Handheld in situ methods for soil organic carbon assessment. Sustainability 16:5592. https://doi.org/10.3390/su16135592

Lotz S, Bucheli TD, Schmidt H-P, Hagemann N (2024) Quantification of soil organic carbon: the challenge of biochar-induced spatial heterogeneity. Frontiers in Climate 6. https://doi.org/10.3389/fclim.2024.1344524

Lu D, Chen Q, Wang G et al (2016) A survey of remote sensing-based aboveground biomass estimation methods in forest ecosystems. Int J Digit Earth 9:63–105. https://doi.org/10.1080/17538947.2014.990526

Luo MR (2015) CIELAB. In: Encyclopedia of color science and technology. Springer, Berlin Heidelberg, Berlin, Heidelberg, pp 1–7

Madugundu R, Al-Gaadi KA, Tola E et al (2022) Estimation of soil organic carbon in agricultural fields: a remote sensing approach. J Environ Biol 43:73–84. https://doi.org/10.22438/jeb/43/1/MRN-1873

Mandal A, Majumder A, Dhaliwal SS et al (2022) Impact of agricultural management practices on soil carbon sequestration and its monitoring through simulation models and remote sensing techniques: a review. Crit Rev Environ Sci Technol 52:1–49. https://doi.org/10.1080/10643389.2020.1811590

Manning DAC, de Azevedo AC, Zani CF, Barneze AS (2024) Soil carbon management and enhanced rock weathering: the separate fates of organic and inorganic carbon. Eur J Soil Sci 75. https://doi.org/10.1111/ejss.13534

Matejovic I (1993) Determination of carbon, hydrogen, and nitrogen in soils by automated elemental analysis (dry combustion method). Commun Soil Sci Plant Anal 24:2213–2222. https://doi.org/10.1080/00103629309368950

Matejovic I (1997) Determination of carbon and nitrogen in samples of various soils by the dry combustion. Commun Soil Sci Plant Anal 28:1499–1511. https://doi.org/10.1080/00103629709369892

McCarty GW, Reeves JB, Reeves VB et al (2002) Mid-infrared and near-infrared diffuse reflectance spectroscopy for soil carbon measurement. Soil Sci Soc Am J 66:640. https://doi.org/10.2136/sssaj2002.0640

McCarty GW, Reeves JB III, Yost R et al (2010) Evaluation of methods for measuring soil organic carbon in West African soils. Afr J Agric Res 5:2169–2177

McKinley DC, Miller-Rushing AJ, Ballard HL et al (2017) Citizen science can improve conservation science, natural resource management, and environmental protection. Biol Conserv 208:15–28. https://doi.org/10.1016/j.biocon.2016.05.015

Meena RS, Kumar S, Yadav GS (2020) Soil carbon sequestration in crop production. In: Nutrient dynamics for sustainable crop production. Springer Singapore, Singapore, pp 1–39

Mikhailova EA, Noble RRP, Post CJ (2003) Comparison of soil organic carbon recovery by Walkley-Black and dry combustion methods in the Russian Chernozem. Commun Soil Sci Plant Anal 34:1853–1860. https://doi.org/10.1081/CSS-120023220

Minasny B, Tranter G, AlexB MB et al (2009) Regional transferability of mid-infrared diffuse reflectance spectroscopic prediction for soil chemical properties. Geoderma 153:155–162. https://doi.org/10.1016/j.geoderma.2009.07.021

Mirzaeitalarposhti R, Demyan MS, Rasche F et al (2016) Overcoming carbonate interference on labile soil organic matter peaks for midDRIFTS analysis. Soil Biol Biochem 99:150–157. https://doi.org/10.1016/j.soilbio.2016.05.010

Mishra U, Lal R, Liu D, Van Meirvenne M (2010) Predicting the spatial variation of the soil organic carbon Pool at a regional scale. Soil Sci Soc Am J 74:906–914. https://doi.org/10.2136/sssaj2009.0158

Morais TG, Teixeira RFM, Domingos T (2019) Detailed global modelling of soil organic carbon in cropland, grassland and forest soils. PLoS One 14:e0222604. https://doi.org/10.1371/journal.pone.0222604

Morgan C, Cappellazzi S, Liptzin D (2020) Assessing soil health: soil carbon cycling and storage. Crops & Soils 53:43–47. https://doi.org/10.1002/crso.20076

Morris AS, Langari R (2012) Calibration of measuring sensors and instruments. In: Measurement and Instrumentation. Elsevier, pp 103–114

Mulder VL, de Bruin S, Schaepman ME, Mayr TR (2011) The use of remote sensing in soil and terrain mapping—a review. Geoderma 162:1–19. https://doi.org/10.1016/j.geoderma.2010.12.018

Naorem A, Jayaraman S, Sinha NK et al (2023) Eight-year impacts of conservation agriculture on soil quality, carbon storage, and carbon emission footprint. Soil Tillage Res 232:105748. https://doi.org/10.1016/j.still.2023.105748

Nawar S, Mouazen AM (2019) On-line vis-NIR spectroscopy prediction of soil organic carbon using machine learning. Soil Tillage Res 190:120–127. https://doi.org/10.1016/j.still.2019.03.006

Nelson DW, Sommers LE (1996) Total carbon, organic carbon, and organic matter. In: Sparks DL, Page AL, Helmke PA et al (eds) Methods of soil analysis part 3 chemical analysis. Soil Science Society of America, American Society of Agronomy, Madison, pp 961–1010

Nevalainen O, Niemitalo O, Fer I et al (2022) Towards agricultural soil carbon monitoring, reporting, and verification through the field observatory network (FiON). Geosci Instrum Method Data Syst 11:93–109. https://doi.org/10.5194/gi-11-93-2022

Niwa K, Yokobori J, Hongo C, Nagata O (2011) Estimating soil carbon stocks in an upland area of Tokachi District, Hokkaido, Japan, by satellite remote sensing. Soil Sci Plant Nutr 57:283–293. https://doi.org/10.1080/00380768.2011.557769

Nocita M, Stevens A, Noon C, van Wesemael B (2013) Prediction of soil organic carbon for different levels of soil moisture using Vis-NIR spectroscopy. Geoderma 199:37–42. https://doi.org/10.1016/j.geoderma.2012.07.020

Nowicki M, DeVries T, Siegel DA (2022) Quantifying the carbon export and sequestration pathways of the ocean's biological carbon pump. Global Biogeochem Cycles 36. https://doi.org/10.1029/2021GB007083

O' Rourke SM, Holden NM (2011) Optical sensing and chemometric analysis of soil organic carbon—a cost effective alternative to conventional laboratory methods? Soil Use Manag 27:143–155. https://doi.org/10.1111/j.1475-2743.2011.00337.x

O'Leary GJ, Liu DL, Ma Y et al (2016) Modelling soil organic carbon 1. Performance of APSIM crop and pasture modules against long-term experimental data. Geoderma 264:227–237. https://doi.org/10.1016/j.geoderma.2015.11.004

Okoli AO, Birkenberg A (2024) Monitoring soil carbon in smallholder carbon projects: insights from Kenya. Clim Chang 177:143. https://doi.org/10.1007/s10584-024-03796-1

Orr R, McBeath AV, Dieleman WIJ et al (2017) Estimating organic carbon content of soil in Papua New Guinea using infrared spectroscopy. Soil Research 55:735. https://doi.org/10.1071/SR16227

Orton TG, Thornton CM, Page KL et al (2023) Evaluation of remotely sensed imagery to monitor temporal changes in soil organic carbon at a long-term grazed pasture trial. Ecol Indic 154:110614. https://doi.org/10.1016/j.ecolind.2023.110614

Pan Y, Fan W, Huang T-H et al (2015) Evaluation of the sinks and sources of atmospheric CO2 by artificial upwelling. Sci Total Environ 511:692–702. https://doi.org/10.1016/j.scitotenv.2014.11.060

Pascucci S, Casa R, Belviso C et al (2014) Estimation of soil organic carbon from airborne hyperspectral thermal infrared data: a case study. Eur J Soil Sci 65:865–875. https://doi.org/10.1111/ejss.12203

Paustian K, Collier S, Baldock J et al (2019) Quantifying carbon for agricultural soil management: from the current status toward a global soil information system. Carbon Manag 10:567–587. https://doi.org/10.1080/17583004.2019.1633231

Pei X, Sudduth K, Veum K, Li M (2019) Improving in-situ estimation of soil profile properties using a multi-sensor probe. Sensors 19:1011. https://doi.org/10.3390/s19051011

Pennock DJ, Frick AH (2001) The role of field studies in landscape-scale applications of process models: an example of soil redistribution and soil organic carbon modeling using CENTURY. Soil Tillage Res 58:183–191. https://doi.org/10.1016/S0167-1987(00)00167-7

Pezzuolo A, Dumont B, Sartori L et al (2017) Evaluating the impact of soil conservation measures on soil organic carbon at the farm scale. Comput Electron Agric 135:175–182. https://doi.org/10.1016/j.compag.2017.02.004

Piekarczyk J, Kaźmierowski C, Królewicz S, Cierniewski J (2016) Effects of soil surface roughness on soil reflectance measured in laboratory and outdoor conditions. IEEE J Sel Top Appl Earth Obs Remote Sens 9:827–834. https://doi.org/10.1109/JSTARS.2015.2450775

Poeplau C, Don A, Six J et al (2018) Isolating organic carbon fractions with varying turnover rates in temperate agricultural soils—a comprehensive method comparison. Soil Biol Biochem 125:10–26. https://doi.org/10.1016/j.soilbio.2018.06.025

Pouladi N, Gholizadeh A, Khosravi V, Borůvka L (2023) Digital mapping of soil organic carbon using remote sensing data: a systematic review. Catena (Amst) 232:107409. https://doi.org/10.1016/j.catena.2023.107409

Pretorius ML, Van Huyssteen CW, Brown LR (2017) Soil color indicates carbon and wetlands: developing a color-proxy for soil organic carbon and wetland boundaries on sandy coastal plains in South Africa. Environ Monit Assess 189:556. https://doi.org/10.1007/s10661-017-6249-z

Pribyl DW (2010) A critical review of the conventional SOC to SOM conversion factor. Geoderma 156:75–83. https://doi.org/10.1016/j.geoderma.2010.02.003

Qin Z, Guan K, Zhou W et al (2023) Assessing long-term impacts of cover crops on soil organic carbon in the central US Midwestern agroecosystems. Glob Chang Biol 29:2572–2590. https://doi.org/10.1111/gcb.16632

Rasel SMM, Groen TA, Hussin YA, Diti IJ (2017) Proxies for soil organic carbon derived from remote sensing. Int J Appl Earth Obs Geoinf 59:157–166. https://doi.org/10.1016/j.jag.2017.03.004

Ravindranath NH, Ostwald M (2008a) Modelling for estimation and projection of carbon stocks in land-use systems. In: Carbon inventory methods handbook for greenhouse gas inventory, carbon mitigation and roundwood production projects. Springer, Netherlands, Dordrecht, pp 201–215

Ravindranath NH, Ostwald M (2008b) Methods for estimating soil organic carbon. In: Carbon inventory methods handbook for greenhouse gas inventory. Carbon Mitigation and Roundwood Production Projects. Springer, Netherlands, Dordrecht, pp 165–180

Raza S, Irshad A, Margenot A et al (2024) Inorganic carbon is overlooked in global soil carbon research: a bibliometric analysis. Geoderma 443:116831. https://doi.org/10.1016/j.geoderma.2024.116831

Rengma NS, Yadav M, Kalambukattu JG, Kumar S (2023) Machine learning-based digital mapping of soil organic carbon and texture in the mid-Himalayan terrain. Environ Monit Assess 195:994. https://doi.org/10.1007/s10661-023-11608-9

Riggers C, Poeplau C, Don A et al (2019) Multi-model ensemble improved the prediction of trends in soil organic carbon stocks in German croplands. Geoderma 345:17–30. https://doi.org/10.1016/j.geoderma.2019.03.014

Römer C, Rodionov A, Behmann J et al (2014) Quantifying moisture and roughness with support vector machines improves spectroscopic soil organic carbon prediction. J Plant Nutr Soil Sci 177:845–847. https://doi.org/10.1002/jpln.201400152

Rubinić V, Pavlović A, Magdić I (2021) Predicting soil organic matter content using soil color at three locations with different land use in Zagreb (Croatia). J Cent Eur Agric 22:646–656. https://doi.org/10.5513/JCEA01/22.3.3193

Saffih-Hdadi K, Mary B (2008) Modeling consequences of straw residues export on soil organic carbon. Soil Biol Biochem 40:594–607. https://doi.org/10.1016/j.soilbio.2007.08.022

Sahwan W, Lucke B, Sprafke T et al (2021) Relationships between spectral features, iron oxides and colours of surface soils in northern Jordan. Eur J Soil Sci 72:80–97. https://doi.org/10.1111/ejss.12986

Salehi MH, Beni OH, Harchegani HB et al (2011) Refining soil organic matter determination by loss-on-ignition. Pedosphere 21:473–482. https://doi.org/10.1016/S1002-0160(11)60149-5

Santi C, Certini G, D'Acqui LP (2006) Direct determination of organic carbon by dry combustion in soils with carbonates. Commun Soil Sci Plant Anal 37:155–162. https://doi.org/10.1080/00103620500403531

Santisteban JI, Mediavilla R, López-Pamo E et al (2004) Loss on ignition: a qualitative or quantitative method for organic matter and carbonate mineral content in sediments? J Paleolimnol 32:287–299. https://doi.org/10.1023/B:JOPL.0000042999.30131.5b

Sato JH, de Figueiredo CC, Marchão RL et al (2014) Methods of soil organic carbon determination in Brazilian savannah soils. Sci Agric 71:302–308. https://doi.org/10.1590/0103-9016-2013-0306

Schipanski ME, McClelland SC, Hughes HM et al (2024) Improving decision support tools for quantifying GHG emissions from organic production systems. Org Agric 14:503–512. https://doi.org/10.1007/s13165-024-00466-5

Schulte EE, Hopkins BG (2015) Estimation of soil organic matter by weight loss-on-ignition. In: Soil organic matter: analysis and interpretation. pp 21–31

Seidel M, Vohland M, Greenberg I et al (2022) Soil moisture effects on predictive VNIR and MIR modeling of soil organic carbon and clay content. Geoderma 427:116103. https://doi.org/10.1016/j.geoderma.2022.116103

Senesi GS, Senesi N (2016) Laser-induced breakdown spectroscopy (LIBS) to measure quantitatively soil carbon with emphasis on soil organic carbon. A review Anal Chim Acta 938:7–17. https://doi.org/10.1016/J.ACA.2016.07.039

Shahrayini E, Shafizadeh-Moghadam H, Noroozi Ali A, Eghbal MK (2022) Multiple-depth modeling of soil organic carbon using visible–near infrared spectroscopy. Geocarto Int 37:1393–1407. https://doi.org/10.1080/10106049.2020.1765887

Shamrikova EV, Kondratenok BM, Tumanova EA et al (2022) Transferability between soil organic matter measurement methods for database harmonization. Geoderma 412:115547. https://doi.org/10.1016/j.geoderma.2021.115547

Shamrikova EV, Vanchikova EV, Lu-Lyan-Min EI et al (2023) Which method to choose for measurement of oranic and inorganic carbon content in carbonate-rich soils? Advantages and disadvantages of dry and wet chemistry. Catena (Amst) 228:107151. https://doi.org/10.1016/j.catena.2023.107151

Shamrikova EV, Vanchikova EV, Kyzyurova EV, Zhangurov EV (2024) Methods for measuring organic carbon content in carbonate-containing soils: a review. Eurasian Soil Sci 57:380–394. https://doi.org/10.1134/S1064229323603104

Sharififar A, Minasny B, Arrouays D, et al (2023) Soil inorganic carbon, the other and equally important soil carbon pool: Distribution, controlling factors, and the impact of climate change. pp 165–231

Shen Z, Ramirez-Lopez L, Behrens T et al (2022) Deep transfer learning of global spectra for local soil carbon monitoring. ISPRS J Photogramm Remote Sens 188:190–200. https://doi.org/10.1016/j.isprsjprs.2022.04.009

Sherrod LA, Dunn G, Peterson GA, Kolberg RL (2002) Inorganic carbon analysis by modified pressure-Calcimeter method. Soil Sci Soc Am J 66:299–305. https://doi.org/10.2136/sssaj2002.2990

Shi J, Jackson T, Tao J et al (2008) Microwave vegetation indices for short vegetation covers from satellite passive microwave sensor AMSR-E. Remote Sens Environ 112:4285–4300. https://doi.org/10.1016/j.rse.2008.07.015

Siegel DA, DeVries T, Cetinić I, Bisson KM (2023) Quantifying the ocean's biological pump and its carbon cycle impacts on global scales. Annu Rev Mar Sci 15:329–356. https://doi.org/10.1146/annurev-marine-040722-115226

Silva JMP, Bomio MRD, Motta FV, Santos RM (2024) Assessment of Calcimetry as a reliable method for monitoring soil inorganic carbon stocks. ACS Agricul Sci Technol 4:781–790. https://doi.org/10.1021/acsagscitech.4c00088

Silva-Olaya AM, Cerri CEP, La Scala JN et al (2013) Carbon dioxide emissions under different soil tillage systems in mechanically harvested sugarcane. Environ Res Lett 8:015014. https://doi.org/10.1088/1748-9326/8/1/015014

Silveira ML, Comerford NB, Reddy KR et al (2008) Characterization of soil organic carbon pools by acid hydrolysis. Geoderma 144:405–414. https://doi.org/10.1016/j.geoderma.2008.01.002

Silvero NEQ, Demattê JAM, Minasny B, et al (2023) Sensing technologies for characterizing and monitoring soil functions: a review. pp 125–168

Sleutel S, De Neve S, Singier B, Hofman G (2007) Quantification of organic carbon in soils: a comparison of methodologies and assessment of the carbon content of organic matter. Commun Soil Sci Plant Anal 38:2647–2657. https://doi.org/10.1080/00103620701662877

Smith JG (2003) Aspects of the loss-on-ignition (loi) technique in the context of clay-rich, glaciolacustrine sediments. Geogr Ann Ser A Phys Geogr 85:91–97. https://doi.org/10.1111/1468-0459.00191

Smith P, Smith JU, Franko U et al (2007) Changes in mineral soil organic carbon stocks in the croplands of European Russia and the Ukraine, 1990–2070; comparison of three models and implications for climate mitigation. Reg Environ Chang 7:105–119. https://doi.org/10.1007/s10113-007-0028-2

Smith WN, Grant BB, Desjardins RL et al (2010) A tool to link agricultural activity data with the DNDC model to estimate GHG emission factors in Canada. Agric Ecosyst Environ 136:301–309. https://doi.org/10.1016/j.agee.2009.12.008

Snæbjörnsdóttir SÓ, Sigfússon B, Marieni C et al (2020) Carbon dioxide storage through mineral carbonation. Nat Rev Earth Environ 1:90–102. https://doi.org/10.1038/s43017-019-0011-8

Sodango T, Sha J, Li X et al (2021) Modeling the spatial dynamics of soil organic carbon using remotely-sensed predictors in Fuzhou City. China Remote Sens (Basel) 13:1682. https://doi.org/10.3390/rs13091682

Soon YK, Abboud S (1991) A comparison of some methods for soil organic carbon determination. Commun Soil Sci Plant Anal 22:943–954. https://doi.org/10.1080/00103629109368465

Stenberg B, Rossel RAV (2010) Diffuse reflectance spectroscopy for high-resolution soil sensing. In: Proximal soil sensing. Springer, Netherlands, Dordrecht, pp 29–47

Stevens A, van Wesemael B, Bartholomeus H et al (2008) Laboratory, field and airborne spectroscopy for monitoring organic carbon content in agricultural soils. Geoderma 144:395–404. https://doi.org/10.1016/j.geoderma.2007.12.009

Stevens A, Udelhoven T, Denis A et al (2010) Measuring soil organic carbon in croplands at regional scale using airborne imaging spectroscopy. Geoderma 158:32–45. https://doi.org/10.1016/J.GEODERMA.2009.11.032

Stevenson A, Zhang Y, Huang J et al (2024) Rates of soil organic carbon change in cultivated and afforested sandy soils. Agric Ecosyst Environ 360:108785. https://doi.org/10.1016/j.agee.2023.108785

Stiglitz R, Mikhailova E, Post C et al (2017) Using an inexpensive color sensor for rapid assessment of soil organic carbon. Geoderma 286:98–103. https://doi.org/10.1016/j.geoderma.2016.10.027

Stockmann U, Adams MA, Crawford JW et al (2013) The knowns, known unknowns and unknowns of sequestration of soil organic carbon. Agric Ecosyst Environ 164:80–99. https://doi.org/10.1016/j.agee.2012.10.001

Strydom T, Riddell ES, Rowe T et al (2019) The effect of experimental fires on soil hydrology and nutrients in an African savanna. Geoderma 345:114–122. https://doi.org/10.1016/j.geoderma.2019.03.027

Stumpe B, Weihermüller L, Marschner B (2011) Sample preparation and selection for qualitative and quantitative analyses of soil organic carbon with mid-infrared reflectance spectroscopy. Eur J Soil Sci 62:849–862. https://doi.org/10.1111/j.1365-2389.2011.01401.x

Sun H, Nelson M, Chen F, Husch J (2009) Soil mineral structural water loss during loss on ignition analyses. Can J Soil Sci 89:603–610. https://doi.org/10.4141/CJSS09007

Sun J, Zhao M, Huang L et al (2023) Recent progress on direct air capture of carbon dioxide. Curr Opin Green Sustain Chem 40:100752. https://doi.org/10.1016/j.cogsc.2023.100752

Swetha RK, Chakraborty S (2021) Combination of soil texture with nix color sensor can improve soil organic carbon prediction. Geoderma 382:114775. https://doi.org/10.1016/J.GEODERMA.2020.114775

Swetha RK, Dasgupta S, Chakraborty S et al (2022) Using Nix color sensor and Munsell soil color variables to classify contrasting soil types and predict soil organic carbon in Eastern India. Comput Electron Agric 199:107192. https://doi.org/10.1016/j.compag.2022.107192

Szava-Kovats R (2009) Re-analysis of the relationship between organic carbon and loss-on-ignition in soil. Commun Soil Sci Plant Anal 40:2712–2724. https://doi.org/10.1080/00103620903173764

Taghizadeh-Toosi A, Olesen JE (2016) Modelling soil organic carbon in Danish agricultural soils suggests low potential for future carbon sequestration. Agric Syst 145:83–89. https://doi.org/10.1016/j.agsy.2016.03.004

Tam NFY, Yao MWY (1998) An accurate, simple and novel analytical method for the determination of Total organic carbon in sediment. Int J Environ Anal Chem 72:137–150. https://doi.org/10.1080/03067319808035885

Tan Z, Lal R, Owens L, Izaurralde R (2007) Distribution of light and heavy fractions of soil organic carbon as related to land use and tillage practice. Soil Tillage Res 92:53–59. https://doi.org/10.1016/j.still.2006.01.003

Teixeira RFM, Morais TG, Domingos T (2021) Global process-based characterization factors of soil carbon depletion for life cycle impact assessment. Sci Data 8:237. https://doi.org/10.1038/s41597-021-01018-2

Tifafi M, Guenet B, Hatté C (2018) Large differences in global and regional total soil carbon stock estimates based on soil grids, HWSD, and NCSCD: Intercomparison and evaluation based on field data from USA, England, Wales, and France. Global Biogeochem Cycles 32:42–56. https://doi.org/10.1002/2017GB005678

Tsouros DC, Bibi S, Sarigiannidis PG (2019) A review on UAV-based applications for precision agriculture. Information 10:349. https://doi.org/10.3390/info10110349

Ulmer MG, Swenson LJ, Patterson DD, Dahnke WC (1992) Organic carbon determination by the Walkley-Black, Udy dye, and dry combustion methods for selected North Dakota soils. Commun Soil Sci Plant Anal 23:417–429. https://doi.org/10.1080/00103629209368599

Ussiri DAN, Jacinthe P-A, Lal R (2014) Methods for determination of coal carbon in reclaimed minesoils: a review. Geoderma 214–215:155–167. https://doi.org/10.1016/j.geoderma.2013.09.015

Vågen T-G, Winowiecki LA, Tondoh JE et al (2016) Mapping of soil properties and land degradation risk in Africa using MODIS reflectance. Geoderma 263:216–225. https://doi.org/10.1016/j.geoderma.2015.06.023

Veum KS, Parker PA, Sudduth KA, Holan SH (2018) Predicting profile soil properties with reflectance spectra via Bayesian covariate-assisted external parameter orthogonalization. Sensors 18:3869. https://doi.org/10.3390/s18113869

Villas-Boas PR, Romano RA, de Menezes Franco MA et al (2016) Laser-induced breakdown spectroscopy to determine soil texture: a fast analytical technique. Geoderma 263:195–202. https://doi.org/10.1016/j.geoderma.2015.09.018

Viscarra Rossel RA, Brus DJ (2018) The cost-efficiency and reliability of two methods for soil organic C accounting. Land Degrad Dev 29:506–520. https://doi.org/10.1002/ldr.2887

Viscarra Rossel RA, Fouad Y, Walter C (2008) Using a digital camera to measure soil organic carbon and iron contents. Biosyst Eng 100:149–159. https://doi.org/10.1016/j.biosystemseng.2008.02.007

Vitti C, Stellacci AM, Leogrande R et al (2016) Assessment of organic carbon in soils: a comparison between the Springer–Klee wet digestion and the dry combustion methods in Mediterranean soils (Southern Italy). Catena (Amst) 137:113–119. https://doi.org/10.1016/J.CATENA.2015.09.001

Vodyanitskii YN, Savichev AT (2017) The influence of organic matter on soil color using the regression equations of optical parameters in the system CIE- L*a*b*. Ann Agrar Sci 15:380–385. https://doi.org/10.1016/j.aasci.2017.05.023

Walkley A, Black IA (1934) An examination of the degtjareff method for determining soil organic matter, and a proposed modification of the chromic acid titration method. Soil Sci 37:29–38. https://doi.org/10.1097/00010694-193401000-00003

Wang D, Anderson DW (1998) Direct measurement of organic carbon content in soils by the Leco CR-12 carbon analyzer. Commun Soil Sci Plant Anal 29:15–21. https://doi.org/10.1080/00103629809369925

Wang X, Smethurst P, Herbert A (1996) Relationships between three measures of organic matter or carbon in soils of eucalypt plantations in Tasmania. Soil Research 34:545. https://doi.org/10.1071/SR9960545

Wang X, Cai D, Hoogmoed W et al (2005) Scenario analysis of tillage, residue and fertilization management effects on soil organic carbon dynamics. Pedosphere 15:473–483

Wang Q, Li Y, Wang Y (2011) Optimizing the weight loss-on-ignition methodology to quantify organic and carbonate carbon of sediments from diverse sources. Environ Monit Assess 174:241–257. https://doi.org/10.1007/s10661-010-1454-z

Wang J-P, Wang X-J, Zhang J (2013) Evaluating loss-on-ignition method for determinations of soil organic and inorganic carbon in arid soils of northwestern China. Pedosphere 23:593–599. https://doi.org/10.1016/S1002-0160(13)60052-1

Wang S, Han P, Song H et al (2019) Application of slope/bias and direct standardization algorithms to correct the effect of soil moisture for the prediction of soil organic matter content based on the near infrared spectroscopy. Spectrosc Spectr Anal 39:1986–1992

Wang X, Zhang Y, Atkinson PM, Yao H (2020) Predicting soil organic carbon content in Spain by combining Landsat TM and ALOS PALSAR images. Int J Appl Earth Obs Geoinf 92:102182. https://doi.org/10.1016/j.jag.2020.102182

Wang J, Liu T, Zhang J et al (2022) Spectral variable selection for estimation of soil organic carbon content using mid-infrared spectroscopy. Eur J Soil Sci 73. https://doi.org/10.1111/ejss.13267

Wang C, Guo D, Zang S (2023a) Thermal remote sensing for mapping the sub-Arctic permafrost and refining its southern limits. Int J Appl Earth Obs Geoinf 118:103235. https://doi.org/10.1016/j.jag.2023.103235

Wang Y, Yang S, Yan X et al (2023b) Evaluation of data pre-processing and regression models for precise estimation of soil organic carbon using Vis–NIR spectroscopy. J Soils Sediments 23:634–645. https://doi.org/10.1007/s11368-022-03337-2

Wangeci A, Adén D, Nikolajsen T et al (2024) Comparing laser-induced breakdown spectroscopy and visible near-infrared spectroscopy for predicting soil properties: a pan-European study. Geoderma 444:116865. https://doi.org/10.1016/j.geoderma.2024.116865

Westman CJ, Hytönen J, Wall A (2006) Loss-on-ignition in the determination of pools of organic carbon in soils of forests and afforested arable fields. Commun Soil Sci Plant Anal 37:1059–1075. https://doi.org/10.1080/00103620600586292

Wielopolski L, Yanai RD, Levine CR et al (2010) Rapid, non-destructive carbon analysis of forest soils using neutron-induced gamma-ray spectroscopy. For Ecol Manag 260:1132–1137. https://doi.org/10.1016/j.foreco.2010.06.039

Wielopolski L, Chatterjee A, Mitra S, Lal R (2011) In situ determination of soil carbon pool by inelastic neutron scattering: comparison with dry combustion. Geoderma 160:394–399. https://doi.org/10.1016/j.geoderma.2010.10.009

Wills SA, Burras CL, Sandor JA (2007) Prediction of soil organic carbon content using field and laboratory measurements of soil color. Soil Sci Soc Am J 71:380–388. https://doi.org/10.2136/sssaj2005.0384

Wright AL, Wang Y, Reddy KR (2008) Loss-on-ignition method to assess soil organic carbon in Calcareous Everglades wetlands. Commun Soil Sci Plant Anal 39:3074–3083. https://doi.org/10.1080/00103620802432931

Wu C, Zheng Y, Yang H et al (2021) Effects of different particle sizes on the spectral prediction of soil organic matter. Catena (Amst) 196:104933. https://doi.org/10.1016/j.catena.2020.104933

Yakubova G, Kavetskiy A, Prior SA, Torbert HA (2017) Measurements of soil carbon by neutron-gamma analysis in static and scanning modes. J Vis Exp. https://doi.org/10.3791/56270

Yang L, He X, Shen F et al (2020) Improving prediction of soil organic carbon content in croplands using phenological parameters extracted from NDVI time series data. Soil Tillage Res 196:104465. https://doi.org/10.1016/j.still.2019.104465

Yang J, Shen F, Wang T et al (2021) Effect of smart phone cameras on color-based prediction of soil organic matter content. Geoderma 402:115365. https://doi.org/10.1016/j.geoderma.2021.115365

Yang P, Wang Y, Hu B et al (2024) Predicting soil organic carbon content using simulated insitu spectra and moisture correction algorithms in southern Xinjiang. China Geoderma Regional 37:e00783. https://doi.org/10.1016/j.geodrs.2024.e00783

Yasuda Y, Ohtani Y, Mizoguchi Y et al (2008) Development of a CO2 gas analyzer for monitoring soil CO_2 concentrations. J For Res 13:320–325. https://doi.org/10.1007/s10310-008-0079-3

Zhang YY, Zhang WH (2014) Automated analyzer, LOI and other methods for determination of soil organic carbon. Appl Mech Mater 580–583:903–906. https://doi.org/10.4028/www.scientific.net/AMM.580-583.903

Zhang M, Li Y, Stoffella PJ (2005) Comparison of analytical methods for organic matter in composts and organic mulches. Commun Soil Sci Plant Anal 36:2587–2599. https://doi.org/10.1080/00103620500257317

Zhang Y, Biswas A, Ji W, Adamchuk VI (2017) Depth-specific prediction of soil properties in situ using Vis-NIR spectroscopy. Soil Sci Soc Am J 81:993–1004. https://doi.org/10.2136/sssaj2016.08.0253

Zhang Y, Zhang T, Li H (2021) Application of laser-induced breakdown spectroscopy (LIBS) in environmental monitoring. Spectrochim Acta Part B At Spectrosc 181:106218. https://doi.org/10.1016/j.sab.2021.106218

Zhang C, Shi T, Liu J et al (2022) Eco-engineering approaches for ocean negative carbon emission. Sci Bull (Beijing) 67:2564–2573. https://doi.org/10.1016/j.scib.2022.11.016

Zhang X, Xie E, Chen J et al (2023) Modelling the spatiotemporal dynamics of cropland soil organic carbon by integrating process-based models differing in structures with machine learning. J Soils Sediments 23:2816–2831. https://doi.org/10.1007/s11368-023-03516-9

Zhang L, Heuvelink GBM, Mulder VL et al (2024) Using process-oriented model output to enhance machine learning-based soil organic carbon prediction in space and time. Sci Total Environ 922:170778. https://doi.org/10.1016/j.scitotenv.2024.170778

Zhou T, Jia C, Zhang K et al (2023) A rapid detection method for soil organic matter using a carbon dioxide sensor in situ. Measurement 208:112471. https://doi.org/10.1016/j.measurement.2023.112471

Chapter 7
Quantifying Soil Organic Carbon Stocks

Abstract This chapter focuses on the assessment of soil organic carbon (SOC) stocks, outlining the key factors that influence accurate estimation, including SOC concentration, bulk density, sampling depth, and gravel content. It discusses the importance of proper sampling protocols and appropriate sampling density to ensure reliable measurements. The chapter also introduces the equivalent soil mass approach to account for variability in bulk density. Common challenges in SOC stock assessments are addressed, highlighting the need for precision in quantifying SOC stocks to support carbon farming initiatives.

7.1 Introduction

Soil organic carbon (C) is critical determinant of soil health affecting soil structure, water retention, nutrient supply, and crop productivity. Soil organic C (SOC) accumulation is influenced by agricultural management practices, such as no-tillage (NT), cover cropping, crop rotation, and the addition of organic inputs like compost and manure (Söderström et al. 2014). These practices not only reduce soil disturbance and erosion but also increase organic matter (OM) input, promoting C sequestration (CS) (Li et al. 2014). Maintaining higher SOC concentration is crucial for achieving climate goals, as it contributes to terrestrial C storage and aligns with global initiatives like the "4 per mille" target, which aims to increase soil C stocks to mitigate climate change (Minasny et al. 2017). While transitioning to sustainable agricultural practices (SAPs) may not greatly reduce greenhouse gas (GHG) emissions at the farm level, increasing SOC stocks on a larger scale is crucial for climate mitigation and adaptation. To ensure these efforts effectively contribute to reducing emissions, C crediting systems must be reformed (Probst et al. 2024). SOC stock measurements are now central to programs like Monitoring, Reporting, and Verification (MRV), which track GHG emission reductions (Badgery et al. 2021). These programs quantify and verify GHG reductions by encouraging agricultural practices that enhance SOC storage, providing financial incentives to farmers through C credits. Each credit represents the removal of one metric ton of CO_2 or its

N. Loria, R. Lal, *Carbon Farming*, https://doi.org/10.1007/978-3-032-00842-8_7

equivalent GHG from the atmosphere, allowing farmers to contribute to climate mitigation while benefiting economically by turning soil C into a valuable commodity in C markets (Nogrady 2024).

SOC stocks are dynamic and exhibit significant variability at spatial and temporal scale, influenced by a multitude of factors, including soil type (Torres-Sallan et al. 2018), climate (Wang et al. 2017), land use practices (Wiesmeier et al. 2015), and OM inputs (Fujisaki et al. 2018). Triantakonstantis et al. (2024) highlighted that various soil properties—including pH, texture, drainage, and slope—impact SOC dynamics in Greek croplands. Although SOC levels decreased from 2010 to 2021, the extent of C loss varied significantly across agroecological contexts, closely linked to land management activities that alter SOC stocks. The temporal variability in SOC was notably reduced under NT practices with residue retention compared to other tillage methods, resulting in enhanced SOC accumulation and stability (Qi et al. 2019). This reduction in variability is associated with fewer fluctuations in heavy and mineral-associated SOC fractions, fostering consistent C accumulation, particularly in the top 0–5 cm soil layer. Key considerations for measuring SOC stocks include using standardized methodologies, establishing pretreatment baselines, and accounting for spatial variability and soil depth (Olson et al. 2014). It is also essential to document temporal changes, track C sources, and be mindful of potential SOC losses in both NT and conventional tillage (CT) systems. Various approaches have been employed to measure and monitor SOC concentration in soils (Conant et al. 2011), with detailed methods already discussed in Chap. 6. This chapter shifts the focus to SOC stock assessments, highlighting key factors to consider and the challenges in ensuring accurate assessments. Accurate SOC stock calculations are crucial for C farming initiatives, as they help quantify the benefits of SAPs, enabling farmers to earn C credits and contribute meaningfully to climate mitigation efforts (Lal and Follett 2009).

7.2 Soil Organic Carbon Stock Assessments

SOC stock refers to the total amount of organic C stored in a given volume of soil, typically expressed in megagrams per hectare (Mg ha^{-1}). These stocks are crucial for monitoring changes in SOC before and after the implementation of conservation practices (Morgan and Ackerson 2022). SOC stock is influenced by various factors, including land use, vegetation, external C inputs, climate, site characteristics such as slope and aspect, and inherent properties like soil type and texture (Jobbágy and Jackson 2000; Chen et al. 2009; Söderström et al. 2014; Fujisaki et al. 2018; Lei et al. 2019; Gocke et al. 2023).

To determine SOC stocks, it is essential to quantify the SOC concentration of fine earth (<2 mm), coarse mineral fraction (>2 mm), and soil bulk density (BD) within a minimum depth of 30 cm (FAO 2019). All samples should be georeferenced, and any associated error or uncertainty must be reported. However, studies may not consider coarse content if its occurrence is rare within the chosen sampling depth (Zhuo et al. 2022). Once all these measurements are obtained, SOC stock is calculated using the recommended formula provided in the IPCC Good Practice Guidance for LULUCF (IPCC 2003; Chen et al. 2009; Loria et al. 2016; FAO 2019):

$$SOC\,stock\left(Mg\,C\,ha^{-1}\right) = SOC\,Concentration\left(g\,Kg^{-1}\right) \times Bulk\ density\left(g\,cm^{-3}\right) \times Sampling\ Depth\left(cm\right) \times\left(1 - Gravel\ content\left[Kg\,Kg^{-1}\right]\right)\times 10$$

The factor of 10 in this formula is used to convert the SOC stock into Mg ha^{-1}. Alternative methods for estimating SOC stock differ based on BD and how rock fragments are treated, which can significantly impact the accuracy estimates. Hailu (2024) reviewed various methods for calculating SOC stocks, with an overview presented in Table 7.1.

Tadiello et al. (2022) outlines a streamlined methodology, including an Excel tool, for calculating SOC stock and its standard deviation, accounting for correlations between SOC concentration and BD across single and multiple soil layers, to streamline C accounting.

Accurate measurement of SOC stocks is challenged by factors such as soil depth, BD, and landscape representation (Vanden Bygaart and Angers 2006). Variability in soil depth can lead to inconsistent sampling results, as different depths may contain varying amounts of SOC (Wang et al. 2022; Dubeux et al. 2024; Odebiri et al. 2024). Additionally, BD is vital for converting SOC concentration into mass per unit area; inaccuracies in this measurement can significantly skew overall estimates (FAO 2019). Landscape representation is another critical aspect, as SOC stocks can vary widely across different landforms, soil types, and land use practices (Chen et al. 2009). To address these challenges, improved experimental designs are necessary, including stratified sampling and compositing methods that account for variations in depth and landscape features, ensuring a representative sample of the area studied (Vanden Bygaart and Angers 2006; Potash et al. 2023). Reporting SOC measurements on an equivalent soil mass (ESM) basis (refer to Sect. 7.5) is essential for detecting small changes in SOC stocks amidst inherent spatial variability (Xiao et al. 2020). The measurement process involves several key steps (Fig. 7.1): site selection of representative areas, developing a robust sampling strategy, careful sample collection, sample processing, laboratory analysis of SOC concentration, calculations and clear reporting of results, and ongoing monitoring to detect changes over time (USDA Natural Resources Conservation Service 2022). By addressing the challenges associated with SOC measurement and following a comprehensive and systematic approach, researchers can improve the reliability of SOC stock assessments and contribute to a better understanding of C dynamics across various ecosystems.

The IPCC guidelines for monitoring SOC stock changes introduce a three-tiered methodology designed to accommodate varying levels of detail and data availability

Table 7.1 Approaches for estimating soil organic carbon (SOC) stocks (Poeplau et al. 2017; FAO 2019; Hailu 2024)

Method	Description	Soil bulk density (BD) calculation	SOC stocks calculation (SCS)	Equations
M1	Samples a known volume of soil, dries it, and weighs it. Does not separate fine soil from rock fragments. Carbon concentration is determined in the fine soil (< 2 mm).	BD is calculated using total mass and total volume without considering rock fragments	SCS is calculated using the total mass of fine soil without accounting for rock fragments	$SOCstock_i = SOC_{confine\ soil} \times BD_{sample} \times depth_i$ (ignores rock fragments)
M2	Similar to CM1, but also measures weight and volume of rock fragments and coarse plant residues.	BD is calculated by including mass and volume of rock fragments.	SCS is calculated using total mass of fine soil, ignoring rock fragments	$SOCstock_i = SOC_{confine\ soil} \times BD_{fine\ soil} \times depth_i$ (ignores coarse roots)
M3	Calculates rock fragment content as a percentage of volume and adjusts soil volume for SCS calculations.	BD is not adjusted for rock fragments; it is based solely on fine soil mass and volume.	SCS is adjusted for volume based on rock fragment content but does not use rock fragments in the BD calculation.	$SOCstock_i = SOC_{confine\ soil} \times BD_{sample} \times depth_i \times (1 - \text{rock fragments fraction})$
M4	Considers rock fragments in both BD and volume calculations for SCS.	BD is calculated by including mass of rock fragments, leading to more accurate estimates.	SCS is calculated by adjusting both volume and mass of fine soil based on rock fragments.	$SOCstock_i = SOC_{confine\ soil} \times BD_{fine\ soil} \times depth_i \times (1 - \text{rock fragments fraction})$

(IPCC 2006). Tier I provides a straightforward approach using default values and response ratios, which facilitates quick assessments but often results in overestimated emissions reductions. Response ratios compare SOC stocks under different management practices to a baseline or conventional practice, quantifying the SOC associated with a specific management strategy relative to a reference condition after a designated period, typically around 20 years (IPCC 2006). Tier II enhances the Tier I approach by incorporating country- or region-specific values and improved data utilization, enabling a more tailored analysis while still relying on predefined

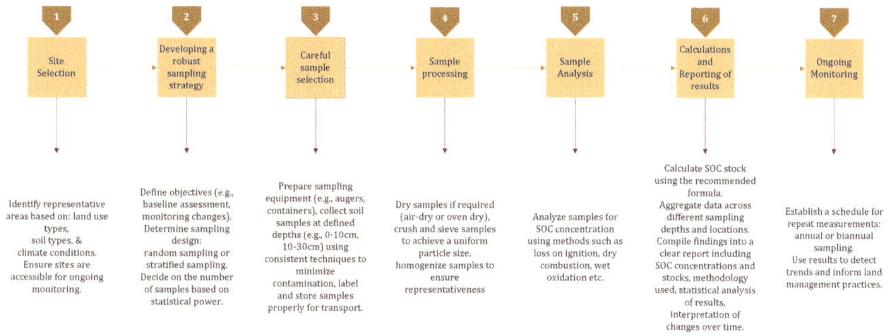

Fig. 7.1 Workflow for assessing soil organic carbon stocks

ratios. This tier allows for potential improvements, such as deriving country-specific emission factors, specifying climate regions better suited to the country's conditions, and creating a more detailed classification of management systems within land-use categories. Finally, Tier III employs detailed modeling and inventory-based methods to track actual SOC changes, yielding robust estimates that consider baseline scenarios and more intricate soil dynamics. This tiered framework allows countries to choose the most appropriate level of sophistication for their specific circumstances, thereby enhancing the accuracy and relevance of SOC monitoring. The Table 7.2 summarizes the key differences, strengths, and limitations of each tier in the three-tiered methodology for monitoring SOC stock changes.

7.2.1 Soil Organic Carbon Concentration

As discussed in Chap. 6 (Sect. 6.3), various methods are available for estimating SOC, including wet combustion (e.g., Walkley-Black titration), dry combustion, spectroscopic, colorimetric, and remote sensing (RS) techniques. The choice of method depends on factors such as available resources, cost, time, and desired level of accuracy (Zhang and Zhang 2014). While wet and dry combustion methods are widely used for their high accuracy, they can be resource intensive (Chatterjee et al. 2009; Loria et al. 2024). Wet combustion, in particular, requires a correction factor based on soil type and land use to account for full recovery of C (Mikhailova et al. 2003; Meersmans et al. 2009; Tivet et al. 2012). A study in Belgium showed that adjusting Walkley-Black measurements for land use, texture, and depth increased SOC estimates by 22%, underscoring the need for region-specific corrections rather than universal ones (Lettens et al. 2007). Spectroscopic and RS methods offer faster, more cost-effective alternatives, though their precision varies depending on the landscape and soil conditions (Das and Ghimire 2024). Comparative studies have shown that the accuracy of these methods can vary significantly across different land uses, soil types, and management practices (Soon and Abboud 1991; Smith

Table 7.2 Comparison of three-tiered methodology for monitoring soil organic carbon stock changes (IPCC 2006; Sanderman and Baldock 2010)

Approach	Description	Strengths	Limitations	Data requirements	Temporal focus	Example use cases
Tier I	Simple accounting using default values and response ratios.	Easy to implement; minimal data required; quick results for initial assessments.	Overestimates emissions reductions; assumes steady-state conditions; does not account for depth beyond 30 cm; does not account inorganic C.	Limited to default values; basic datasets; no depth or detailed local information.	Typically considers changes over a set period (e.g., last 20 years).	Initial assessments, quick estimates for policymaking, preliminary evaluations of management practices.
Tier II	Incorporates country/region-specific values and improved data utilization.	More tailored; incorporates local data and management practices; allows for estimating SOC storage to greater depth beyond 30 cm if sufficient data is available.	Still relies on response ratios; potential inaccuracies in local data; may not fully capture the complexity of soil dynamics.	Requires regional data, improved inventories, and greater detail in local management practices.	Considers longer periods but still focuses on recent management impacts.	National inventories, comparative studies among regions, more detailed assessments of management impacts.
Tier III	Detailed modeling and/or inventory-based approach tracking actual SOC changes.	Robust estimates of net emissions/removals; considers baseline scenarios; can account for inorganic C changes.	More complex; requires extensive data and long-term monitoring; higher resource intensity.	Extensive datasets; requires time series data, modeling tools, and potential spatial data.	Capable of capturing longer-term legacy effects of land use and management.	Research studies; comprehensive national GHG accounting; policy development based on rigorous data.

et al. 1997; Da Silva Dias et al. 2013; Davis et al. 2017; Roper et al. 2019; Ciric et al. 2021; Burgos Hernández et al. 2023). In agricultural lands, where SOC distribution may be more uniform, methods like dry combustion are often highly reliable (Matejovic 1997; Paustian et al. 2019). In contrast, RS and spectroscopic approaches may perform better in large, heterogeneous landscapes such as forests or grasslands, where rapid and broad-scale assessment is required (Wang et al. 2020; Gholizadeh et al. 2023). Each method's applicability and precision must be evaluated within the specific context of the study to ensure reliable SOC stock estimates.

7.2.2 *Bulk Density*

BD is a critical parameter in calculating SOC stocks because it links the concentration of organic C to the mass of soil per unit volume (Lal et al. 2001). Accurate BD measurements are essential for converting SOC concentrations from a percentage or mass per unit mass (e.g., g kg^{-1}) to a stock value per unit area (e.g., Mg C ha^{-1}). Key challenges in measuring soil BD include ensuring the sample's representativeness of the target layer, avoiding modifications to soil structure during sampling (which can lead to compaction or cracking), and selecting an appropriate sample size (Timm et al. 2005). There are two main approaches to measuring BD: direct methods and indirect methods (Al-Shammary et al. 2018). Table 7.3 provides a detailed comparison of these methods, highlighting their key characteristics.

Direct methods include the core, clod, and excavation methods (Blake and Hartge 1986). The core method, which uses a volumetric cylinder, is the most common. It involves extracting a soil core, drying it, and calculating BD based on its mass and the core volume (Blake and Hartge 1986). This method is simple and cost-effective but can be influenced by soil type, depth, and field conditions, with reduced accuracy in sandy or gravelly soils (Al-Shammary et al. 2018). Additionally, its time-consuming nature and diminishing accuracy with depth can limit its applicability in large-scale studies (Gross and Harrison 2018). The clod method measures BD by coating soil clods in paraffin, then weighing them in air and water to calculate volume (Blake and Hartge 1986). Although this method can yield accurate results, it's labor-intensive and requires stable clods between 4–10 cm in diameter (Al-Shammary et al. 2018). Additionally, inconsistencies in the wax coating can introduce errors, and it tends to overestimate BD because it doesn't account for inter-aggregate pores in the soil (Casanova et al. 2016). The excavation or pit method, used for stony soils or on sloped terrain, involves excavating a hole and measuring its volume by filling it with sand or water (Donald C. Erbach 1987). While generally accurate, this method is time-consuming and may suffer from errors caused by non-uniformity in the soil or surface slope (Grossman and Reinsch 2002; Vadeboncoeur et al. 2012). According to FAO (2019), soil BD should be measured in the same core as SOC concentration using direct methods like the undisturbed core or excavation methods, as these provide the most accurate estimates; the clod method is not recommended since it does not adequately represent BD across soil layers or horizons.

Table 7.3 Overview of bulk density (BD) measurement methods (Grossman and Reinsch 2002; Gross and Harrison 2018; Al-Shammary et al. 2018)

	Direct			Indirect	
Method	Core	Clod	Excavation	Radiation	Regression
Sample collection	Cylinder inserted into the soil, sample taken within cylinder.	Natural clod cut/broken from soil mass, coated with resin	Involves digging an irregularly shaped hole and collecting the soil sample	No physical sample collection; uses radioactive isotopes or gamma-ray attenuation	No physical sample; uses predictive models based on indirect measurements
Volume determination	Volume of soil equals volume of cylinder.	Volume measured by water displacement or weight in water	Volume of cavity filled by sand, spheres, water, etc.	Radiation attenuation is used to estimate BD.	Calculated using a regression equation based on relationships between soil properties and BD.
Moisture dependency	Requires moist soil for insertion without disruption.	Can measure at different water tensions; moisture dependent.	Not highly dependent on moisture; soil can be oven-dried.	Water content must be measured separately for dry BD calculation.	May require moisture as a variable in the regression model.
Suitability	Ideal for soft, cohesive soils like clay loam; less suitable for non-cohesive, fracturing soils like sand or gravel.	Not suitable for coarse-textured soils, as clods may not form or break during transport.	Suitable for stony or gravelly soils and for slopy terrain; unsuitable for soils with large pores	Useful for soils with rough stones; requires calibration and depends on soil texture and mineral content.	Effective for predicting BD based on SOC and texture
Impact on BD	Possible compaction during insertion affects BD	Higher BD due to absence of interclod spaces.	No compaction; more accurate for layers with rock fragments or varying thickness.	Nondestructive and does not affect soil structure	BD predicted based on input factors
Accuracy	High for cohesive soils but generally underestimates BD due to sampling bias, volume measurement errors, and potential moisture loss or gain.	Moderate, depends on clod stability, user experience, calibration	Moderate, depends on uniformity of fill material	High for surface layers, decreases with depth	Accuracy depends on the quality of input data and models, indirect estimation.
Challenges	May underestimate rock content and BD, risk of soil compaction, operator-dependent	Cumbersome transport, wax coating can introduce errors	Destructive, time-intensive, requires precise measurement of fill material	Requires calibration, costly equipment, regulatory requirements	Time-consuming data collection, influenced by soil properties

Indirect methods include the radiation and regression methods (Al-Shammary et al. 2018). The radiation method measures BD indirectly by-passing gamma rays through the soil. The attenuation of these rays correlates with soil density (Donald C. Erbach 1987). This non-destructive technique is fast and preserves the soil structure, making it useful for applications such as road construction and civil engineering (Al-Shammary et al. 2018). However, it's expensive, requires calibration, and becomes less accurate with increasing soil depth. Operator experience and soil properties like texture and mineral content also affect accuracy. Regression methods, such as Pedotransfer Functions (PTFs), estimate BD based on readily available data like SOC concentration, texture, structure, and environmental conditions (Casanova et al. 2016). These methods have evolved, with advanced machine learning techniques such as artificial neural networks (ANN) and genetic programming improving accuracy (Padarian et al. 2012; Choudhury et al. 2023). However, PTFs are sensitive to soil depth, texture, and environmental variables, which can influence their precision (Qiao et al. 2019). Combining soil data with climate information can enhance predictions (Barros and de Jong van Lier 2014). Uncertainty in SOC stock estimates can increase when using PTFs if associated errors are not adequately addressed, leading to potential bias and significant variability at regional levels (Schrumpf et al. 2011).

Various studies demonstrate that the choice of BD measurement methods significantly impacts the accuracy of SOC stock estimates. One study compared three common methods—clod, core, and excavation—in a managed forest ecosystem, finding that the core method underestimated both BD and SOC stocks by up to 36% at depths below 20 cm compared to excavation (Gross and Harrison 2018). It emphasized that these methods are not interchangeable and recommended sampling to at least 1 meter in depth, using clod or excavation methods as standard references for BD measurements in non-rocky soils. Coring methods often underestimate rock fragment (RF) content and BD, leading to biased estimates, while mass-based calculations can reduce these biases, with quantitative pits identified as the best method for establishing reference soil masses (Beem-Miller et al. 2016). Walter et al. (2016) showed that direct BD measurements with sample rings were more accurate than those derived from PTFs, which introduced significant errors. In challenging climates like southern Australia, Holmes et al. (2011) showed that SOC stock estimates remained consistent despite BD variations when coarse fragment volume exceeded 20%. Clancy et al. (2023) proposed a novel "core-scaling" method that combines core and pit sampling to account for large RFs, achieving strong correlation with pit estimates and avoiding inaccuracies in RF-rich soils. Singla and Dhingra (2024) utilized gamma-ray spectrometry to determine the mass attenuation coefficient and soil BD in different soil textures and depths, revealing that both gamma-ray energy and soil composition affect attenuation. The method showed promising accuracy for estimating BD and could be applied to assess other soil properties like porosity and moisture content. Pires et al. (2009) demonstrated that both the volumetric ring method and gamma-ray attenuation effectively captured soil spatial variability, confirming the reliability of these techniques for BD assessments.

7.2.3 Soil Depth

Soil depth is crucial for accurately evaluating SOC stock, as C concentration varies across depths due to management practices (Follett et al. 2015). Proper assessment of SOC distribution is essential for understanding how these practices impact CS. For instance, transitioning from CT to NT alters C distribution, with CT promoting a more uniform mixing of OM throughout the topsoil, while NT allows OM to accumulate in surface layers due to reduced disturbance (Luo et al. 2010). However, the climate change mitigation potential of NT may be overestimated if deeper soil layers are not considered (Meurer et al. 2018). Studies suggest that standard depth intervals, such as 10 cm, provide more reliable results than pedogenetic horizons, as they reduce variation and prevent overestimation of SOC stocks (Dincă et al. 2015). Grüneberg et al. (2010) found that sampling by depth increments is preferable for accurately assessing regional SOC stocks, whereas sampling by soil horizons is better suited for investigating pedogenic processes that shape organic C distribution in soil.

Shallow sampling can lead to inconsistent SOC findings, as it may miss C redistribution across the soil profile (Olson and Al-Kaisi 2015). To fully capture root zone C, sampling should extend beyond the tillage zone (0–30 cm) to the entire root zone, ideally up to 1–2 meters. Deep soil sampling is especially important, as studies show that reduced tillage in organic farming can increase SOC stocks by 3.6% at depths up to 100 cm, highlighting the often-overlooked role of subsoil C storage (Krauss et al. 2022). In a study comparing SOC stocks in afforested mineral soils, it was found that soils below 30 cm contributed over 22% of the total SOC within a 0–60 cm profile, underscoring the necessity of considering deeper soil layers in SOC assessments (Premrov et al. 2017). Moreover, significant differences in SOC stocks were observed between soil types. Gleysols had substantially higher mineral SOC stocks than Podzols, particularly in the upper 30 cm, indicating the need for tailored C management strategies based on soil type and depth. Wiesmeier et al. (2012) noted that cropland soils in southeast Germany had lower SOC concentrations in their A horizons, but their overall SOC stocks were comparable to those of other land uses, primarily due to a deeper topsoil layer. Grassland soils demonstrated higher SOC levels in the B horizon, largely attributable to the presence of C-rich Gleysols, emphasizing the importance of accounting for soil depth in SOC assessments rather than relying solely on fixed depth (FD) increments.

Recent studies emphasize the importance of considering soil depth in SOC assessments. Research in the alpine grasslands of the Peruvian Andes found that total SOC averaged 215 ± 21 Mg ha^{-1}, with limited surface sampling leading to significant underestimations of total SOC stocks (Yang et al. 2018). This study also highlighted the risks of limited soil sampling, as focusing only on surface layers can lead to significant underestimations of total SOC stocks and misrepresentation of the effects of soil formation and environmental factors. A study in KwaZulu-Natal, South Africa, further emphasizes the importance of soil depth in predicting SOC stocks, revealing that prediction accuracy declines with increasing depth, with root mean square error (RMSE) ranging from 8.34 Mg ha^{-1} to 11.97 Mg ha^{-1} (Odebiri et al. 2024). This indicates a weakening relationship between environmental factors and SOC stocks at greater depths. Additionally, Xiao et al. (2020) investigated the controversial effectiveness of NT in enhancing SOC stocks, revealing that while NT significantly alters SOC distribution in the top 5 cm of soil, changes at greater depths are minimal. This emphasizes the need for deeper sampling to clarify NT's role in CS and climate mitigation. Lastly, Ge et al. (2023) highlighted the critical role of soil depth in shrublands, revealing those subsoils (30–100 cm) store about 53.30% of total SOC, with edaphic properties significantly influencing SOC stocks in this layer. Overall, these findings illustrate that to fully capture SOC dynamics and assess total C storage potential, a comprehensive understanding of soil depth and its implications for SOC assessment is essential.

7.2.4 *Gravel Content*

Gravel or RFs, typically defined as particles larger than 2 mm in diameter, is commonly found in many soils, especially in arid and semi-arid regions, forests and mountainous areas, and soils with parent materials that produce coarse fragments (Lai et al. 2022). Traditionally, soil scientists have excluded the gravel fraction from SOC calculations, assuming it does not significantly contribute to organic C storage (Zabowski et al. 2011). However, recent studies indicate that gravel influences SOC stock in several ways: it occupies space that could support fine mineral soil, thereby reducing nutrient availability (Qin et al. 2015); it creates micro-conditions that enhance OM decay and alter soil moisture and temperature, influencing enzyme activity essential for nutrient cycling (Certini et al. 2004); and it can host significant amounts of OM, either bound to gravel surfaces, present as fine organic coatings, or in the form of microbial biomass (Agnelli et al. 2001; Mehler et al. 2014). Additionally, RFs affect various physical properties, such as BD and water infiltration, impacting crop yield by enhancing macro-pore formation that improves infiltration and reduces runoff (Sekucia et al. 2020). Increased gravel content has been shown to reduce dry matter accumulation and grain yield in crops like durum wheat, primarily due to restrictions on root growth (Ercoli et al. 2006). Qin et al. (2015) reported negative effects of gravel on SOC, total nitrogen (N), fractional vegetation cover, and aboveground biomass, particularly in the 2 to 20 mm size range, highlighting its adverse impact on nutrient sequestration and vegetation growth in the semi-arid and arid regions of the Qinghai-Tibetan Plateau. In contrast, Danalatos et al. (1995) emphasize the crucial role of RFs in enhancing SOC storage and moisture conservation in Mediterranean conditions, ultimately supporting plant growth and helping to mitigate desertification, especially in dry years.

In soils with high gravel content, coarse fragments can represent a significant portion of total SOC. For example, a study of Red Chromosol soils in Eastern Australia found that the gravel fraction accounted for 19% and 23% of the total C and N stocks, respectively, to a depth of 180 cm. The gravel content varied from approximately 5% in surface layers to 45% at deeper levels, with gravel contributing up to 40% of total SOC below 100 cm (Kirkby et al. 2021). Additionally, Rytter (2012) examined the impact of stones and gravel on nutrient stock estimates in arable soils across five Swedish sites, revealing that neglecting their volume led to an 8–9% overestimation of C and N stocks. This underscores the importance of including RFs when comparing nutrient stocks across sites or assessing land-use changes at larger scales. Zheng et al. (2021) emphasizes the important role of RFs in shaping soil properties and nutrient dynamics across various grassland ecosystems. While certain physical characteristics showed no significant differences, factors like number density, volumetric content, and surface area density were significantly related to SOC and nutrient levels. These findings suggest that RFs enhance soil health by improving nutrient accumulation and water retention, which supports plant growth (Zheng et al. 2021). Future research should focus on the interactions between RFs and soil-vegetation dynamics, as well as their implications for hydrological processes and nutrient cycling, especially in the context of climate change.

7.3 Sampling Protocol

Soil spatial and temporal variability is a well-documented phenomenon that highlights the importance of carefully planned sampling strategies to accurately represent the entire area of interest (Rodeghiero and Cescatti 2008; Zhang et al. 2011; Francaviglia et al. 2017; Welsch et al. 2019). Consistency in soil sampling across sites is essential for making SOC stock calculations comparable across sites and regions (Davis et al. 2017). A robust sampling design is crucial for accurately assessing SOC stock changes, requiring careful consideration of sample size, precision, and the ability to detect temporal trends across diverse landscapes (Chang et al. 2016). In studies focused on SOC stocks, two sampling approaches are commonly employed: the design-based and the model-based sampling. Design-based sampling relies on random selection, assuming sample representativeness for simple average calculations, making it suitable for capturing broad SOC variability with minimal assumptions (de Gruijter et al. 2006). This approach is often applied in regional or global SOC estimates. Conversely, model-based sampling uses purposeful selection and incorporates auxiliary data, allowing for spatially detailed estimates and robust trend analysis (de Gruijter et al. 2006). By employing spatial models, model-based approaches improve site-specific estimates, facilitating refined monitoring in areas with complex variability (Somarathna et al. 2017; Sharma et al. 2024). A review by Lawrence et al. (2020) suggests that while model-based sampling shows potential in addressing spatial variability and site-specific management options, design-based methods remain more practical and beneficial for farmers by reducing sample sizes, minimizing bias, and supporting more efficient nutrient applications. These two approaches include multiple sampling types, each with distinct strengths and limitations (Table 7.4).

Table 7.4 Sampling approaches for soil organic carbon stock (de Gruijter et al. 2006; Abbott 2010; Allen et al. 2010)

Approach	Sampling type	Description	Purpose	Advantages	Limitations
Design-based sampling	Simple random sampling	Randomly selects sampling locations across the study area.	Unbiased mean estimation.	Easy to implement; unbiased.	Can result in clustered samples, leaving areas unsampled.
	Stratified random sampling	Divides area into strata based on geography or ancillary variables, sampling randomly within each stratum.	Improved representation of spatial variability.	Reduces bias; improves coverage of spatial heterogeneity.	Requires knowledge of relevant strata; may be complex to implement.
	Systematic grid sampling	Places sampling points in a grid with random starting position.	Uniform coverage across study area.	Reduces clustering; covers large areas evenly.	Risk of bias if periodic features match grid spacing.
	Cluster sampling	Divides population into clusters (groups), and a random selection of these clusters is made	To reduce costs and simplify the sampling process, especially in geographically dispersed populations.	Cost-effective; simplifies data collection; useful when a complete list of the population is difficult to obtain.	High sampling error due to potential similarity within clusters; risk of bias if clusters are not representative of the entire population.
	Nested sampling	Hierarchical sampling with increasing density around initial points.	Captures multi-scale spatial variability.	Allows analysis at multiple spatial scales.	Complex; requires large sample size for full replication.
Model-based sampling	Purposive sampling	Selects locations based on expected SOC variability, informed by prior data or models.	Spatial optimization of samples.	Efficient for targeted studies; fewer samples needed.	May introduce bias if model assumptions are incorrect.
	Geostatistical sampling (kriging)	Uses variograms to select points that capture spatial autocorrelation, optimizing sample location for interpolation.	Generates spatial predictions (e.g., SOC maps).	Produces spatially detailed estimates; minimizes kriging variance.	Requires variogram estimation; complex analysis.
	Cokriging (multivariate sampling)	Uses paired variables (e.g., SOC concentration and bulk density) to optimize sample locations.	Simultaneous estimation of multiple variables.	Reduces sampling effort for correlated variables.	Requires correlation models between variables; complex to implement.
	Adaptive sampling	Iterative approach that adjusts sample locations based on initial SOC data.	Focuses resources on areas of high variance.	Increases efficiency; refines sampling in high-variance areas.	Requires multiple sampling stages; suitable only for dynamic studies.

Choosing a particular sampling type depends on several key factors, including the specific objectives of the study, the spatial and temporal variability of the target population, and the available resources (Potash et al. 2023). The objectives of the research determine the level of detail and accuracy required, which influences the choice of sampling method (Zhang and Hartemink 2017). Additionally, understanding the natural variability within the area of interest is crucial, as it can impact the representativeness of the samples (Zhang et al. 2011). Practical considerations, such as budget constraints, time availability, and the accessibility of the study area, also play a significant role in selecting an appropriate sampling type (Abbott 2010; Chang et al. 2014; Wang et al. 2023b). Ultimately, the chosen method should strike a balance between enhanced accuracy, minimized bias, and increased efficiency, aligning with both the scientific goals and logistical realities of the study (Lawrence et al. 2020).

Several studies have proposed various sampling approaches based on specific study goals, site conditions, and available resources. For instance, Zhang and Hartemink (2017) identified four sampling designs for assessing soil profile SOC stocks, highlighting that horizon-based sampling is effective when depths are accurately measured and representative samples are collected. Vertical transect sampling, requiring two to three cores, captures horizontal variation, while depth-wise stratified random sampling improves accuracy in areas where SOC concentrations decrease with depth. Fuzzy c-means sampling can also stratify the soil profile but requires additional samples and specialized equipment, suggesting that a combination of vertical transects or depth-based methods may be optimal for SOC estimation (Zhang and Hartemink 2017). Rodeghiero and Cescatti (2008) demonstrated that stratified sampling reduced estimation errors by 12% in grasslands and 55–57% in forests, effectively accounting for spatial variability in soil respiration. Systematic sampling, as shown by Sherpa et al. (2016), is another valuable method, especially for covering geographic and attribute variations in SOC estimation. Stratified sampling, based on parent material, has also been emphasized by Heim et al. (2009) as an essential tool for reducing uncertainty in SOC stock estimates, which can significantly improve detection limits for changes in SOC over time. Peltoniemi et al. (2007) further showed that incorporating model-predicted variables into stratification enhances sampling efficiency and reduces standard error in SOC stock estimates, particularly when uncertainties are controlled.

In addition to these traditional methods, more advanced techniques have shown promising results in improving the accuracy of SOC stock estimates. Potash et al. (2023) demonstrated that doubly balanced sampling using readily available data reduces the number of samples needed for SOC estimation by 32%, outperforming stratified sampling, which reduces sample sizes by 17%. This method provides a more efficient way to monitor SOC in agricultural fields, improving both cost-effectiveness and accuracy. Furthermore, advancements like conditioned Latin hypercube sampling (cLHS) are transforming predictive soil mapping by ensuring better characterization of covariate feature space (Malone et al. 2019; Sorenson et al. 2024). This approach helps to enhance spatial predictions and ensures a more representative distribution of samples, which is particularly beneficial for localized

C stock assessments and precision agriculture (Minasny and McBratney 2006). Despite these improvements, challenges persist in determining the optimal sampling density, especially for small-scale assessments critical to C offset markets, where high precision is essential for financial incentives and regulatory compliance. Striking a balance between sampling density, cost, and accuracy is key to improving SOC estimate reliability, ultimately aiding in better soil management and climate mitigation efforts.

7.4 Sampling Density

The precision of SOC stock estimates is heavily influenced by sample size, with small samples potentially leading to large errors, while excessively large samples waste time and resources (Somarathna et al. 2017; Wu et al. 2022; Safaee et al. 2024). Multiple factors, including cost, desired accuracy, SOC variability, and sampling method, determine the required sample size for SOC stock assessment (Chaudhuri et al. 2011; Malone et al. 2019; Lai et al. 2021; Bettigole et al. 2023). Large sample sizes are particularly necessary in areas with high SOC and BD variability, but prior knowledge of the site can help tailor these requirements (Chaudhuri et al. 2011). Research underscores the role of sampling density in improving SOC estimation accuracy, with higher densities improving the detection of spatial variations and impacts of human management (Sun et al. 2012). For example, in Guangdong, China, a sample size of 800 with regression kriging was considered optimal for digital soil mapping of soil OM accurately (Lai et al. 2021), while in Xi'an City, high-density sampling revealed significant spatial variations in SOC density due to urban factors (Zhang et al. 2021a). Further evidence of the importance of optimal sample size comes from Zhang et al. (2021b), who found that as sampling distance increased, SOC stock estimates became smoother but less detailed. They highlighted that a sampling density of 150 × 150 m provided reliable estimates, with 357 points at 450 × 450 m spacing capturing surface SOC variability (0–20 cm) effectively, and 200 points at 600 × 600 m sufficient for deeper layers (100 cm). Long et al. (2018) demonstrated that higher sampling densities improved prediction accuracy in regions with complex topography, with accuracy being more sensitive in flat areas than in slope-rich terrains. These findings underscore the need for tailored sampling schemes based on desired accuracy, topographic complexity, and the balance between cost and precision for efficient SOC stock assessments.

Statistical methods such as power analysis are commonly used to determine the minimum sample size needed to achieve a specified level of confidence and precision (Lark 2009; Heikkinen et al. 2021; Stanley et al. 2023). This calculation typically considers factors like expected variance, desired margin of error, and the size of the population or area. Saurette et al. (2024) proposed a technique to determine the minimum sample size for the cLHS algorithm by analyzing histograms of predictor variables using the Freedman-Diaconis rule. This method adjusts for multimodal and non-normally distributed data and calculates sample size based on the selected quantile range (QR), with the sample size increasing as the QR expands. Morgan et al. (2003) demonstrated that decision-tree model accuracy improves with increasing sample size, but at a decreasing rate, with 10,000 records achieving accuracy within 0.5% of the theoretical maximum. This trend suggests that fitting a power curve to sample data can optimize sample sizes for large datasets. While model-based sampling often relies heavily on sample size, increasing training samples does not always guarantee improved model performance (Lai et al. 2021; Safaee et al. 2024). For example, models like Artificial Neural Networks (ANN) perform better with larger sample sizes, whereas methods like Random Forest are generally less sensitive to sample size (Khaledian and Miller 2020). Additionally, Lagacherie et al. (2020) demonstrated that balanced spatial sampling achieved through dense, evenly distributed samples optimizes digital soil mapping model accuracy and uncertainty estimation, especially for clay content prediction using hyperspectral imagery in Tunisia. These methods for sample size determination are crucial for achieving accurate and efficient SOC stock assessments, considering factors such as variability, topography, and model performance, while being tailored to site conditions, budget, and available resources.

7.5 Equivalent Soil Mass

To accurately estimate SOC changes, SOC concentrations should be converted to area-based stocks, typically by using soil BD. However, BD can vary due to factors like land use, elevation, soil type, and OM changes across different soil depths, which may distort results if using a FD (Li et al. 2019). When BD changes, FD methods can falsely indicate SOC gains or losses due to altered soil mass at the same depth, skewing SOC and ratio-based measurements (von Haden et al. 2020). To avoid this, it's better to express SOC stock on an ESM basis, adjusting depth to ensure consistent dry mass across samples (Wendt and Hauser 2013). The ESM method is preferred over the FD method for tracking changes in SOC concentration over time or comparing effects of land-use changes on its distribution (Heikkinen et al. 2021). Peng et al. (2024) study compared different methods for estimating SOC stocks in a 24-year agricultural experiment, revealing that the ESM approach is more accurate than the FD method, which overestimated SOC under NT practices by about 15%. Similarly, the study by Raffeld et al. (2024) found that FD estimates can overestimate SOC by more than 100% due to BD changes, while the ESM method offers more accurate measurements. It recommends using ESM with 60 cm sampling for reliable SOC assessments and C credits in annual row crop systems. The ESM method should be the default for assessing changes in SOC stocks, mass percentages, and stable isotope values, as it improves accuracy by accounting for BD changes and reduces sampling errors, providing better insights into ecosystem responses (von Haden et al. 2020).

The ESM method enhances the accuracy of SOC stock estimations, but its application is not without challenges. These include difficulties in calculating soil mass at varying depths, accounting for coarse materials and RFs, addressing heterogeneity between pedogenic horizons, and managing the complexities of diverse landscapes (Poeplau and Gregorich 2022). Despite these challenges, the ESM approach remains a valuable tool for improving the accuracy of SOC measurements and should be considered the preferred method for monitoring SOC dynamics in many contexts, particularly in studies focused on long-term land-use changes and climate change mitigation strategies. The ESM corrective approach can incentivize SAPs, reward farmers, improve SOC stock estimates, and validate C market accounting, supporting GHG reduction efforts and enhancing environmental quality (Fowler et al. 2023). The study by Rovira et al. (2022) acknowledges that while the FD approach may introduce minor biases in SOC sequestration estimates, these do not necessitate major revisions to global estimates, and both the FD and ESM approaches should be used complementarily to ensure accurate SOC comparisons across varied conditions.

In the given example (Fig. 7.2), SOC stocks are calculated using both FD and ESM methods for 30 cm sampling depth. The baseline SOC concentration is 2%, with a BD of 1.2 g cm^{-3}. Using the FD method, the soil mass for the 0–30 cm depth is 36 g cm^{-2} (1.2 g cm^{-3}× 30 cm). The baseline SOC stock is then calculated as 0.72 g cm^{-2} (36 g cm^{-2} × 0.02), converted to 72 Mg ha^{-1} with a factor of 100. This initial value serves as a reference SOC stock before any management practices alter soil structure.

Two scenarios are explored, simulating changes in BD due to different tillage practices—CT that decompacts the soil, and NT that compacts it. SOC concentration is assumed to remain constant, while BD changes in each scenario.

1. **Scenario 1: Conventional Tillage (Decompaction)**
   CT reduces BD to 1.1 g cm^{-3}, decreasing soil mass in the 0–30 cm depth to 33 g cm^{-2} (1.1 g cm^{-3}× 30 cm), which results in an apparent SOC stock of

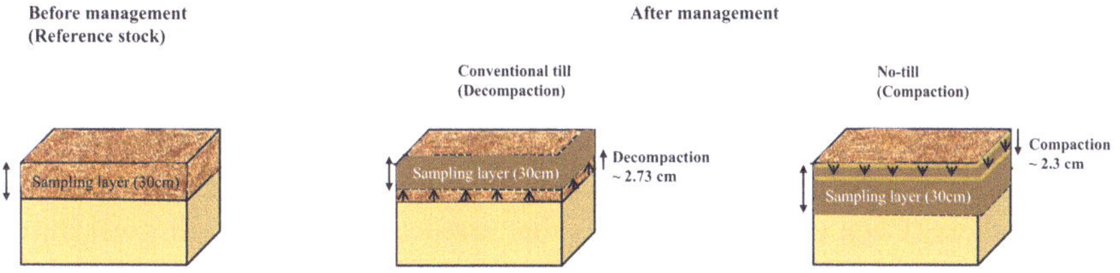

FD method	
Depth (cm)	30
BD (g cm^{-3})	1.2
Soil mass (g cm^{-2})	36
SOC concentration (%)	2
SOC stock(Mg ha^{-1})	72

Method	FD	ESM
Depth (cm)	30	
BD (g cm^{-3})	1.1	1.1
Soil mass (g cm^{-2})	33	33
Initial soil mass (g cm^{-2})	-	36
Adjusted sampling depth (cm)	-	36/1.1 = 32.73
SOC concentration (%)	2	2
SOC stock(Mg ha^{-1})	66	72
Stock change (Mg ha^{-1})	72-66 = 6 **(decrease)**	No change

	FD	ESM
Depth (cm)	30	-
BD (g cm^{-3})	1.3	1.3
Soil mass (g cm^{-2})	39	39
Initial soil mass (g cm^{-2})	-	36
Adjusted sampling depth (cm)	-	36/1.3 = 27.7
SOC concentration (%)	2	2
SOC stock(Mg ha^{-1})	78	72
Stock change (Mg ha^{-1})	78-72 = 6 **(increase)**	No change

Fig. 7.2 Soil organic carbon stock comparison using fixed-depth and equivalent soil mass method

66 Mg ha^{-1} (33 g cm^{-2} × 0.02 × 100) under the FD method. This decline, however, is artificial, as it reflects only the decreased BD from decompaction, not an actual loss of SOC. To correct this, the ESM method maintains a consistent soil mass by adjusting the sampling depth. Dividing the target mass of 36 g cm^{-2} by the new BD (1.1 g cm^{-3}) gives an adjusted depth of approximately 32.73 cm. Calculating SOC stock using the ESM-adjusted depth confirms that it remains at 72 Mg ha^{-1}, accurately reflecting no real change in SOC stock.

2. **Scenario 2: No-Tillage (Compaction)**

   In the NT scenario, BD increases to 1.3 g/cm^3, increasing soil mass to 39 g cm^{-2} (1.3 g cm^{-3}× 30 cm), giving an apparent SOC stock of 78 Mg ha^{-1} (39 g cm^{-2} × 0.02 × 100) with the FD method. Again, this apparent increase is misleading and reflects only the BD increase from compaction. To correct this, the ESM method adjusts the depth, dividing the target soil mass (36 g cm^{-2}) by the new BD (1.3 g cm^{-3}) to reach an adjusted depth of 27.7 cm. Using this depth, the SOC stock is 72 Mg ha^{-1}, showing no real change in SOC stock.

This example illustrates how the FD method can yield misleading results due to variations in BD, often overestimating SOC in NT systems. In contrast, the ESM method adjusts for these density differences, providing a more reliable SOC assessment by comparing consistent soil mass across different management practices. By accounting for soil mass consistency, ESM enhances the accuracy of SOC stock comparisons across different land management practices, supporting better-informed decisions on SOC dynamics and land use impacts.

7.6 Challenges in Soil Organic Carbon Stock Assessments

Accurately measuring SOC stocks presents several interrelated challenges that arise from both technical and environmental complexities, beginning with understanding C saturation in soils. Soils have a finite capacity for C storage, influenced by factors such as soil texture, structure, biochemical composition, and C input levels (Stewart et al. 2007). However, determining these saturation levels remains a significant challenge in soil C research (McBratney et al. 2014), as soils across different sites exhibit varying C saturation points, with some having a higher capacity for C storage than others (Chen et al. 2019). Soils that are near saturation will not exhibit significant changes in SOC stock, regardless of additional C inputs, rendering it an inefficient use of time and resources to focus on these areas for CS (Stewart et al. 2007). In the context of C farming, targeting soils that have reached or are near their saturation point would be counterproductive, as no substantial increase in SOC could be achieved. Identifying soils with lower saturation points or those not yet saturated is essential for optimizing C farming strategies, ensuring that resources are allocated to regions where CS potential remains (Wang et al. 2023a). Additionally, spatial and temporal variability in SOC levels further complicates accurate measurements. Factors such as soil properties (Vos et al. 2019; Fu et al. 2023), climate, land use histories, topography, and management practices create significant variations in SOC across landscapes (Liu et al. 2011; Vos et al. 2019; Yu et al. 2020; Paltineanu et al. 2024). Temporal variability is particularly challenging, as SOC changes occur over long periods, making it difficult to assess long-term impacts on SOC levels, especially in the absence of extensive monitoring data (Triantakonstantis et al. 2024).

Other factors, such as BD estimation and sampling depth, introduce additional uncertainty in SOC stock calculations. BD, which is critical for converting SOC concentrations into stock estimates, can vary due to soil texture and compaction (Li et al. 2019). Inaccurate BD measurements can lead to significant errors in SOC stock calculations. Furthermore, sampling depth is a crucial consideration, as C is not uniformly distributed throughout the soil profile. Deeper soils or soils with significant amounts of buried OM may require deeper sampling to capture the full extent of SOC, but determining the optimal depth for sampling is challenging and can vary depending on the soil type and land management practices (Olson and Al-Kaisi 2015; Xiao et al. 2020). Shallow or inconsistent sampling can therefore lead to underestimation of SOC stocks, particularly in areas with deep C reserves. In addition to these challenges, lack of long-term monitoring data poses a significant obstacle to accurately assessing SOC dynamics. Since SOC changes occur over long periods, often taking decades or centuries to manifest, there is a significant need for long-term studies to understand how various factors—such as land use change, climate variability, and soil management practices—impact SOC levels (Conant and Paustian 2002). However, such studies are expensive, resource-intensive, and often not feasible in regions with limited research infrastructure or funding (Smith and Smith 2004; Tate et al. 2005). Finally, modeling SOC dynamics across diverse landscapes is hindered by uncertainty, as models must account for spatial and temporal variations (Yigini and Panagos 2016). The lack of validated, context-specific models, especially in the Global South, further undermines the reliability of C credit systems (Garsia et al. 2023).

7.7 Conclusion

Accurate SOC stock measurements are critical for programs focused on MRV, which incentivize farmers to adopt C farming practices and generate C credits. The process of estimating SOC stock involves several key steps, including site selection, the development of a robust sampling strategy, sample collection, laboratory analysis, and ongoing monitoring. The IPCC-recommended formula for calculating SOC stocks incorporates key variables such as SOC concentration, BD, coarse soil fraction, and sampling depth. Each of these components can be affected by edaphic and environmental factors, land use, and management history. The choice of method for determining SOC concentration and BD depends on available resources and the required level of measurement accuracy. Wet chemistry and dry combustion are commonly used for measuring SOC, while BD is typically measured using core sampling, though alternative methods, like clod or excavation, may be applied in certain contexts. To reduce biases in SOC estimates, the ESM method, which adjusts for variations in BD, is preferred over the FD method.

Reliable SOC stock estimates require consistent sampling protocols and optimal sample sizes, determined through statistical methods such as power analysis. The calculation process is inherently complex, requiring careful consideration of factors such as C saturation limits, the dynamics of SOC sequestration and stabilization, and the influence of environmental and management factors on SOC storage capacity. Temporal variability, resulting from decomposition, mineralization, and C inputs from vegetation, further complicates SOC assessment. To address these challenges and improve the accuracy of SOC stock assessments, there is a growing need for enhanced methodologies, more robust data collection protocols, and advanced models that can account for the complexities of SOC dynamics. These models should integrate multiple sources of data, including remote sensing, field observations, and long-term soil monitoring, to provide more accurate and scalable estimates of SOC stocks. Additionally, advancing the understanding of the interactions between soil, climate, and land use is critical to improving C farming programs and ensuring that they deliver the intended climate benefits.

References

Abbott AM (2010) Statistical sampling methods for soils monitoring. In: Page-Dumroese D, Neary D, Trettin C (eds) Scientific background for soil monitoring on National Forests and Rangelands: workshop proceedings, April 2008. Denver, pp 109–120

Agnelli A, Ugolini FC, Corti G, Pietramellara G (2001) Microbial biomass-C and basal respiration of fine earth and highly altered rock fragments of two forest soils. Soil Biol Biochem 33:613–620. https://doi.org/10.1016/S0038-0717(00)00203-0

Allen DE, Pringle MJ, Page KL, Dalal RC (2010) A review of sampling designs for the measurement of soil organic carbon in Australian grazing lands. The Rangeland Journal 32:227. https://doi.org/10.1071/RJ09043

Al-Shammary AAG, Kouzani AZ, Kaynak A et al (2018) Soil bulk density estimation methods: a review. Pedosphere 28:581–596. https://doi.org/10.1016/S1002-0160(18)60034-7

Badgery W, Murphy B, Cowie A et al (2021) Soil carbon market-based instrument pilot—the sequestration of soil organic carbon for the purpose of obtaining carbon credits. Soil Research 59:12. https://doi.org/10.1071/SR19331

Barros AHC, de Jong Van Lier Q (2014) Pedotransfer functions for Brazilian soils. In: Application of soil physics in environmental analyses. Springer International Publishing, Cham, pp 131–162

Beem-Miller JP, Kong AYY, Ogle S, Wolfe D (2016) Sampling for soil carbon stock assessment in rocky agricultural soils. Soil Sci Soc Am J 80:1411–1423. https://doi.org/10.2136/sssaj2015.11.0405

Bettigole C, Hanle J, Kane DA et al (2023) Optimizing sampling strategies for near-surface soil carbon inventory: one size doesn't fit all. Soil Syst 7:27. https://doi.org/10.3390/soilsystems7010027

Blake GR, Hartge KH (1986) Bulk density. In: Klute A (ed) Methods of soil analysis. pp 363–375

Burgos Hernández TD, Slater BK, Shaffer JM, Basta N (2023) Comparison of methods for determining organic carbon content of urban soils in Central Ohio. Geoderma Reg 34:e00680. https://doi.org/10.1016/j.geodrs.2023.e00680

Casanova M, Tapia E, Seguel O, Salazar O (2016) Direct measurement and prediction of bulk density on alluvial soils of Central Chile. Chil J Agric Res 76:105–113. https://doi.org/10.4067/S0718-58392016000100015
Certini G, Campbell CD, Edwards AC (2004) Rock fragments in soil support a different microbial community from the fine earth. Soil Biol Biochem 36:1119–1128. https://doi.org/10.1016/j.soilbio.2004.02.022
Chang X, Wang S, Cui S et al (2014) Alpine grassland soil organic carbon stock and its uncertainty in the three Rivers source region of the Tibetan plateau. PLoS One 9:e97140. https://doi.org/10.1371/journal.pone.0097140
Chang X, Bao X, Wang S et al (2016) Exploring effective sampling design for monitoring soil organic carbon in degraded Tibetan grasslands. J Environ Manag 173:121–126. https://doi.org/10.1016/j.jenvman.2016.03.010
Chatterjee A, Lal R, Wielopolski L et al (2009) Evaluation of different soil carbon determination methods. CRC Crit Rev Plant Sci 28:164–178. https://doi.org/10.1080/07352680902776556
Chaudhuri S, Pena-Yewtukhiw EM, McDonald LM et al (2011) Land use effects on sample size requirements for soil organic carbon stock estimations. Soil Sci 176:110–114. https://doi.org/10.1097/SS.0b013e31820a0fe2
Chen H, Marhan S, Billen N, Stahr K (2009) Soil organic-carbon and total nitrogen stocks as affected by different land uses in Baden-Württemberg (Southwest Germany). J Plant Nutr Soil Sci 172:32–42. https://doi.org/10.1002/jpln.200700116
Chen S, Arrouays D, Angers DA et al (2019) Soil carbon stocks under different land uses and the applicability of the soil carbon saturation concept. Soil Tillage Res 188:53–58. https://doi.org/10.1016/j.still.2018.11.001
Choudhury BU, Santra P, Singh N, Chakraborty P (2023) Development of land-use-specific pedotransfer functions for predicting bulk density of acidic topsoil in eastern Himalayas (India). Geoderma Reg 34:e00671. https://doi.org/10.1016/j.geodrs.2023.e00671
Ciric V, Manojlovic M, Svarc-Gajic J, Kumar Sitaula B (2021) Comparison, limitations and uncertainty of wet chemistry techniques, loss on ignition and dry combustion in soil organic carbon analysis. Agrochimica 65:295–314. https://doi.org/10.12871/00021857202136
Clancy MA, Ruffing BE, Jarmain C, Byrne KA (2023) A novel method incorporating large rock fragments for improved soil bulk density and carbon stock estimation. Soil Sci Soc Am J 87:1029–1041. https://doi.org/10.1002/saj2.20562
Conant RT, Paustian K (2002) Spatial variability of soil organic carbon in grasslands: implications for detecting change at different scales. Environ Pollut 116:S127–S135. https://doi.org/10.1016/S0269-7491(01)00265-2
Conant RT, Ogle SM, Paul EA, Paustian K (2011) Measuring and monitoring soil organic carbon stocks in agricultural lands for climate mitigation. Front Ecol Environ 9:169–173. https://doi.org/10.1890/090153
Da Silva Dias R, De Abreu CA, De Abreu MF et al (2013) Comparison of methods to quantify organic carbon in soil samples from São Paulo state, Brazil. Commun Soil Sci Plant Anal 44:429–439. https://doi.org/10.1080/00103624.2013.742345
Danalatos NG, Kosmas CS, Moustakas NC, Yassoglou N (1995) Rock fragments IL their impact on soil physical properties and biomass production under Mediterranean conditions. Soil Use Manag 11:121–126. https://doi.org/10.1111/j.1475-2743.1995.tb00509.x
Das S, Ghimire D (2024) Soil organic carbon: measurement and monitoring using remote sensing data. In: Remote Sensing of Soils. Elsevier, pp 395–409
Davis M, Alves B, Karlen D et al (2017) Review of soil organic carbon measurement protocols: a US and Brazil comparison and recommendation. Sustainability 10:53. https://doi.org/10.3390/su10010053
de Gruijter JJ, Bierkens MFP, Brus DJ, Knotters M (2006) Sampling for natural resource monitoring. Springer, Berlin Heidelberg
Dincă L, Dincă M, Vasile D et al (2015) Calculating organic carbon Stock from Forest soils. Not Bot Horti Agrobot Cluj Napoca 43:568–575. https://doi.org/10.15835/nbha43210141

Donald C. Erbach (1987) Measurement of soil bulk density and moisture. Transactions of the ASAE 30:0922–0931. https://doi.org/10.13031/2013.30500

Dubeux JCB, de Lira Junior M A, Simili FF et al (2024) Deep soil organic carbon: a review. CABI Reviews. https://doi.org/10.1079/cabireviews.2024.0024

Ercoli L, Masoni A, Mariotti M, Arduini I (2006) Dry matter accumulation and remobilization of durum wheat as affected by soil gravel content. Cereal Res Commun 34:1299–1306. https://doi.org/10.1556/CRC.34.2006.4.272

FAO (2019) Measuring and modelling soil carbon stocks and stock changes in livestock production systems—Guidelines for assessment. Version 1—Advanced copy. 152

Follett RF, Kimble JM, Pruessner EG, et al (2015) Soil Organic Carbon Stocks with Depth and Land Use at Various U.S. Sites. In: Lal R, Follett R (eds) Soil Carbon Sequestration and the Greenhouse Effect, Second. SSSA Special Publications, Madison, WI, USA, pp 29–46

Fowler AF, Basso B, Millar N, Brinton WF (2023) A simple soil mass correction for a more accurate determination of soil carbon stock changes. Sci Rep 13:2242. https://doi.org/10.1038/s41598-023-29289-2

Francaviglia R, Renzi G, Doro L et al (2017) Soil sampling approaches in Mediterranean agro-ecosystems. Influence on soil organic carbon stocks. Catena (Amst) 158:113–120. https://doi.org/10.1016/j.catena.2017.06.014

Fu Z, Hu W, Beare M et al (2023) Response of soil organic carbon stock to land use is modulated by soil hydraulic properties. Soil Tillage Res 233:105793. https://doi.org/10.1016/j.still.2023.105793

Fujisaki K, Chevallier T, Chapuis-Lardy L et al (2018) Soil carbon stock changes in tropical croplands are mainly driven by carbon inputs: a synthesis. Agric Ecosyst Environ 259:147–158. https://doi.org/10.1016/j.agee.2017.12.008

Garsia A, Moinet A, Vazquez C et al (2023) The challenge of selecting an appropriate soil organic carbon simulation model: a comprehensive global review and validation assessment. Glob Chang Biol 29:5760–5774. https://doi.org/10.1111/gcb.16896

Ge J, Xu W, Xiong G et al (2023) Depth-dependent controls over soil organic carbon stock across Chinese Shrublands. Ecosystems 26:277–289. https://doi.org/10.1007/s10021-022-00757-6

Gholizadeh A, Saberioon M, Pouladi N, Ben-Dor E (2023) Quantification and depth distribution analysis of carbon to nitrogen ratio in forest soils using reflectance spectroscopy. Int Soil Water Conserv Res 11:112–124. https://doi.org/10.1016/j.iswcr.2022.06.004

Gocke MI, Guigue J, Bauke SL et al (2023) Interactive effects of agricultural management on soil organic carbon accrual: a synthesis of long-term field experiments in Germany. Geoderma 438:116616. https://doi.org/10.1016/j.geoderma.2023.116616

Gross CD, Harrison RB (2018) Quantifying and comparing soil carbon stocks: underestimation with the Core sampling method. Soil Sci Soc Am J 82:949–959. https://doi.org/10.2136/sssaj2018.01.0015

Grossman RB, Reinsch TG (2002) Bulk density and linear extensibility. In: Methods of Soil Analysis, pp 201–228

Grüneberg E, Schöning I, Kalko EKV, Weisser WW (2010) Regional organic carbon stock variability: a comparison between depth increments and soil horizons. Geoderma 155:426–433. https://doi.org/10.1016/j.geoderma.2010.01.002

Hailu H (2024) A review of possible calculation methods (CMs) for soil organic carbon stock estimates and evaluating their methodological biases. In: Haddout S, Priya K, Hoguane AM (eds) Proceedings of the 2nd International Conference on Climate Change and Ocean Renewable Energy. Springer, Cham, pp 315–323

Heikkinen J, Keskinen R, Regina K et al (2021) Estimation of carbon stocks in boreal cropland soils—methodological considerations. Eur J Soil Sci 72:934–945. https://doi.org/10.1111/ejss.13033

Heim A, Wehrli L, Eugster W, Schmidt MWI (2009) Effects of sampling design on the probability to detect soil carbon stock changes at the Swiss CarboEurope site Lägeren. Geoderma 149:347–354. https://doi.org/10.1016/j.geoderma.2008.12.018

Holmes KW, Wherrett A, Keating A, Murphy DV (2011) Meeting bulk density sampling requirements efficiently to estimate soil carbon stocks. Soil Research 49:680. https://doi.org/10.1071/SR11161

IPCC (2003) Good practice guidance for land use, Land-Use Change and Forestry

IPCC (2006) Agriculture, Forestry and Other Land Use. Institute for Global Environmental Strategies, Hayama

Jobbágy EG, Jackson RB (2000) The vertical distribution of soil organic carbon and its relation to climate and vegetation. Ecol Appl 10:423–436

Khaledian Y, Miller BA (2020) Selecting appropriate machine learning methods for digital soil mapping. Appl Math Model 81:401–418. https://doi.org/10.1016/j.apm.2019.12.016

Kirkby CA, Kirkegaard JA, Richardson AE (2021) Gravel-associated organic material is important to quantify soil carbon and nitrogen stocks to depth in an agricultural cropping soil. Soil Research 60:224–233. https://doi.org/10.1071/SR21140

Krauss M, Wiesmeier M, Don A et al (2022) Reduced tillage in organic farming affects soil organic carbon stocks in temperate Europe. Soil Tillage Res 216:105262. https://doi.org/10.1016/j.still.2021.105262

Lagacherie P, Arrouays D, Bourennane H et al (2020) Analysing the impact of soil spatial sampling on the performances of digital soil mapping models and their evaluation: a numerical experiment on quantile random forest using clay contents obtained from Vis-NIR-SWIR hyperspectral imagery. Geoderma 375:114503. https://doi.org/10.1016/j.geoderma.2020.114503

Lai Y-Q, Wang H-L, Sun X-L (2021) A comparison of importance of modelling method and sample size for mapping soil organic matter in Guangdong, China. Ecol Indic 126:107618. https://doi.org/10.1016/j.ecolind.2021.107618

Lai X, Zhu Q, Castellano MJ, Liao K (2022) Soil rock fragments: unquantified players in terrestrial carbon and nitrogen cycles. Geoderma 406:115530. https://doi.org/10.1016/j.geoderma.2021.115530

Lal R, Kimble JM, Follett RF, Stewart BA (eds) (2001) Assessment Methods for Soil Carbon. CRC Press, Boca Raton, FL

Lal R, Follett RF (eds) (2009) Soil Carbon Sequestration and the Greenhouse Effect, Second. SSSA Special Publications, Madison, WI, USA

Lark RM (2009) Estimating the regional mean status and change of soil properties: two distinct objectives for soil survey. Eur J Soil Sci 60:748–756. https://doi.org/10.1111/j.1365-2389.2009.01156.x

Lawrence PG, Roper W, Morris TF, Guillard K (2020) Guiding soil sampling strategies using classical and spatial statistics: a review. Agron J 112:493–510. https://doi.org/10.1002/agj2.20048

Lei Z, Yu D, Zhou F et al (2019) Changes in soil organic carbon and its influencing factors in the growth of Pinus sylvestris var. mongolica plantation in Horqin Sandy land, Northeast China. Sci Rep 9:16453. https://doi.org/10.1038/s41598-019-52945-5

Lettens S, De Vos B, Quataert P et al (2007) Variable carbon recovery of Walkley-Black analysis and implications for national soil organic carbon accounting. Eur J Soil Sci 58:1244–1253. https://doi.org/10.1111/j.1365-2389.2007.00916.x

Li Y, Yu H, Chappell A et al (2014) How much soil organic carbon sequestration is due to conservation agriculture reducing soil erosion? Soil Research 52:717. https://doi.org/10.1071/SR14078

Li S, Li Q, Wang C et al (2019) Spatial variability of soil bulk density and its controlling factors in an agricultural intensive area of Chengdu plain, Southwest China. J Integr Agric 18:290–300. https://doi.org/10.1016/S2095-3119(18)61930-6

Liu Z, Shao M, Wang Y (2011) Effect of environmental factors on regional soil organic carbon stocks across the Loess Plateau region, China. Agric Ecosyst Environ 142:184–194. https://doi.org/10.1016/j.agee.2011.05.002

Long J, Liu Y, Xing S et al (2018) Effects of sampling density on interpolation accuracy for farmland soil organic matter concentration in a large region of complex topography. Ecol Indic 93:562–571. https://doi.org/10.1016/j.ecolind.2018.05.044

Loria N, Bhardwaj SK, Ndungu CK (2016) Impact of cropping systems on soil properties, nutrient availability, and their carbon sequestration potential in Shivalik hills of Himachal Pradesh. J Appl Nat Sci 8:1479–1484

Loria N, Lal R, Chandra R (2024) Handheld in situ methods for soil organic carbon assessment. Sustainability 16:5592. https://doi.org/10.3390/su16135592

Luo Z, Wang E, Sun OJ (2010) Can no-tillage stimulate carbon sequestration in agricultural soils? A meta-analysis of paired experiments. Agric Ecosyst Environ 139:224–231. https://doi.org/10.1016/j.agee.2010.08.006

Malone BP, Minansy B, Brungard C (2019) Some methods to improve the utility of conditioned Latin hypercube sampling. PeerJ 7:e6451. https://doi.org/10.7717/peerj.6451

Matejovic I (1997) Determination of carbon and nitrogen in samples of various soils by the dry combustion. Commun Soil Sci Plant Anal 28:1499–1511. https://doi.org/10.1080/00103629709369892

McBratney AB, Stockmann U, Angers DA et al (2014) Challenges for soil organic carbon research. In: Soil Carbon. Springer International Publishing, Cham, pp 3–16

Meersmans J, Van Wesemael B, Van Molle M (2009) Determining soil organic carbon for agricultural soils: a comparison between the Walkley & Black and the dry combustion methods (north Belgium). Soil Use Manag 25:346–353. https://doi.org/10.1111/j.1475-2743.2009.00242.x

Mehler K, Schöning I, Berli M (2014) The importance of rock fragment density for the calculation of soil bulk density and soil organic carbon stocks. Soil Sci Soc Am J 78:1186–1191. https://doi.org/10.2136/sssaj2013.11.0480

Meurer KHE, Haddaway NR, Bolinder MA, Kätterer T (2018) Tillage intensity affects total SOC stocks in boreo-temperate regions only in the topsoil—a systematic review using an ESM approach. Earth Sci Rev 177:613–622. https://doi.org/10.1016/j.earscirev.2017.12.015

Mikhailova EA, Noble RRP, Post CJ (2003) Comparison of soil organic carbon recovery by Walkley-Black and dry combustion methods in the Russian Chernozem. Commun Soil Sci Plant Anal 34:1853–1860. https://doi.org/10.1081/CSS-120023220

Minasny B, McBratney AB (2006) A conditioned Latin hypercube method for sampling in the presence of ancillary information. Comput Geosci 32:1378–1388. https://doi.org/10.1016/j.cageo.2005.12.009

Minasny B, Malone BP, McBratney AB et al (2017) Soil carbon 4 per mille. Geoderma 292:59–86. https://doi.org/10.1016/j.geoderma.2017.01.002

Morgan J, Daugherty R, Hilchie A, Carey B (2003) Sample size and modelling accuracy of decision tree based data mining tools. Academy of Information and Management Sciences Journal 6:77–91

Morgan CLS, Ackerson JP (2022) Sampling Design for Quantifying Soil Organic Carbon Stock in production Ag fields. Crops & Soils 55:28–33. https://doi.org/10.1002/crso.20156

Nogrady B (2024) How farming could become the ultimate climate-change tool. Nature 630:S23–S25. https://doi.org/10.1038/d41586-024-02036-x

Odebiri O, Mutanga O, Odindi J et al (2024) Remote sensing of depth-induced variations in soil organic carbon stocks distribution within different vegetated landscapes. Catena (Amst) 243:108216. https://doi.org/10.1016/j.catena.2024.108216

Olson KR, Al-Kaisi MM (2015) The importance of soil sampling depth for accurate account of soil organic carbon sequestration, storage, retention and loss. Catena (Amst) 125:33–37. https://doi.org/10.1016/j.catena.2014.10.004

Olson KR, Al-Kaisi MM, Lal R, Lowery B (2014) Experimental consideration, treatments, and methods in determining soil organic carbon sequestration rates. Soil Sci Soc Am J 78:348–360. https://doi.org/10.2136/sssaj2013.09.0412

Padarian J, Minasny B, McBratney A (2012) Using genetic programming to transform from Australian to USDA/FAO soil particle-size classification system. Soil Research 50:443. https://doi.org/10.1071/SR12139

Paltineanu C, Dumitru S, Vizitiu O et al (2024) Soil organic carbon and total nitrogen stocks related to land use and basic environmental properties—assessment of soil carbon sequestration potential in different ecosystems. Catena (Amst) 246:108435. https://doi.org/10.1016/j.catena.2024.108435

Paustian K, Collier S, Baldock J et al (2019) Quantifying carbon for agricultural soil management: from the current status toward a global soil information system. Carbon Manag 10:567–587. https://doi.org/10.1080/17583004.2019.1633231

Peltoniemi M, Heikkinen J, Mäkipää R (2007) Stratification of regional sampling by model-predicted changes of carbon stocks in forested mineral soils. Silva Fennica 41. https://doi.org/10.14214/sf.287

Peng Y, Chahal I, Hooker DC, Van Eerd LL (2024) Comparison of equivalent soil mass approaches to estimate soil organic carbon stocks under long-term tillage. Soil Tillage Res 238:106021. https://doi.org/10.1016/j.still.2024.106021

Pires LF, Rosa JA, Pereira AB et al (2009) Gamma-ray attenuation method as an efficient tool to investigate soil bulk density spatial variability. Ann Nucl Energy 36:1734–1739. https://doi.org/10.1016/j.anucene.2009.08.016

Poeplau C, Gregorich E (2022) Advances in measuring soil organic carbon stocks and dynamics at the profile scale. In: Understanding and fostering soil carbon sequestration, pp 323–350

Poeplau C, Vos C, Don A (2017) Soil organic carbon stocks are systematically overestimated by misuse of the parameters bulk density and rock fragment content. Soil 3:61–66. https://doi.org/10.5194/soil-3-61-2017

Potash E, Guan K, Margenot AJ et al (2023) Multi-site evaluation of stratified and balanced sampling of soil organic carbon stocks in agricultural fields. Geoderma 438:116587. https://doi.org/10.1016/j.geoderma.2023.116587

Premrov A, Cummins T, Byrne KA (2017) Assessing fixed depth carbon stocks in soils with varying horizon depths and thicknesses, sampled by horizon. Catena (Amst) 150:291–301. https://doi.org/10.1016/j.catena.2016.11.030

Probst BS, Toetzke M, Kontoleon A et al (2024) Systematic assessment of the achieved emission reductions of carbon crediting projects. Nat Commun 15:9562. https://doi.org/10.1038/s41467-024-53645-z

Qi J, Wang X, Zhao X et al (2019) Temporal variability of soil organic carbon in paddies during 13-year conservation tillage. Land Degrad Dev 30:1840–1850. https://doi.org/10.1002/ldr.3384

Qiao J, Zhu Y, Jia X et al (2019) Development of pedotransfer functions for predicting the bulk density in the critical zone on the Loess Plateau, China. J Soils Sediments 19:366–372. https://doi.org/10.1007/s11368-018-2040-1

Qin Y, Yi S, Chen J et al (2015) Effects of gravel on soil and vegetation properties of alpine grassland on the Qinghai-Tibetan plateau. Ecol Eng 74:351–355. https://doi.org/10.1016/j.ecoleng.2014.10.008

Raffeld AM, Bradford MA, Jackson RD et al (2024) The importance of accounting method and sampling depth to estimate changes in soil carbon stocks. Carbon Balance Manag 19:2. https://doi.org/10.1186/s13021-024-00249-1

Rodeghiero M, Cescatti A (2008) Spatial variability and optimal sampling strategy of soil respiration. For Ecol Manag 255:106–112. https://doi.org/10.1016/j.foreco.2007.08.025

Roper WR, Robarge WP, Osmond DL, Heitman JL (2019) Comparing four methods of measuring soil organic matter in North Carolina soils. Soil Sci Soc Am J 83:466–474. https://doi.org/10.2136/sssaj2018.03.0105

Rovira P, Sauras-Yera T, Romanyà J (2022) Equivalent-mass versus fixed-depth as criteria for quantifying soil carbon sequestration: how relevant is the difference? Catena (Amst) 214:106283. https://doi.org/10.1016/j.catena.2022.106283

Rytter R-M (2012) Stone and gravel contents of arable soils influence estimates of C and N stocks. Catena (Amst) 95:153–159. https://doi.org/10.1016/j.catena.2012.02.015

Safaee S, Libohova Z, Kladivko EJ et al (2024) Influence of sample size, model selection, and land use on prediction accuracy of soil properties. Geoderma Reg 36:e00766. https://doi.org/10.1016/j.geodrs.2024.e00766

Sanderman J, Baldock JA (2010) Accounting for soil carbon sequestration in national inventories: a soil scientist's perspective. Environ Res Lett 5:034003. https://doi.org/10.1088/1748-9326/5/3/034003

Saurette DD, Biswas A, Heck RJ et al (2024) Determining minimum sample size for the conditioned Latin hypercube sampling algorithm. Pedosphere 34:530–539. https://doi.org/10.1016/j.pedsph.2022.09.001

Schrumpf M, Schulze ED, Kaiser K, Schumacher J (2011) How accurately can soil organic carbon stocks and stock changes be quantified by soil inventories? Biogeosciences 8:1193–1212. https://doi.org/10.5194/bg-8-1193-2011

Sekucia F, Dlapa P, Kollár J et al (2020) Land-use impact on porosity and water retention of soils rich in rock fragments. Catena (Amst) 195:104807. https://doi.org/10.1016/j.catena.2020.104807

Sharma R, Levi MR, Ricker MC et al (2024) Scaling of soil organic carbon in space and time in the Southern Coastal Plain, USA. Sci Total Environ 933:173060. https://doi.org/10.1016/j.scitotenv.2024.173060

Sherpa SR, Wolfe DW, van Es HM (2016) Sampling and data analysis optimization for estimating soil organic carbon stocks in agroecosystems. Soil Sci Soc Am J 80:1377–1392. https://doi.org/10.2136/sssaj2016.04.0113

Singla A, Dhingra N (2024) Investigation of impact of soil texture, depth and gamma ray energy on the mass attenuation coefficient and determination of soil bulk density using gamma-ray spectrometry. Radiat Phys Chem 216:111400. https://doi.org/10.1016/j.radphyschem.2023.111400

Smith P, Smith P (2004) Monitoring and verification of soil carbon changes under Article 3.4 of the Kyoto protocol. Soil Use Manag 20:264–270. https://doi.org/10.1079/SUM2004239

Smith P, Smith JU, Powlson DS et al (1997) A comparison of the performance of nine soil organic matter models using datasets from seven long-term experiments. Geoderma 81:153–225. https://doi.org/10.1016/S0016-7061(97)00087-6

Söderström B, Hedlund K, Jackson LE et al (2014) What are the effects of agricultural management on soil organic carbon (SOC) stocks? Environ Evid 3:2. https://doi.org/10.1186/2047-2382-3-2

Somarathna PDSN, Minasny B, Malone BP (2017) More data or a better model? Figuring out what matters most for the spatial prediction of soil carbon. Soil Sci Soc Am J 81:1413–1426. https://doi.org/10.2136/sssaj2016.11.0376

Soon YK, Abboud S (1991) A comparison of some methods for soil organic carbon determination. Commun Soil Sci Plant Anal 22:943–954. https://doi.org/10.1080/00103629109368465

Sorenson PT, Kiss J, Bedard-Haughn A (2024) A proposed methodology for determining the economically optimal number of sample points for carbon stock estimation in the Canadian prairies. Land (Basel) 13:114. https://doi.org/10.3390/land13010114

Stanley P, Spertus J, Chiartas J et al (2023) Valid inferences about soil carbon in heterogeneous landscapes. Geoderma 430:116323. https://doi.org/10.1016/j.geoderma.2022.116323

Stewart CE, Paustian K, Conant RT et al (2007) Soil carbon saturation: concept, evidence and evaluation. Biogeochemistry 86:19–31. https://doi.org/10.1007/s10533-007-9140-0

Sun W, Zhao Y, Huang B et al (2012) Effect of sampling density on regional soil organic carbon estimation for cultivated soils. J Plant Nutr Soil Sci 175:671–680. https://doi.org/10.1002/jpln.201100181

Tadiello T, Perego A, Valkama E et al (2022) Computation of total soil organic carbon stock and its standard deviation from layered soils. MethodsX 9:101662. https://doi.org/10.1016/j.mex.2022.101662

Tate KR, Wilde RH, Giltrap DJ et al (2005) Soil organic carbon stocks and flows in New Zealand: system development, measurement and modelling. Can J Soil Sci 85:481–489. https://doi.org/10.4141/S04-082

Timm LC, Pires LF, Reichardt K et al (2005) Soil bulk density evaluation by conventional and nuclear methods. Soil Research 43:97. https://doi.org/10.1071/SR04054

Tivet F, de Moraes C, Sá J, Borszowskei PR et al (2012) Soil carbon inventory by wet oxidation and dry combustion methods: effects of land use, soil texture gradients, and sampling depth on

the linear model of C-equivalent correction factor. Soil Sci Soc Am J 76:1048–1059. https://doi.org/10.2136/sssaj2011.0328
Torres-Sallan G, Creamer RE, Lanigan GJ et al (2018) Effects of soil type and depth on carbon distribution within soil macroaggregates from temperate grassland systems. Geoderma 313:52–56. https://doi.org/10.1016/j.geoderma.2017.10.012
Triantakonstantis D, Batsalia M, Lolos N (2024) Spatio-temporal dynamics of soil organic carbon stock in Greek croplands: a Long-term assessment. Sustainability 16:7984. https://doi.org/10.3390/su16187984
USDA Natural Resources Conservation Service (2022) Soil Organic Carbon Stock Monitoring
Vadeboncoeur MA, Hamburg SP, Blum JD et al (2012) The quantitative soil pit method for measuring belowground carbon and nitrogen stocks. Soil Sci Soc Am J 76:2241–2255. https://doi.org/10.2136/sssaj2012.0111
Vanden Bygaart AJ, Angers DA (2006) Towards accurate measurements of soil organic carbon stock change in agroecosystems. Can J Soil Sci 86:465–471. https://doi.org/10.4141/S05-106
von Haden AC, Yang WH, DeLucia EH (2020) Soils' dirty little secret: depth-based comparisons can be inadequate for quantifying changes in soil organic carbon and other mineral soil properties. Glob Chang Biol 26:3759–3770. https://doi.org/10.1111/gcb.15124
Vos C, Don A, Hobley EU et al (2019) Factors controlling the variation in organic carbon stocks in agricultural soils of Germany. Eur J Soil Sci 70:550–564. https://doi.org/10.1111/ejss.12787
Walter K, Don A, Tiemeyer B, Freibauer A (2016) Determining soil bulk density for carbon stock calculations: a systematic method comparison. Soil Sci Soc Am J 80:579–591. https://doi.org/10.2136/sssaj2015.11.0407
Wang M, Xu S, Zhao Y, Shi X (2017) Climatic effect on soil organic carbon variability as a function of spatial scale. Arch Agron Soil Sci 63:375–387. https://doi.org/10.1080/03650340.2016.1213812
Wang S, Gao J, Zhuang Q et al (2020) Multispectral remote sensing data are effective and robust in mapping regional Forest soil organic carbon stocks in a northeast Forest region in China. Remote Sens 12:393. https://doi.org/10.3390/rs12030393
Wang M, Guo X, Zhang S et al (2022) Global soil profiles indicate depth-dependent soil carbon losses under a warmer climate. Nat Commun 13:5514. https://doi.org/10.1038/s41467-022-33278-w
Wang S, Wang Z, Zhuang Q et al (2023a) Assessing the carbon sequestration potential and identifying influential factors of cultivated soils in Northeast China. Geoderma Reg 33:e00655. https://doi.org/10.1016/j.geodrs.2023.e00655
Wang Y, Qi Q, Bao Z et al (2023b) A novel sampling design considering the local heterogeneity of soil for farm field-level mapping with multiple soil properties. Precis Agric 24:1–22. https://doi.org/10.1007/s11119-022-09926-y
Welsch J, Songling C, Buckley HL et al (2019) How many samples? Soil variability affects confidence in the use of common agroecosystem soil indicators. Ecol Indic 102:401–409. https://doi.org/10.1016/j.ecolind.2019.02.065
Wendt JW, Hauser S (2013) An equivalent soil mass procedure for monitoring soil organic carbon in multiple soil layers. Eur J Soil Sci 64:58–65. https://doi.org/10.1111/ejss.12002
Wiesmeier M, Spörlein P, Geuß U et al (2012) Soil organic carbon stocks in Southeast Germany (Bavaria) as affected by land use, soil type and sampling depth. Glob Chang Biol 18:2233–2245. https://doi.org/10.1111/j.1365-2486.2012.02699.x
Wiesmeier M, von Lützow M, Spörlein P et al (2015) Land use effects on organic carbon storage in soils of Bavaria: the importance of soil types. Soil Tillage Res 146:296–302. https://doi.org/10.1016/j.still.2014.10.003
Wu T, Wu Q, Zhuang Q et al (2022) Optimal sample size for SOC content prediction for mapping using the random forest in cropland in northern Jiangsu, China. Eurasian Soil Sci 55:1689–1699. https://doi.org/10.1134/S1064229322600816

Xiao L, Zhou S, Zhao R et al (2020) Evaluating soil organic carbon stock changes induced by no-tillage based on fixed depth and equivalent soil mass approaches. Agric Ecosyst Environ 300:106982. https://doi.org/10.1016/j.agee.2020.106982

Yang S, Cammeraat E, Jansen B et al (2018) Soil organic carbon stocks controlled by lithology and soil depth in a Peruvian alpine grassland of the Andes. Catena (Amst) 171:11–21. https://doi.org/10.1016/j.catena.2018.06.038

Yigini Y, Panagos P (2016) Assessment of soil organic carbon stocks under future climate and land cover changes in Europe. Sci Total Environ 557–558:838–850. https://doi.org/10.1016/j.scitotenv.2016.03.085

Yu H, Zha T, Zhang X et al (2020) Spatial distribution of soil organic carbon may be predominantly regulated by topography in a small revegetated watershed. Catena (Amst) 188:104459. https://doi.org/10.1016/j.catena.2020.104459

Zabowski D, Whitney N, Gurung J, Hatten J (2011) Total soil carbon in the coarse fraction and at depth. For Sci 57:11–18. https://doi.org/10.1093/forestscience/57.1.11

Zhang Y, Hartemink AE (2017) Sampling designs for soil organic carbon stock assessment of soil profiles. Geoderma 307:220–230. https://doi.org/10.1016/j.geoderma.2017.08.013

Zhang YY, Zhang WH (2014) Automated analyzer, LOI and other methods for determination of soil organic carbon. Appl Mech Mater 580–583:903–906. https://doi.org/10.4028/www.scientific.net/AMM.580-583.903

Zhang W, Weindorf DC, Zhu Y (2011) Soil organic carbon variability in croplands. Soil Sci 176:367–371. https://doi.org/10.1097/SS.0b013e31821eb7d2

Zhang P, Wang Y, Sun H et al (2021a) Spatial variation and distribution of soil organic carbon in an urban ecosystem from high-density sampling. Catena (Amst) 204:105364. https://doi.org/10.1016/j.catena.2021.105364

Zhang Z, Zhou Y, Huang X (2021b) Exploring the optimal sampling density to characterize spatial heterogeneity of soil carbon stocks in a Karst Region. Agron J 113:99–110. https://doi.org/10.1002/agj2.20467

Zheng Y, Chen N, Zhang C et al (2021) Effects of rock fragments on the soil physicochemical properties and vegetation on the Northeastern Tibetan plateau. Front Environ Sci 9. https://doi.org/10.3389/fenvs.2021.693769

Zhuo Z, Chen Q, Zhang X et al (2022) Soil organic carbon storage, distribution, and influencing factors at different depths in the dryland farming regions of northeast and North China. Catena (Amst) 210:105934. https://doi.org/10.1016/j.catena.2021.105934

Chapter 8
Carbon Trading

Abstract This chapter explores agricultural carbon (C) trading systems, highlighting their societal value in addressing climate change and promoting sustainable land use. It discusses the potential of C markets to encourage C farming through financial incentives and market-based solutions. Core elements of the C credit process such as project design, measurement, verification, and issuance are outlined, along with the distinction between regulatory and voluntary C markets. Challenges including low credit prices, additionality, permanence, leakage, and verification issues are critically examined. The chapter concludes with strategies such as Payment for Ecosystem Services, supportive policy frameworks, and innovation to enhance farmer participation and advance both environmental and societal goals.

8.1 Introduction

Carbon (C) has become an increasingly valuable commodity as markets for C credits and emissions trading (ET) systems grow, offering financial incentives for industries and farmers to reduce their greenhouse gas (GHG) emissions and invest in sustainable agricultural practices (SAPs). C markets began to take shape in the early 1990s when a network of economists and policymakers promoted ET as an efficient mechanism to reduce GHG emissions (Paterson 2012). The concept gained momentum through key reports and discussions, culminating in the Kyoto Protocol of 1997, which formally introduced ET, along with mechanisms like the Clean Development Mechanism (CDM) and Joint Implementation (JI) (UNFCCC 1997). After Kyoto, the European Union launched the European Union ET System (EU ETS) in 2005, establishing the largest C market globally (Wettestad 2011). By the mid-2000s, C markets expanded globally, with countries and regions like New Zealand, California, and various US states implementing their own ET system (ETS), positioning C markets as central to international climate policy (Paterson 2012; Bullock 2012). In North America, regional initiatives like the Western Climate Initiative (WCI) reflect economic ties but face integration challenges due to differing caps, compliance periods, and cost-containment provisions (Haites and Mehling

N. Loria, R. Lal, *Carbon Farming*, https://doi.org/10.1007/978-3-032-00842-8_8

2009). Policymakers can address GHG emissions through three main strategies: (1) capping emissions for companies, (2) imposing a C tax to charge for emissions, or (3) establishing an ETS, allowing companies to trade pollution.

C trading offers a powerful framework for integrating environmental sustainability with economic growth, emphasizing the societal value of managing CO_2 emissions and incentivizing actions to address climate change (Hunjra et al. 2024). Central to C trading is the monetary valuation of C, which incorporates the benefits of ecosystem services (ESs) and the economic costs associated with GHG emissions that affect public health, ecosystems, and the broader economy (Lal 2014). This valuation is fundamental to the creation of C markets, both voluntary and regulatory, where C credits or allowances can be bought and sold, allowing governments, businesses, and individuals to trade the right to release C. One innovative aspect of C trading is the monetization of C sequestration (CS) in agricultural soils. Practices like no-tillage and permanent soil cover, key components of Conservation Agriculture (CA), can generate C credits by enhancing soil organic C (SOC) stocks and reducing CO_2 emissions. In regions like India, such practices can provide farmers with an additional income stream, with potential earnings of up to $30 per hectare through the sale of C credits (Cariappa et al. 2024). Beyond the direct economic benefits, CA practices offer multiple environmental co-benefits, including enhanced soil fertility, increased biodiversity, reduced soil erosion, and minimized chemical runoff, contributing to long-term environmental health and agricultural productivity (Uri et al. 1999).

However, the adoption of SAPs for CS varies widely due to differing regional conditions, costs, and local challenges (Derpsch and Friedrich 2009), limiting broad participation in C markets, especially for smallholder farmers. In the U.S., agriculture presents significant opportunities for GHG reduction through C offset projects, such as adjustments in rice cultivation, compost additions to rangelands, and the avoidance of grassland conversion (Proville et al. 2020). However, high transaction and verification costs remain key challenges. To address these, future research should focus on reducing barriers and improving market designs (Proville et al. 2020). A blended approach combining private sector investment with public support is crucial to ensure equitable access, especially for smallholder farmers facing financial and technical challenges (Li and Martino 2024). Financial incentives, subsidies, and targeted solutions can help overcome these barriers, while promoting both environmental sustainability and economic resilience in rural communities (Men et al. 2024).

This chapter provides a comprehensive overview of C trading systems, identifying the challenges and opportunities within C markets. It highlights pathways for enhancing participation, driving development, and contributing to global climate goals, thus advancing both sustainability and economic growth. This chapter is based on two hypothesis: Hypothesis one posits that agricultural C markets have the potential to drive the widespread adoption of C farming practices among farmers by providing financial incentives and fostering market-based solutions for CS. Hypothesis two suggests that implementing targeted incentives, such as Payment for ESs (PES) schemes, alongside strengthened policy support, can effectively address farmers' concerns and barriers to participation in agricultural C markets, thus enhancing their engagement and long-term commitment. By leveraging opportunities in both voluntary and regulatory C markets, farmers can be incentivized to implement sustainable land management (SLM) practices that sequester C. This will enhance agricultural resilience, and promote long-term environmental sustainability, advancing the global shift toward more sustainable farming practices.

8.1.1 Societal Value of Carbon

The societal value of soil C represents the economic worth of ESs provided by a unit of SOC, including enhancing crop productivity, improving water retention, reducing erosion and pollution, climate regulation, and supporting biodiversity (Lal 2014). These services are often overlooked and undervalued in traditional economic models, resulting in misallocation of resources and inadequate investment in conservation practices (Costanza et al. 1997). Economic valuation of these services is critical for recognizing SOC's true value. The value of 17 key ESs for the entire biosphere, including GHG regulation, erosion control, and food production, has been estimated at $16–54 trillion annually, with an average of $33 trillion (Costanza et al. 1997). More recent estimates for global ESs, including those linked to SOC, place their value at $125–145 trillion annually (in 2007 $US), while the loss of these services due to land use changes between 1997 and 2011 was estimated at $4.3–20.2 trillion per year (Costanza et al. 2014). These figures highlight the critical role of ESs in global welfare and environmental health. As scarcity increases, the value of these services is likely to rise, making it essential to incorporate this into land use and resource management decisions. However, factors such as market power, food price fluctuations, and the complexity of non-market goods like biodiversity are frequently left out of economic valuations (Graham et al. 2013). To fully integrate SOC into policy frameworks, more robust scientific evidence and improved economic accounting are necessary. Graham et al. (2013) emphasized the need to compensate farmers more fairly for their contributions to ESs, particularly considering climate change uncertainties and extreme weather events. Balancing ESs and their economic valuation is crucial for ensuring that both environmental benefits and economic value are considered in decision-making (Fig. 8.1). Quantifying these benefits enables policymakers and businesses to make more informed decisions that reflect long-term environmental costs and economic gains. Figure 8.2 illustrates the benefits of enhancing soil C and the costs associated with its loss, emphasizing the critical role these factors play in determining the overall value of soil C for agriculture, global warming mitigation, and ecosystem health.

While it is difficult to place a precise value on SOC, its inherent value can be recognized through a broader understanding of its contribution to various ESs and its role in mitigating climate change (Lal 2014). A study in Western Australia estimated the marginal value of SOC at $7.1–8.7 per Mg C ha^{-1} year^{-1}, with sequestration benefits comprising 75% of this value (Petersen and Hoyle 2016). Over a

Fig. 8.1 Balancing ecosystem services and their economic value in decision-making

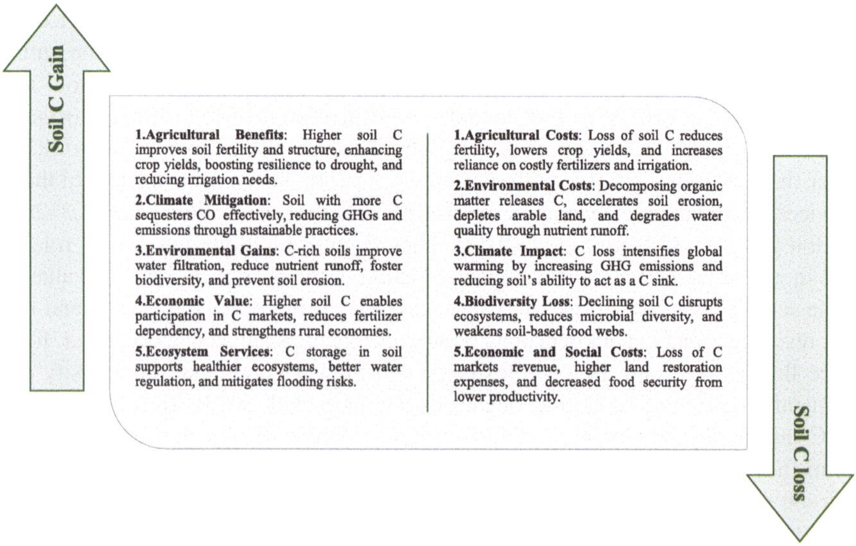

Fig. 8.2 Impact of soil carbon (C) changes on benefits and costs

50-year period, the present value of SOC increases ranged from \$130 to \$160 per Mg C ha^{-1}, illustrating the long-term economic benefits of SOC management in agriculture. The cost of sequestering 10,000 kg of C is estimated at around \$1327, with the inherent value of SOC calculated to be approximately \$0.13 per kg of C (Lal 2014). However, these estimates reflect only the direct costs related to the sequestration process, such as input costs for fertilizers, crop residues, and labor, and do not fully encompass the long-term benefits SOC provides to the environment and society at large. A key challenge in valuing SOC lies in the difficulty of accurately assessing the wide range of ESs it provides, as well as the variability in these values based on factors like geographic location, market conditions, and time (Lal 2014). For instance, Mikhailova et al. (2019) estimated that the value of SOC storage in U.S. soils ranges from \$4.64 trillion to \$23.1 trillion, with certain soil types such as Mollisols, Histosols, and Alfisols having higher sequestration potential. This illustrates the complexity of valuing SOC and the importance of considering regional differences in SOC management. In another example, Kristanto et al. (2024) found that cacao agroforestry systems (AFS) have higher indirect use values (IUVs) for soil health, ranging from \$633 to \$723, compared to monoculture systems, which have IUVs between \$415 and \$575. These values reflect the contribution of SOC and nitrogen (N) to ESs like water regulation, nutrient retention, and climate regulation. This emphasizes the need for targeted policies that fully recognize SOC's contributions and support SLM practices. Furthermore, research by Gascoigne et al. (2011) on land-use trade-offs in the Prairie Pothole Region found that increasing Conservation Reserve Program (CRP) and Wetlands Reserve Program (WRP) lands, while preserving native prairie, offers significant societal benefits, including a \$4 billion welfare loss from converting native prairie to cropland. This highlights how land-use policies that prioritize SOC conservation can provide long-term societal benefits, including enhancing water quality, reducing erosion, and supporting biodiversity.

Policies like Payment for ESs (PES) can incentivize SAPs by redistributing part of SOC's societal value to farmers and landowners investing in long-term soil and land management (Pascual et al. 2015). However, many PES schemes face challenges, including design flaws, scientific uncertainties, and unintended negative impacts on other ESs (Kinzig et al. 2011). To effectively account for SOC's true societal value, policies must consider regional variations, market dynamics, and the long-term benefits of conservation. Frameworks such as C credits, agricultural productivity improvements, and ES assessments can help enhance understanding of SOC's value and encourage SLM. Ultimately, the full potential of SOC as a key environmental and agricultural resource can be only realized through comprehensive, evidence-based policies that balance ecological, economic, and social considerations.

8.2 Carbon Credits

C credits are financial instruments that represent the reduction or sequestration of one metric ton of CO_2 equivalent (CO_2e) through activities like afforestation, cover cropping, and practices such as CA that prevent the release of stored C (Oldfield et al. 2022; Brummitt et al. 2024). These credits are generated through a multi-stage process that includes the following key stages:

8.2.1 Project Design and Baseline Establishment

The process begins with designing a project aimed at reducing or sequestering emissions (Chandra and Ruffalo 2024). This includes defining the project scope, objectives, and methodologies while establishing a baseline to represent the C stocks that would exist without the project. The baseline ensures that the reductions are additional, achieved solely due to the project's intervention. Accurate baseline setting is critical for assessing additionality, validating the project's impact, and ensuring the integrity of the C credits generated (Bento et al. 2016). For example, a lenient baseline may lead to over-crediting, issuing unjustified credits, while an overly stringent baseline could result in under-crediting, jeopardizing the project's financial viability despite achieving genuine emissions reductions. California's forest C offsets program illustrates these risks. Using regional C stock averages, the program awarded approximately 30 million excess credits valued at $410 million (Badgley et al. 2022b), undermining the program's goal of generating real climate benefits. Privately defined baselines in C trading also leads to risk manipulation, particularly by high emitters (Liu and Cui 2017). Policymakers can mitigate this by restricting baseline development to low-emission entities to ensure market integrity.

8.2.2 Measurement, Reporting, and Verification

Measurement, Reporting, and Verification (MRV) systems track emissions reductions by collecting data, quantifying CS or emission reductions, and verifying estimates over time. It involves measuring, monitoring, and reporting GHG emissions and sinks through direct measurement or modeling (USDA 2023). For example, Indigo Ag's MRV pipeline quantified emissions reductions from regenerative practices on 553,743 hectares of U.S. cropland, generating 296,662 tCO_2e in C credits (Brummitt et al. 2024). After quantifying C savings, third-party verifiers, independent of the project, C registries, and programs, verify that the project meets the requirements of C offset protocols (USDA 2023). This ensures that the reductions are real, additional, permanent, and enforceable (CAR 2021). Verified C credits are then issued by recognized C standards like the Climate Action Reserve (CAR), Gold Standard (GS), American C Registry (ACR), or Verra's Verified C Standard (VCS), ensuring compliance with methodologies for C reduction, avoidance, or removal (Spilker and Nugent 2022).

8.2.3 Issuance of Credits

Upon successful verification, C credits are issued based on the verified amount of CO_2e sequestered or reduced (Kollmuss et al. 2010). These credits can be sold or traded in C markets, with issuance typically tied to defined monitoring periods. Challenges in this phase are exemplified by Australia's 2023 soil C credits, where sequestration estimates were inflated due to short-term climatic variability (Mitchell et al. 2024). High rainfall during the measurement period led to SOC gains far exceeding scientific norms, which were unsustainable under average conditions. To enhance credit integrity, extending measurement periods to at least five years and establishing science-based limits for expected long-term SOC gains are recommended, ensuring credited sequestration reflects management-driven outcomes rather than temporary climate effects.

8.2.4 Retiring Credits

The final stage is credit retirement, where credits are permanently removed from the market and cannot be reused or resold (Chandra and Ruffalo 2024). This ensures the associated C reductions or sequestration are counted toward achieving emissions reduction goals. The retirement of credits is typically recorded in a registry to maintain transparency and prevent double counting (Blaufelder et al. 2020). This stage is essential for ensuring that the credits provide genuine environmental benefits and contribute to mitigating climate change. Once retired, the credits can no longer be utilized for offsetting emissions by other parties.

C credit calculation involves determining the difference in GHG emissions and sequestration between a baseline scenario (what would have occurred without the project) and a project scenario (with the mitigation activity in place). The basic formula for calculating C credits is (USDA 2023):

$$C\,credits = \begin{pmatrix} Incremental\ Project\ sequestration \\ -Incremental\ Baseline\ Sequestration \end{pmatrix} + \left(Baseline\ Emissions - Project\ Emissions\right)$$

For projects that focus only on emissions reductions, without sequestration, the formula simplifies to:

$$C\,credits = Baseline\ Emissions - Project\ Emissions$$

This process requires accurate data on emissions, sequestration, and baseline conditions, and often includes addressing uncertainties and factors like leakage, permanence, and additionality to ensure the environmental integrity of the credits generated (USDA 2023).

8.3 Carbon Markets

C markets are economic systems that facilitate the trading of C credits. These markets can be classified into two main types: regulatory and voluntary (USDA 2023).

8.3.1 Regulatory Carbon Markets

Regulatory or compliance C markets, such as the EU ETS, Regional Greenhouse Gas Initiative (RGGI), and California Air Resources Board's (CARB) cap-and-trade program, are mandatory systems where businesses are required to reduce their GHG emissions. Non-compliance with these regulations results in penalties, and the prices of credits in these markets tend to be higher due to the strict requirements (Musier and Adib 2010; Climate Seed 2023). These markets operate through a cap-and-trade system, where a cap is set on total emissions, and businesses can trade allowances, thus incentivizing decarbonization (Mnif and Davison 2011). The Kyoto Protocol, signed in 1997, played a pivotal role in establishing the first global framework for regulating CO_2 emissions (UNFCCC 2002). It introduced three regulatory mechanisms: CDM, JI, and ET (Table 8.1), which laid the foundation for C trading. These mechanisms facilitated international cooperation on climate change mitigation and enabled countries meet emission reduction targets more efficiently (Hepburn 2007). Subsequent agreements, such as the Paris Agreement, have further expanded these regulatory frameworks to support global emissions reduction strategies (UNFCCC 2016).

Table 8.1 Carbon crediting mechanisms under Kyoto Protocol (UNFCCC 2002; Hepburn 2007)

	Clean Development Mechanism (CDM)	Joint Implementation (JI)	Emissions Trading (ET)
Article	Article 12	Article 6	Article 17
Project location	Non-Annex I parties (developing nations)	Between Annex I parties (industrialized/ developed nations)	Between Annex I parties
Purpose	To allow Annex I parties to implement projects in non-Annex I parties	To allow Annex I parties to trade emission units with each other	To allow Annex I parties to trade emission units to meet their targets
Trading units	CERs (certified emission reductions)	ERUs (emission reduction units)	AAUs (assigned amount units)
Main mechanism	Project-based emissions reductions in developing countries	Transfer of emission reduction units between developed countries	Trade of surplus emission allowances or units

8.3.2 *Voluntary Carbon Markets*

Voluntary C markets (VCMs) allow businesses and individuals to purchase C credits from certified projects to offset their GHG emissions voluntarily (Climate Seed 2023). These markets offer more flexibility and lower costs compared to regulatory markets but lack the strict accountability mechanisms seen in mandatory systems. VCMs incentivize SAPs, where C credits are generated by farmers adopting practices that avoid or remove CO_2 emissions (Cammarata et al. 2024). Participation in these programs has surged, with over 10,000 farmers across 35 states managing ten million acre (four million hectare) through initiatives by organizations such as the North Dakota National Farmers Union and Iowa Farm Bureau, promoting practices like no-till farming and forest restoration (Johnson 2009). Despite this growth, challenges persist, including concerns over effectiveness, accountability, financial risks, complexity, and the potential for greenwashing (Dibattista 2024). The difficulty in verifying emissions reductions and the high costs of implementation deter small-scale farmers from fully participating in these markets (Proville et al. 2020). Furthermore, the risk of C credit misuse and low-quality credits threatens the credibility of VCMs. To address these challenges, several initiatives have been established to promote transparency and integrity in VCMs. These include the Science Based Targets Initiative (SBTi), the Voluntary C Markets Integrity Initiative (VCMI), and the Integrity Council for the Voluntary C Market (ICVCM), which have introduced guidelines to enhance the reliability of C credits (Pineda et al. 2020; ICVCM 2024; VCMI 2025). One significant development is the introduction of the Ten Core C Principles (CCPs) by the ICVCM (Fig. 8.3), endorsed by multiple organizations to ensure the credibility and effectiveness of voluntary C credits, supporting their role in global warming mitigation (ICVCM 2024).

Agricultural C markets are pivotal in reducing GHG emissions through a range of practices, such as methane capture from livestock manure, optimized nitrogen management, rice management, and soil CS. Table 8.2 provides an overview of various agricultural C projects aimed at generating C credits through these sustainable practices, outlining the relevant standards and examples of project developers and third-party verifiers involved. The success of these markets relies heavily on factors like the quality and credibility of C credits, effective MRV systems, and consistent demand for credits (Wongpiyabovorn et al. 2023). Reliable MRV systems are essential to avoid fraud, overstatement of C offsets, or the issuance of low-quality credits that fail to deliver real environmental benefits (Trencher et al. 2024). High demand for credits, often driven by regulatory requirements or voluntary corporate sustainability goals, is crucial for maintaining market stability and providing farmers with strong incentives (Schofield et al. 2024; Cammarata et al. 2024). Conversely, low or unpredictable demand may destabilize the market and dissuade farmer participation. As noted by Wongpiyabovorn et al. (2023), scenarios in which demand for credits fluctuates or becomes uncertain can lead to market inefficiencies and reduced confidence in the system (Fig. 8.4). Moreover, achieving a successful C farming market requires balancing high-quality credits, strong demand, and supportive

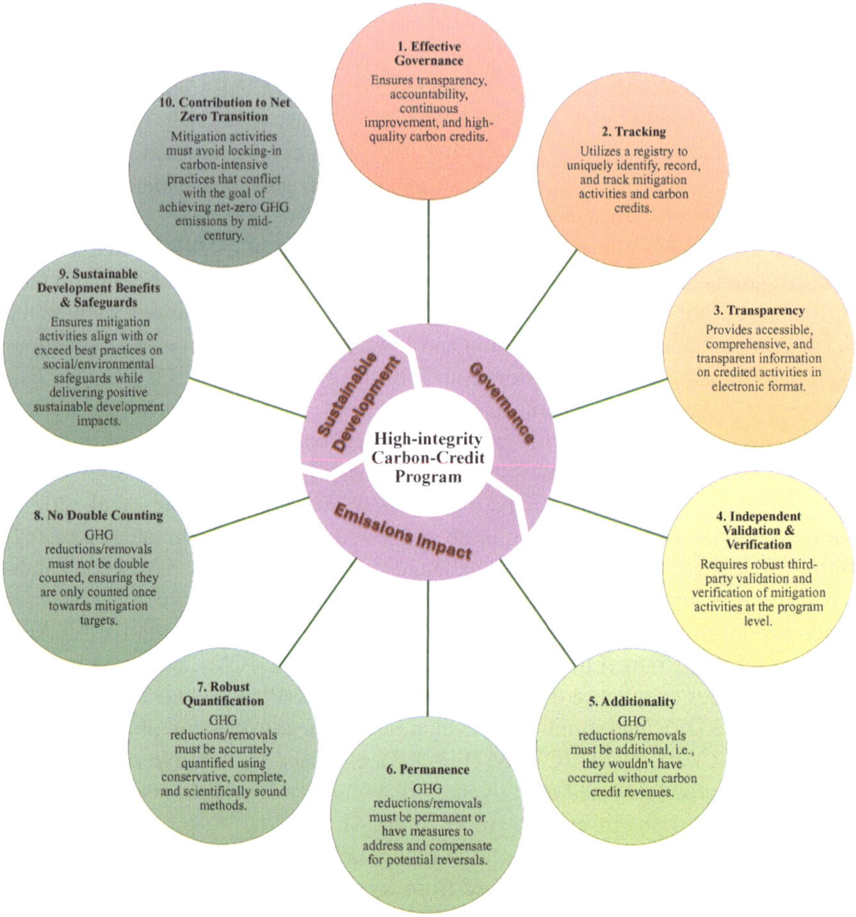

Fig. 8.3 Ten core carbon principles of a high-integrity carbon crediting program, based on the information from ICVCM (2024)

regulatory frameworks that encourage long-term farmer participation aligned with global climate goals. Businesses must collaborate with governments to secure access to high-quality credits and meet climate targets (Kizzier et al. 2021).

8.4 Carbon Pricing

C pricing, which includes mechanisms such as C taxes and cap-and-trade systems, has become a central policy tool towards efforts to mitigate GHG emissions. C taxes directly charge emitters based on the amount of C they produce, providing clear price signals to encourage reduction in emissions. This approach is favored for its

Table 8.2 Overview of agricultural carbon projects (USDA 2023; Haya et al. 2024)

Project type	Focus/Activity	Project developers	Protocols / standards	Verifier
Livestock Manure Methane Digester	Digesters capture methane from manure, which is either flared, used for electricity generation, or processed as renewable natural gas.	Camco offsets I, LLC; 3Degrees Group, Inc.; Ag Methane Advisors, LLC; Profit Carbon Environmental Energy Technology (Shanghai) Co. Ltd.; Solika Energy Pvt. Ltd.; Muyuan Foods Co., Ltd.; Animalia Genetics Ltd.; The Climate Trust	CARB, CAR, ACR, VCS, GS	Agri-Waste Technology, Inc.; Dillon Consulting Ltd.; First Environment of California, Inc.; SCS Global Services
Enteric Fermentation	Reduction of enteric methane emissions from ruminants through the use of feed Ingredients	Blue Ocean Barns; Symbrosia; Carbon Scale; Agri Beef Co.; Sea Forest Ltd.; Feedworks Pty Ltd.	GS, VCS	-
Optimized Nitrogen Management	Reduction of N_2O emissions from nitrogen fertilizer use	Nutrien Ag Solutions, Inc.; Pivot Bio, Inc.	CAR, VCS, ACR	Aster Global Environmental Solutions, Inc.
Improved Rice Management	Reduction of methane emissions from rice cultivation (e.g., dry seeding, AWD, early drainage)	Terra Global Capital; Suhas Joshi; Kosher Climate India Pvt Ltd.; Nurture Agtech Private Limited; Core CX Solutions Pvt. Ltd.	CARB, ACR, GS, VCS	SCS (Scientific Certification Systems) Global Services
Improved Irrigation Management	Reducing GHG emissions in irrigated crops through micro irrigation or solar powered irrigation services	International Development Enterprises (India); Climate Impact Partners Ltd.; SunCulture Kenya Ltd.	GS, VCS	–
Sustainable agriculture	Practices like no-till, strip-till, and cover crops to sequester carbon in soil, improved cropland and grassland management	One Carbon World Ltd.; Boomitra Inc.; Bayer Crop Science LP; Climate Neutral Group; Indigo Ag; Truterra LLC	CAR, GS, VCS	Aster Global Environmental Solutions, Inc.; TUV SUD America, Inc
Solid waste separation	Methane avoidance through separation of solids from wastewater or manure treatment systems	3Degrees Group, Inc.; Willow Creek; Mercer Vu Farms, Inc.; Bos Dairy Farm Project	VCS	–

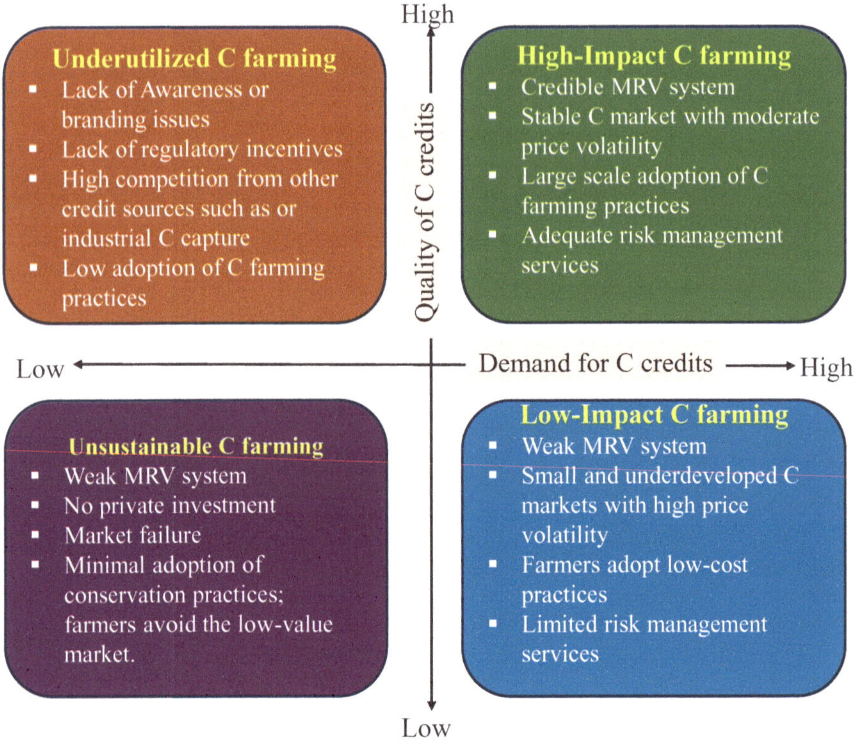

Fig. 8.4 Four agricultural carbon farming scenarios (This matrix is based on the three scenarios outlined in Wongpiyabovorn et al. (2023), with the addition of a new scenario, "Underutilized C Farming")

simplicity and efficiency, although it can face political resistance due to its potential impact on consumer prices (Rabe 2018). On the other hand, cap-and-trade systems set a cap on total emissions and allow businesses to trade allowances, offering flexibility to meet reduction goals. While cap-and-trade systems ensure specific emissions cuts, they can suffer from price volatility and complex administration (Rabe 2018). Despite these challenges, C pricing remains one of the most effective tools for driving emissions reductions, although misconceptions persist. A common myth is that C pricing will significantly harm the economy, but studies show it has minimal GDP impact and can even foster positive effects through biofuel and energy efficiency promotion (Metcalf 2023). Similarly, C pricing is not job-killing; it can create employment in cleaner sectors, though support for displaced workers is necessary. Additionally, while C taxes are often viewed as regressive, redistributing revenue through rebates or tax cuts can make them progressive, benefiting lower-income households (Metcalf 2023).

Global studies on C pricing in agriculture demonstrate its potential for significant emissions reductions. In the U.S., for instance, a C tax can raise input costs, especially for fertilizers, but it could also drive-up commodity prices, partially offsetting losses for farmers (Dumortier and Elobeid 2021). In the EU, integrating agriculture into a C pricing scheme could reduce emissions by 10% in Germany and 14% in the broader EU by 2030, with technological innovations contributing significantly to these reductions (Stepanyan et al. 2023). Similarly, in Nepal, where the agriculture, forestry, and other land use (AFOLU) sectors account for over 80% of GHG emissions, C pricing could mitigate up to 41.5% of emissions from 2016 to 2050 (Pradhan et al. 2017). In China, C pricing in the agricultural sector could reduce emissions by 554.1 $MtCO_2e$ by 2060, with nearly 39% of this reduction being cost-negative (Deng et al. 2024). In Scotland, an ET scheme in agriculture, based on a case study of 192 farmers, showed that up to a 37% reduction in emissions could be achieved at a reasonable cost, provided farmers adopted cost-effective on-farm abatement measures (Bakam et al. 2012). Additionally, mechanisms like the Agricultural C Border Adjustment Mechanism (Ag-CBAM) could play a critical role in preventing C leakage by ensuring that agricultural products imported from countries with lower environmental standards face C fees, thereby aligning global C pricing efforts (Fournier Gabela et al. 2024).

The potential for C pricing to drive emissions reductions is further emphasized in global policy frameworks. The Commission tasked with supporting the Paris Agreement's target of limiting global warming to well below 2 °C recommended C prices ranging from $40–80 per ton of CO_2 by 2020 and $50–100 per ton by 2030, contingent on technological developments and national circumstances (Stiglitz et al. 2017). However, C pricing policies must be carefully designed to balance environmental benefits with potential food security concerns, such as slight reductions in global calorie consumption (Jansson et al. 2024). Revenues from C taxes should be used to support vulnerable populations and promote agricultural efficiency (Schneider and McCarl 2005). Studies on ESs management have stressed the importance of cost-effective assessment methods and the trade-off between cost and accuracy (Richter et al. 2021). Various ES management tools, such as cap-and-trade, direct payments, and self-regulation, are best suited to specific geographical scales, socio-economic conditions, and political environments (Kinzig et al. 2011). Economic incentives, policy coordination, and participation from farmers are key to optimizing CS and reducing emissions (Bamière et al. 2021). Factors such as C prices, transaction costs, local conditions, resource access, and institutional support significantly influence farmers' participation (Antle and Stoorvogel 2009). To strengthen rural C trading, lessons from past projects should inform emission-reducing practices, supported by incentives, governance, and local community empowerment through decentralized resource management and capacity building (Ekka et al. 2016). Policymakers must ensure that C pricing frameworks are flexible, responsive, and consider broader environmental and economic impacts (Schneider and McCarl 2005).

8.5 Challenges in Soil Carbon Sequestration Credit Programs

Soil CS programs face several challenges that limit their broader adoption and full climate potential. These challenges include issues with additionality, permanence, leakage, and accurate measurement and verification (Carpenter 2010; Liu 2022; USDA 2023), as outlined below.

8.5.1 Low Carbon Credit Prices

Agricultural C credit programs have struggled with an oversupply of credits, leading to a significant price drop and undermining financial incentives for farmers to adopt SAPs (Andrew 2021). This issue often arises from weak demand and inadequate regulatory frameworks, as exemplified by the failure of the Chicago Climate Exchange (CCX) in 2010 (Gronewold 2011; Sabbaghi and Sabbaghi 2017; Andrew 2021). The CCX, the U.S.'s first cap-and-trade program, saw the price of C credits plummet from $7.50 to just 5 cents per ton, primarily due to the voluntary nature of the program and the absence of strong regulatory support (Gronewold 2011). This allowed certain participants, such as those already below the emissions cap or easily able to implement compliant technologies, to benefit, while farmers were left exposed to volatile prices (DiPerna 2018). This instability left farmers with worthless credits, eroding their trust in C markets as a reliable source of income.

Despite the potential for C markets to incentivize sustainable farming practices, issues such as low prices, inconsistent compensation, and regulatory uncertainty continue to hinder their effectiveness (USDA 2023). For example, research suggests that farmers require $10–$40 per acre to adopt no-till practices, with government payments being preferred over C market payments (Gramig and Widmar 2018). Additionally, participation in C programs is highly sensitive to price levels, with higher prices like $70/Mg C driving more interest, while lower prices such as $30/Mg C result in a sharp decline in enrollment (Wang et al. 2024). To ensure the success of these programs, it is crucial for policymakers to address uncertainties, enhance perceived benefits, and create stronger regulatory frameworks that improve the economic viability of C offset initiatives.

8.5.2 Additionality

To meet the key criteria of a C offset program, a project must demonstrate additionality, meaning its activities must result in GHG emission reductions beyond what would have occurred without the project. This involves considering legal, financial, and practice-related factors, as well as common practices (USDA 2023). While many farmers already implement SAPs to reduce agrochemical use and improve soil fertility (Phelan et al. 2024), they are not motivated by C payments, seeing them more as supplementary income for practices they already adopt (Barbato and Strong 2023). As a result, credits are often awarded for non-additional practices, undermining the potential for global warming mitigation. Defining an accurate baseline to determine what would have occurred without the project is challenging, as it requires projecting counterfactual scenarios and factoring in subjective elements like social and cultural barriers (Thamo and Pannell 2016). Approaches like the common-practice method, which sets adoption thresholds to determine additionality, have limitations. These methods fail to account for fluctuating adoption rates, may overlook opportunities to scale up practices, and risk excluding small-scale expansions or crediting already common practices, thereby reducing the program's effectiveness (Liu 2022). Additionally, as time progresses, practices once considered additional may no longer meet this criterion, requiring ongoing adjustments to maintain fairness and accuracy (Murray et al. 2007).

8.5.3 Permanence

The permanence of CS is another significant concern for soil C offset programs. While practices like no-till farming and cover cropping can sequester C in soils, they are vulnerable to reversal due to changes in farming practices driven by economic or environmental factors (USDA 2023). These factors include crop rotations, market fluctuations, and climate-related events such as droughts, fires or floods, all of which can lead to the release of stored C back into the atmosphere (Liu 2022; USDA 2023). Advances in remote sensing (RS) technologies may help reduce monitoring costs and improve enforcement of permanence, but current C prices in markets like EU-ETS remain too low to incentivize widespread adoption of continuous no-till farming (Gramig and Widmar 2018). The economic viability of such practices remains limited unless C prices increase or policies related to permanence, penalties for reversals, and reserve pool contributions become more favorable to landowners (Gramig and Widmar 2018). The length of the permanence period in C offset programs is another critical factor in shaping incentives for land-use change and ensuring long-term C storage. For example, a study comparing scenarios with a 25-year crediting period and a 100-year permanence period found that a shorter crediting period led to a 69% reduction in the economically viable supply of C offsets (Regan et al. 2020). This highlights the importance of providing adequate compensation to landholders for the temporal opportunity costs associated with long-term commitments. Ultimately, ensuring permanence in agricultural CS programs requires robust monitoring systems, appropriate compensation mechanisms, and long-term policies that account for the risks and challenges of agricultural systems (Wongpiyabovorn et al. 2023). Moreover, flexibility in the design of these programs, such as the use of reserve pools or buffer credits to mitigate risks, will be crucial in maintaining their environmental integrity and encouraging participation from farmers (Gren and Aklilu 2016; Badgley et al. 2022a).

8.5.4 Leakage

Leakage refers to the risk that CS efforts in one area may lead to emissions or land-use changes outside the project's spatial and temporal boundaries, undermining net reductions (Murray et al. 2004; USDA 2023). For example, leakage could occur if farmers adopt C-sequestering practices but then shift other land areas or activities to practices that release C, such as deforestation or intensified farming (Thamo and Pannell 2016). In Ireland, the growing bovine population is making it hard to reduce GHG emissions enough to meet climate targets (Adenaeuer et al. 2023). Voluntary measures to reduce emissions haven't been very effective. As a result, if stricter rules in Ireland push farming practices to other countries with weaker regulations, emissions could rise in those countries, increasing the risk of C leakage (Adenaeuer et al. 2023). Similarly, policies like AFS in tropical regions, which aim to reduce deforestation and enhance C sinks, can lead to leakage if intensified livestock production results in cattle being moved to new pastures, sometimes at the forest edge, contributing to further deforestation (Castro-Nunez et al. 2021). C leakage risks have also been highlighted in EU agricultural policies, where increased imports and reduced exports can shift emissions to non-EU countries with weaker environmental regulations (Fellmann et al. 2018). To address these risks, policies like C Border Adjustment Mechanisms (CBAM) could be implemented to regulate emissions from high-leakage agricultural imports, ensuring that domestic mitigation efforts are not undermined by displaced emissions (Fournier Gabela et al. 2024). Effective C leakage management in agriculture demands comprehensive policies addressing both production and consumption, along with robust monitoring of emissions and land-use changes across borders to ensure genuine reductions (Fellmann et al. 2018).

8.5.5 Measurement and Verification

Measurement and verification (M&V) are fundamental to the credibility and accuracy of soil CS programs. However, accurately measuring CS in soils, particularly on smallholdings, remains a significant challenge due to the complexity of soil dynamics and the variability in C storage across different soil types and climates (USDA 2023; Okoli and Birkenberg 2024). While traditional methods, such as soil sampling, are reliable, they are costly and time-consuming, limiting large-scale monitoring and smallholder participation (Loria et al. 2024). The high spatial variability of soil C stocks further complicates C quantification, as a small number of samples may not adequately represent C content across a large area (Conant and Paustian 2002).

While model-based and RS approaches offer more cost-effective alternatives, they lack precision and face challenges in meeting regulatory standards for certification (Paul et al. 2023). Disagreements among soil scientists about the duration and extent of C storage, especially regarding practices like no-till farming, add to the uncertainty. This variability, driven by soil heterogeneity and local conditions, makes developing reliable CS models more challenging (Liu 2022).

One promising solution to enhance M&V is a two-tiered approach, combining intensive field-level sampling at specific sites with diffuse, geo-referenced regional or national monitoring (Conant and Paustian 2002). This strategy offers a scalable, cost-effective way to track soil C changes while maintaining accuracy. However, the high costs of verification and certification remain a significant barrier (Proville et al. 2020), especially for smallholder farmers. The financial burden of upfront verification costs and penalties for not maintaining C farming practices may deter participation in C offset programs. These high fees can reduce the economic incentives for small-scale farmers, who are often operating on narrow profit margins (Wongpiyabovorn et al. 2023). Therefore, developing cost-effective and scalable M&V systems that balance accuracy with affordability is crucial for the success and growth of agricultural C offset markets.

8.6 Path Forward

For agricultural C markets to reach their full potential, addressing the key challenges identified in C farming schemes and improving their design is essential. The following steps are crucial to advancing C trading in the agricultural sector:

8.6.1 Integrating Payment for Ecosystem Services into Agricultural Carbon Markets

PES is a system where farmers or landowners are paid for providing environmental benefits, such as CS, improved water quality or biodiversity conservation, but only if they adopt practices that ensure these benefits are delivered (Wunder et al. 2008). Participation is voluntary, with landowners or farmers choosing to engage in the program and receiving payments for achieving specified environmental outcomes. Unlike traditional C markets, which primarily focus on GHG emission reductions, PES schemes provide more tailored and context-specific incentives for land-use practices that benefit a broader range of ESs (Supriyanto et al. 2024). For example, research from South Carolina, USA, shows that PES can generate significant community benefits by improving ESs, with residents expressing a willingness to pay up to $6.39 per month for these improvements (Ureta et al. 2022). In Michigan, Environmental Quality Incentives Program (EQIP) payments for cover crops have been closely linked to increased adoption, showing that financial support reduces risks and encourages sustainable practices (Surdoval et al. 2024).

8.6.1.1 Localized Approaches and Tailored Incentives for Effective Participation

The success of PES programs depends on tailoring them to local social, economic, and cultural contexts goals (Reed et al. 2022; Schofield et al. 2024). Research in Vermont found that while farmers are driven by stewardship, financial pressures and a lack of trust in government can limit their participation in PES programs (White et al. 2022). In Ethiopia, a study on climate-smart agroforestry highlighted that smallholder preferred low numbers of trees, short-term contracts, and up-front payments, with food as the preferred payment type over cash (Haile et al. 2019), emphasizing the importance of understanding local preferences for effective PES design. Likewise, PES incentives in Lake Rawa Pening, Indonesia, were structured to align with local ecological needs, varying based on water quality and pollution levels (Supriyanto et al. 2024). These studies underscore the need for context-specific information and mechanisms that reflect the realities of local farming systems to drive higher participation and adoption rates.

8.6.1.2 Overcoming Barriers and Expanding Participation

Barriers to participation in PES programs are especially challenging for smallholder farmers, who often face constraints related to land tenure and economic constraints (Lansing 2017). The Kansas study found that farmers with strong environmental stewardship attitudes and larger operations were more likely to enroll in PES programs like the Conservation Stewardship Program (CSP). In contrast, those with fixed cash leases and business-only relationships with landowners were less likely to participate (Palm-Forster et al. 2023). To increase participation, PES programs need to address issues like land tenure and financial incentives while fostering better tenant-landowner relationships. Simplifying contract structures and reducing administrative costs can also help overcome barriers and enhance engagement (Palm-Forster et al. 2023).

Farmers' reluctance to engage with agricultural C markets often stems from concerns about market rules, long-term commitments, and additionality criteria (Barbato and Strong 2023; Phelan et al. 2024). Motivation factors such as knowledge, sustainability attitudes, available resources, and incentives from government programs are significant for adoption (Swinton et al. 2015). For example, a study on low-C crop production in China found that farmers were most motivated by knowledge and market opportunities, rather than government regulations (Zhu and Chen 2024). Similarly, farmers' willingness to participate in PES programs is influenced by farm characteristics and payment offers, suggesting that tiered strategies could increase enrollment (Ma et al. 2012).

In the Tropical Andes, a survey of PES programs revealed that most programs focus on ecological and social goals but rarely provide cash payments, instead facing behavioral barriers to participation (Bauchet et al. 2020). The study emphasized the importance of in-kind incentives, dual goals, and compliance enforcement, underscoring the need for better alignment between program design and implementation. A bottom-up approach, adapting PES theory based on practical implementation, and fostering collaboration between academics and practitioners could help bridge the gap between theoretical design principles and on-the-ground practices (Bauchet et al. 2020). This approach would ensure that PES programs are more responsive to local farmer preferences and needs, ultimately increasing participation in low-C agricultural practices.

8.6.1.3 Refining Payment for Ecosystem Services Mechanisms for Improved Effectiveness

To improve PES program effectiveness, incentive mechanisms must be refined. A study by Raina et al. (2024) suggests that combining result-based and action-based incentives to offer low-risk, guaranteed payments, enhancing farmers' confidence in long-term participation. For example, MEKA-B4 in Baden-Württemberg was a result-based agri-environmental measure that compensated farmers for preserving species-rich grasslands based on outcomes rather than actions (Russi et al. 2016). Giger (1998) stressed the importance of subsidies to support economically viable technologies and improve market access. However, El Bakali et al. (2023) cautioned that while PES and input subsidies can promote CA, excessive reliance on subsidies, particularly for practices like minimum tillage, may have negative long-term effects. Furthermore, Piñeiro et al. (2020) emphasized the importance of balancing short-term economic incentives with long-term environmental goals, noting that voluntary incentives need to be adequately compensated to drive adoption, while compulsory ones provide more certainty for participants.

8.6.1.4 Financing and Policy Support

Financing for agricultural C management should be sourced from diverse channels, including government funding, international sources, and sustainability-focused policies. Such financing can integrate C mitigation into broader sustainability frameworks and eliminate subsidies for unsustainable practices (Muller 2012). Farmers have suggested that policy changes and greater resources for PES programs could further drive the adoption of diversified, climate-smart farming systems (Surdoval et al. 2024). Ward (2020) further demonstrated that PES programs supported by financial incentives and social networks could drive CA adoption, especially by offsetting transition costs. These mechanisms, when carefully designed and combined, can enhance the reach and success of PES programs, ensuring that they support both environmental and economic sustainability in agriculture.

8.6.2 Addressing Key Barriers

Addressing barriers such as permanence, leakage, and additionality (PLA) is vital for advancing C trading in agricultural C markets. To ensure CS efforts are both permanent and additional, robust protocols and standards are needed. This includes long-term monitoring, enforcement mechanisms, and enhanced MRV systems for accurate C tracking and market integrity (USDA 2023). For example, the Soil Enrichment Protocol (SEP) by the CAR offers scalable, high-quality agricultural C credits through flexible modeling, hybrid credit generation, and advanced uncertainty quantification, which could enhance soil analysis, improve data accuracy, and drive more impactful climate action (Jackson Hammond et al. 2021). Adjusting for PLA factors is key to enhancing the profitability of Agricultural Soil CS (ASCS) projects, even in the face of high discount rates (Murray et al. 2007). By focusing on activities with high additionality and low leakage, project developers can mitigate risks and design long-term commitments to manage reversal risks. Furthermore, addressing challenges like landowner heterogeneity and uncertainty, particularly in forest CS, requires continued development of efficient policy tools such as optimal contracts and emissions baselines (Gren and Aklilu 2016). The Australian Government's C Farming Initiative illustrated how barriers such as undeveloped methodologies, complex approval processes, long-term permanence obligations, and the additionality requirement limit participation (Rochecouste et al. 2017).

With high transaction costs and low profitability, C offset production is unappealing to farmers (Swinton et al. 2015). To advance C trading in agriculture, simplifying processes, reducing transaction costs, and enhancing incentives for farmers is essential to unlock the full potential of C markets. An aggregate approach, which combines multiple projects and uses default values for large areas, effectively harnesses the mitigation potential of agricultural land (Muller 2012). Aggregating smaller landowners reduces transaction costs and enhances access to C markets, especially for underrepresented farmers. Successful examples include the ACR Grassland Protocol, CAR SEP, and Verra's Family Forest C Program (USDA 2023). However, while aggregation reduces verification costs, it does not eliminate the substantial costs of quantifying GHG reductions.

8.6.3 Policy and Governance Support

Governments and international bodies should implement favorable policies to support C farming, ensure regulatory consistency, and incentivize participation in C markets (van Kooten and Zanello 2023). Key policy tools such as direct regulations, subsidy programs, C taxes, and C offset markets can encourage the adoption of SAPs that sequester C in soils and reduce GHG emissions (Aldy and Stavins 2012). While direct regulations mandate specific practices, voluntary approaches like subsidies and tax incentives, such as the CRP, are more commonly used but may face challenges in widespread adoption due to sociocultural factors and the financial burden on farmers (Ogle et al. 2023). C taxes, which internalize the cost of GHG emissions, are effective but require careful design due to complexities in tracing agricultural product footprints and addressing potential inequities (Ogle et al. 2023). Clear regulations and guidelines are essential to reduce uncertainty and increase market stability, which will attract both farmers and investors (USDA 2023).

PES can incentivize the adoption of SAPs in Mozambique, with a study showing that 60% adoption over 20 years resulted in double the revenue compared to disadoption (Simone et al. 2017). However, establishing a sustainable PES system requires significant investment, institutional support, and addressing market fluctuations. Similarly, C trading policies in China have significantly reduced land-use C emissions by at least four million tons annually, primarily through changes in the energy structure, as found by Xia et al. (2021). The study highlighted the importance of optimizing the energy structure and considering regional differences for effective policy expansion. A grassroots-driven approach is essential to overcome challenges and ensure the success of C trading policies in agriculture (Simone et al. 2017).

8.6.4 Research and Innovation

Research and innovation in agricultural C markets are pivotal for advancing SAPs and improving CS efforts. The integration of advanced technologies like RS, drones, and artificial intelligence (AI) into MRV systems is crucial for improving the efficiency, accuracy, and cost-effectiveness of monitoring C offsets (Mmbando 2025). These technologies enable more precise tracking of CS in agricultural landscapes, making C markets more accessible, particularly for smallholder farmers. RS tools, such as satellite imagery and drones, provide real-time data on soil health, crop conditions, and land use changes, which can be incorporated into AI algorithms to predict and optimize C storage over time (Inoue 2020). Additionally, AI solutions improve energy efficiency and reduce C emissions in controlled agricultural environments, such as greenhouses. Through smart systems, including sensors, IoT devices, and intelligent shading, AI helps regulate environmental factors to reduce energy consumption, contributing to C offset efforts (Maraveas et al. 2023).

Additionally, research highlighted the importance of integrating social-ecological systems (SES) into soil C management. A review in Australia calls for a more holistic approach, incorporating farmers' perspectives alongside ecological factors to support sustainable practices and contribute to global sustainability goals (Amin et al. 2020). Despite these advances, challenges persist in the widespread adoption of new research and technologies, especially in developing regions. High costs, data quality concerns, and limited infrastructure are significant barriers that must be addressed to ensure the scalability of AI and RS in agriculture (Mmbando 2025). Overcoming these barriers requires strong institutional support, including policies that encourage public-private partnerships, provide funding for research and development, and address the financial barriers faced by smallholder farmers (Mmbando 2025). Training programs to overcome technophobia (fear or distrust of technology), improve digital literacy, and bridge skill gaps are essential for increasing awareness and fostering the adoption of these technologies (Boniface 2019). By addressing these challenges, agricultural C markets can be expanded, leading to enhanced sustainability, higher productivity, and increased C offset contributions in agriculture.

8.7 Conclusion

This chapter revolves around two main hypotheses: Hypothesis one suggests that agricultural C markets can drive the adoption of C farming practices by offering financial incentives and market-based solutions. Hypothesis two proposes that targeted incentives, such as PES and stronger policy support, can address farmers' concerns and increase their participation. These hypotheses guide the exploration of agricultural C markets and underscore the critical need for well-designed systems to overcome barriers like low credit prices, additionality concerns, permanence risks, leakage, and challenges with measurement and verification. Through practices such as no-tillage, cover cropping, and SLM, C farming offers significant opportunities to sequester C in soils, generate C credits, and contribute to global GHG emission reduction goals. To unlock this potential, it is essential to recognize the societal value of soil C, integrate SOC into decision-making, and incentivize farmers via mechanisms like PES and C credits. This approach will not only enhance agricultural resilience but also ensure long-term environmental benefits.

Despite promising opportunities, agricultural C markets face several hurdles that need to be addressed. For these markets to thrive, it is crucial to refine baseline definitions for additionality, enhance transparency in MRV, and establish more reliable, cost-effective monitoring solutions. Effective policy frameworks and clear regulations will create an environment where C farming practices are financially viable, especially for smallholder farmers who may struggle with the complexity and cost of verification processes.

A comprehensive approach is needed to improve market designs, build trust, and foster innovation. Strengthening incentive structures, integrating PES schemes, and simplifying processes will drive broader adoption of C farming practices, helping to tackle challenges such as high transaction costs and inequitable participation. By supporting technological advancements and offering targeted financial incentives, agricultural C markets can become more accessible and effective. As these markets evolve, C farming can become a cornerstone of both environmental sustainability and economic growth, benefiting farmers, ecosystems, and society as a whole.

References

Adenaeuer L, Breen J, Witzke P et al (2023) The potential impacts of an EU-wide agricultural mitigation target on the Irish agriculture sector. Clim Pol 23:495–508. https://doi.org/10.1080/14693062.2022.2105791

Aldy JE, Stavins RN (2012) The promise and problems of pricing carbon. J Environ Dev 21:152–180. https://doi.org/10.1177/1070496512442508

Amin MN, Hossain MS, Lobry de Bruyn L, Wilson B (2020) A systematic review of soil carbon management in Australia and the need for a social-ecological systems framework. Sci Total Environ 719:135182. https://doi.org/10.1016/j.scitotenv.2019.135182

Andrew H (2021) Agricultural carbon markets: how might they work?

Antle JM, Stoorvogel JJ (2009) Payments for ecosystem services, poverty and sustainability: the case of agricultural soil carbon sequestration. In: Payment for environmental services in agricultural landscapes. Springer, New York, pp 133–161

Badgley G, Chay F, Chegwidden OS et al (2022a) California's forest carbon offsets buffer pool is severely undercapitalized. Frontiers in Forests and Global Change 5. https://doi.org/10.3389/ffgc.2022.930426

Badgley G, Freeman J, Hamman JJ et al (2022b) Systematic over-crediting in California's forest carbon offsets program. Glob Chang Biol 28:1433–1445. https://doi.org/10.1111/gcb.15943

Bakam I, Pajot G, Matthews RB (2012) Estimating carbon price from a closed emissions trading scheme in the agricultural sector. J Land Use Sci 7:221–238. https://doi.org/10.1080/1747423X.2011.563324

Bamière L, Jayet P, Kahindo S, Martin E (2021) Carbon sequestration in French agricultural soils: a spatial economic evaluation. Agric Econ 52:301–316. https://doi.org/10.1111/agec.12619

Barbato CT, Strong AL (2023) Farmer perspectives on carbon markets incentivizing agricultural soil carbon sequestration. npj Climate Action 2:26. https://doi.org/10.1038/s44168-023-00055-4

Bauchet J, Asquith N, Ma Z et al (2020) The practice of payments for ecosystem services (PES) in the tropical Andes: evidence from program administrators. Ecosyst Serv 45:101175. https://doi.org/10.1016/j.ecoser.2020.101175

Bento A, Kanbur R, Leard B (2016) On the importance of baseline setting in carbon offsets markets. Clim Chang 137:625–637. https://doi.org/10.1007/s10584-016-1685-2

Blaufelder C, Katz J, Levy C et al (2020) How the voluntary carbon market can help address climate change. McKinsey. Sustainability

Boniface PJ (2019) Constraints faced by farmers and agricultural extensionists in using selected information technology enabled Systems for Agriculture*. J Soc Sci 58. https://doi.org/10.31901/24566756.2019/58.1-3.2234

Brummitt CD, Mathers CA, Keating RA et al (2024) Solutions and insights for agricultural monitoring, reporting, and verification (MRV) from three consecutive issuances of soil carbon credits. J Environ Manag 369:122284. https://doi.org/10.1016/j.jenvman.2024.122284

Bullock D (2012) Emissions trading in New Zealand: development, challenges and design. Env Polit 21:657–675. https://doi.org/10.1080/09644016.2012.688359

Cammarata M, Scuderi A, Timpanaro G, Cascone G (2024) Factors influencing farmers' intention to participate in the voluntary carbon market: an extended theory of planned behavior. J Environ Manag 369:122367. https://doi.org/10.1016/j.jenvman.2024.122367

CAR (2021) Climate Action Reserve Offset Program Manual. https://www.climateactionreserve.org/wp-content/uploads/2021/03/Reserve_Offset_Program_Manual_March_2021.pdf. Accessed 5 Jan 2025

Cariappa AAG, Konath NC, Sapkota TB, Krishna VV (2024) Evaluating the potential and eligibility of conservation agriculture practices for carbon credits. Sci Rep 14:9193. https://doi.org/10.1038/s41598-024-59262-6

Carpenter RJ (2010) Implementation of biological sequestration offsets in a carbon reduction policy: Answers to Key Questions for a Successful Domestic Offset Program

Castro-Nunez A, Buriticá A, Gonzalez C et al (2021) The risk of unintended deforestation from scaling sustainable livestock production systems. Conserv Sci Pract 3. https://doi.org/10.1111/csp2.495

Chandra A, Ruffalo W (2024) Understanding carbon credits in agricultural voluntary carbon markets: ecosystem actors and their roles. CSA News 69:3–7. https://doi.org/10.1002/csan.21428

Climate Seed (2023) Voluntary carbon market vs. regulated carbon market: key differences. In: Climate seed. Voluntary Carbon Market vs. Regulated Carbon Market: key differences. Accessed 5 Jan 2025

Conant RT, Paustian K (2002) Spatial variability of soil organic carbon in grasslands: implications for detecting change at different scales. Environ Pollut 116:S127–S135. https://doi.org/10.1016/S0269-7491(01)00265-2

Costanza R, d'Arge R, de Groot R et al (1997) The value of the world's ecosystem services and natural capital. Nature 387:253–260. https://doi.org/10.1038/387253a0

Costanza R, de Groot R, Sutton P et al (2014) Changes in the global value of ecosystem services. Glob Environ Chang 26:152–158. https://doi.org/10.1016/j.gloenvcha.2014.04.002

Deng Y, Liu J-Y, Xie W et al (2024) Impact of carbon pricing on mitigation potential in Chinese agriculture: a model-based multi-scenario analysis at provincial scale. Environ Impact Assess Rev 105:107409. https://doi.org/10.1016/j.eiar.2023.107409

Derpsch R, Friedrich T (2009) Development and current status of no-till adoption in the world. In: Proceedings on CD, 18th Triennial Conference of the International Soil Tillage Research Organization (ISTRO). Izmir, Turkey

Dibattista I (2024) Agricultural voluntary carbon markets: challenges and recommendations. The Florence School of Regulation

DiPerna P (2018) Pricing carbon: integrating promise, practice and lessons learned from the Chicago climate exchange. In: Walker T, Kibsey SD, Crichton R (eds) Designing a sustainable financial system. Palgrave Macmillan, Cham, pp 115–148

Dumortier J, Elobeid A (2021) Effects of a carbon tax in the United States on agricultural markets and carbon emissions from land-use change. Land Use Policy 103:105320. https://doi.org/10.1016/j.landusepol.2021.105320

Ekka A, Aftabuddin M, Pandit A (2016) Effective carbon management for carbon market compliance by the rural sector in India. Curr Sci 111:1780. https://doi.org/10.18520/cs/v111/i11/1780-1786

El Bakali I, Ait El Mekki A, Maatala N, Harbouze R (2023) A systematic review on the impact of incentives on the adoption of conservation agriculture: new guidelines for policymakers and researchers. Int J Agric Sustain 21. https://doi.org/10.1080/14735903.2023.2290415

Fellmann T, Witzke P, Weiss F et al (2018) Major challenges of integrating agriculture into climate change mitigation policy frameworks. Mitig Adapt Strateg Glob Chang 23:451–468. https://doi.org/10.1007/s11027-017-9743-2

Fournier Gabela JG, Spiegel A, Stepanyan D et al (2024) Carbon leakage in agriculture: when can a carbon border adjustment mechanism help? Clim Pol 24:1410–1425. https://doi.org/10.1080/14693062.2024.2387237

Gascoigne WR, Hoag D, Koontz L et al (2011) Valuing ecosystem and economic services across land-use scenarios in the prairie pothole region of the Dakotas, USA. Ecol Econ 70:1715–1725. https://doi.org/10.1016/j.ecolecon.2011.04.010

Giger M (1998) Using incentives and subsidies for sustainable management of agricultural soils—a challenge for projects and policy-makers. In: Blume H, Eger H, Fleischhauer E et al (eds) Towards sustainable land use. Catena Verlag, pp 1405–1411

Graham A, Ferrier H, Mitchell D et al (2013) Ecosystem services: the farmers' challenge. Science (1979) 342:420–421. https://doi.org/10.1126/science.342.6157.420-b

Gramig BM, Widmar NJO (2018) Farmer preferences for agricultural soil carbon sequestration schemes. Appl Econ Perspect Policy 40:502–521. https://doi.org/10.1093/aepp/ppx041

Gren I-M, Aklilu AZ (2016) Policy design for forest carbon sequestration: a review of the literature. Forest Policy Econ 70:128–136. https://doi.org/10.1016/j.forpol.2016.06.008

Gronewold N (2011) Chicago climate exchange closes nation's first cap-and-trade system but keeps eye to the future. The New York Times
Haile KK, Tirivayi N, Tesfaye W (2019) Farmers' willingness to accept payments for ecosystem services on agricultural land: the case of climate-smart agroforestry in Ethiopia. Ecosyst Serv 39:100964. https://doi.org/10.1016/j.ecoser.2019.100964
Haites E, Mehling M (2009) Linking existing and proposed GHG emissions trading schemes in North America. Clim Pol 9:373–388. https://doi.org/10.3763/cpol.2009.0622
Haya BK, Abayo A, Rong X et al (2024) Voluntary registry offsets database v2024–10. Berkeley Carbon Trading Project, University of California, Berkeley
Hepburn C (2007) Carbon trading: a review of the Kyoto mechanisms. Annu Rev Environ Resour 32:375–393. https://doi.org/10.1146/annurev.energy.32.053006.141203
Hunjra AI, Bouri E, Azam M et al (2024) Economic growth and environmental sustainability in developing economies. Res Int Bus Finance 70:102341. https://doi.org/10.1016/j.ribaf.2024.102341
ICVCM (2024) Core carbon principles, assessment framework and assessment procedure. The Integrity Council for the Voluntary Carbon Market
Inoue Y (2020) Satellite- and drone-based remote sensing of crops and soils for smart farming—a review. Soil Sci Plant Nutr 66:798–810. https://doi.org/10.1080/00380768.2020.1738899
Jackson Hammond AA, Motew M, Brummitt CD et al (2021) Implementing the soil enrichment protocol at scale: opportunities for an agricultural carbon market. Frontiers in Climate 3. https://doi.org/10.3389/fclim.2021.686440
Jansson T, Malmström N, Johansson H, Choi H (2024) Carbon taxes and agriculture: the benefit of a multilateral agreement. Clim Pol 24:13–25. https://doi.org/10.1080/14693062.2023.2171355
Johnson R (2009) Provisions supporting ecosystem services markets in U.S. Farm Bill Legislation
Kinzig AP, Perrings C, Chapin FS et al (2011) Paying for ecosystem services—promise and peril. Science (1979) 334:603–604. https://doi.org/10.1126/science.1210297
Kizzier K, Hanafi A, Ogata C et al (2021) Mobilizing voluntary carbon markets to drive climate action: trends in the voluntary carbon markets. EDF and ENGIE Impact
Kollmuss A, Lee C, Lazarus M (2010) How offset programs assess and approve projects and credits. Carbon Manag 1:119–134. https://doi.org/10.4155/cmt.10.6
Kristanto Y, Tarigan S, June T et al (2024) Indirect use value of improved soil health as natural capital that supports essential ecosystem services: a case study of cacao agroforestry. Agric Econ (Zemědělská ekonomika) 70:137–154. https://doi.org/10.17221/281/2023-AGRICECON
Lal R (2014) Societal value of soil carbon. J Soil Water Conserv 69:186A–192A. https://doi.org/10.2489/jswc.69.6.186A
Lansing DM (2017) Understanding smallholder participation in payments for ecosystem services: the case of Costa Rica. Hum Ecol 45:77–87. https://doi.org/10.1007/s10745-016-9886-x
Li X, Martino S (2024) Assessing the economic feasibility of voluntary carbon markets in land use management scenarios for Scottish saltmarshes. Ocean Coast Manag 251:107099. https://doi.org/10.1016/j.ocecoaman.2024.107099
Liu S (2022) Restoring carbon sinks in agriculture through carbon credits. Sustainable Finance Initiative
Liu X, Cui Q (2017) Baseline manipulation in voluntary carbon offset programs. Energy Policy 111:9–17. https://doi.org/10.1016/j.enpol.2017.09.014
Loria N, Lal R, Chandra R (2024) Handheld in situ methods for soil organic carbon assessment. Sustainability 16:5592. https://doi.org/10.3390/su16135592
Ma S, Swinton SM, Lupi F, Jolejole-Foreman C (2012) Farmers' willingness to participate in payment-for-environmental-services programmes. J Agric Econ 63:604–626. https://doi.org/10.1111/j.1477-9552.2012.00358.x
Maraveas C, Karavas C-S, Loukatos D et al (2023) Agricultural greenhouses: resource management technologies and perspectives for zero greenhouse gas emissions. Agriculture 13:1464. https://doi.org/10.3390/agriculture13071464

Men P, Hok L, Seeniang P et al (2024) Assessment of influencing factors and mechanisms to engage the private sector in promoting conservation agriculture production systems in Cambodia. Int J Agric Sustain 22. https://doi.org/10.1080/14735903.2024.2399398
Metcalf GE (2023) Five myths about carbon pricing. Oxf Rev Econ Policy 39:680–693. https://doi.org/10.1093/oxrep/grad042
Mikhailova EA, Groshans GR, Post CJ et al (2019) Valuation of soil organic carbon stocks in the contiguous United States based on the avoided social cost of carbon emissions. Resources 8:153. https://doi.org/10.3390/resources8030153
Mitchell E, Takeda N, Grace L et al (2024) Making soil carbon credits work for climate change mitigation. Carbon Manag 15:2430780. https://doi.org/10.1080/17583004.2024.2430780
Mmbando GS (2025) Harnessing artificial intelligence and remote sensing in climate-smart agriculture: the current strategies needed for enhancing global food security. Cogent Food Agric 11. https://doi.org/10.1080/23311932.2025.2454354
Mnif W, Davison M (2011) Carbon emission markets. In: Quantitative Financial Risk Management. Springer, Berlin Heidelberg, pp 95–108
Muller A (2012) Agricultural land management, carbon reductions and climate policy for agriculture. Carbon Manag 3:641–654. https://doi.org/10.4155/cmt.12.64
Murray BC, McCarl BA, Lee H-C (2004) Estimating leakage from Forest Carbon Sequestration Programs. Land Econ 80:109–124. https://doi.org/10.2307/3147147
Murray BC, Sohngen B, Ross MT (2007) Economic consequences of consideration of permanence, leakage and additionality for soil carbon sequestration projects. Clim Chang 80:127–143. https://doi.org/10.1007/s10584-006-9169-4
Musier R, Adib P (2010) Emerging carbon markets and fundamentals of tradable permits. In: Generating electricity in a carbon-constrained world. Elsevier, pp 57–85
Ogle SM, Conant RT, Fischer B et al (2023) Policy challenges to enhance soil carbon sinks: the dirty part of making contributions to the Paris agreement by the United States. Carbon Manag 14. https://doi.org/10.1080/17583004.2023.2268071
Okoli AO, Birkenberg A (2024) Monitoring soil carbon in smallholder carbon projects: insights from Kenya. Clim Chang 177:143. https://doi.org/10.1007/s10584-024-03796-1
Oldfield EE, Eagle AJ, Rubin RL et al (2022) Crediting agricultural soil carbon sequestration. Science (1979) 375:1222–1225. https://doi.org/10.1126/science.abl7991
Palm-Forster LH, Taylor M, Banerjee S, Xie L (2023) Factors influencing enrollment of leased cropland in the conservation stewardship program in Kansas. Land Use Policy 135:106956. https://doi.org/10.1016/j.landusepol.2023.106956
Pascual U, Termansen M, Abson DJ (2015) The economic value of soil carbon. In: Soil carbon: science, management and policy for multiple benefits. CABI, pp 179–187
Paterson M (2012) Who and what are carbon markets for? Politics and the development of climate policy. Clim Pol 12:82–97. https://doi.org/10.1080/14693062.2011.579259
Paul C, Bartkowski B, Dönmez C et al (2023) Carbon farming: are soil carbon certificates a suitable tool for climate change mitigation? J Environ Manag 330:117142. https://doi.org/10.1016/j.jenvman.2022.117142
Petersen EH, Hoyle FC (2016) Estimating the economic value of soil organic carbon for grains cropping systems in Western Australia. Soil Research 54:383. https://doi.org/10.1071/SR15101
Phelan L, Chapman PJ, Ziv G (2024) The emerging global agricultural soil carbon market: the case for reconciling farmers' expectations with the demands of the market. Environ Dev 49:100941. https://doi.org/10.1016/j.envdev.2023.100941
Pineda AC, Chang A, Faria P (2020) Foundations for science-based net-zero target setting in the corporate sector
Piñeiro V, Arias J, Dürr J et al (2020) A scoping review on incentives for adoption of sustainable agricultural practices and their outcomes. Nat Sustain 3:809–820. https://doi.org/10.1038/s41893-020-00617-y
Pradhan BB, Shrestha RM, Hoa NT, Matsuoka Y (2017) Carbon prices and greenhouse gases abatement from agriculture, forestry and land use in Nepal. Glob Environ Chang 43:26–36. https://doi.org/10.1016/j.gloenvcha.2017.01.005

Proville J, Parkhurst R, Koller S et al (2020) Agricultural offset potential in the United States, New York

Rabe BG (2018) Why carbon pricing is appealing. In: can we Price carbon? The MIT Press, pp 1–12

Raina N, Zavalloni M, Viaggi D (2024) Incentive mechanisms of carbon farming contracts: a systematic mapping study. J Environ Manag 352:120126. https://doi.org/10.1016/j.jenvman.2024.120126

Reed MS, Curtis T, Gosal A et al (2022) Integrating ecosystem markets to co-ordinate landscape-scale public benefits from nature. PLoS One 17:e0258334. https://doi.org/10.1371/journal.pone.0258334

Regan CM, Connor JD, Summers DM et al (2020) The influence of crediting and permanence periods on Australian forest-based carbon offset supply. Land Use Policy 97:104800. https://doi.org/10.1016/j.landusepol.2020.104800

Richter F, Jan P, El Benni N et al (2021) A guide to assess and value ecosystem services of grasslands. Ecosyst Serv 52:101376. https://doi.org/10.1016/j.ecoser.2021.101376

Rochecouste J-F, Dargusch P, King C (2017) Farmer perceptions of the opportunities and constraints to producing carbon offsets from Australian dryland grain cropping farms. Aust J Environ Manag 24:441–452. https://doi.org/10.1080/14486563.2017.1379037

Russi D, Margue H, Oppermann R, Keenleyside C (2016) Result-based agri-environment measures: market-based instruments, incentives or rewards? The case of Baden-Württemberg. Land Use Policy 54:69–77. https://doi.org/10.1016/j.landusepol.2016.01.012

Sabbaghi O, Sabbaghi N (2017) The Chicago climate exchange and market efficiency: an empirical analysis. Environ Econ Policy Stud 19:711–734. https://doi.org/10.1007/s10018-016-0171-4

Schneider UA, McCarl BA (2005) Implications of a carbon-based energy tax for U.S. Agriculture Agric Resour Econ Rev 34:265–279. https://doi.org/10.1017/S1068280500008418

Schofield LR, Pearson ME, Newell S et al (2024) Why aren't more landowners enrolling in land-based carbon credit exchanges? Rangelands 46:117–131. https://doi.org/10.1016/j.rala.2024.05.004

Simone TE, Lambert DM, Cuvaca I, Eash NS (2017) Soil carbon sequestration, carbon markets, and conservation agriculture practices: a hypothetical examination in Mozambique. Int Soil Water Conserv Res 5:167–179. https://doi.org/10.1016/j.iswcr.2017.06.001

Spilker G, Nugent N (2022) Voluntary carbon market derivatives: Growth, innovation and usage. Borsa Istanbul Rev 22:S109–S118. doi:https://doi.org/10.1016/j.bir.2022.11.008

Stepanyan D, Heidecke C, Osterburg B, Gocht A (2023) Impacts of national vs European carbon pricing on agriculture. Environ Res Lett 18:074016. https://doi.org/10.1088/1748-9326/acdcac

Stiglitz JE, Stern N, Duan M et al (2017) Report of the high-level commission on carbon prices. DC, USA, Washington

Supriyanto S, Nowo Martono D, Sari Hasibuan H, Mulyo Hartono D (2024) Assessing payment for ecosystem services to improve lake water quality using the InVEST model. Environ Econ 15:149–173. https://doi.org/10.21511/ee.15(1).2024.12

Surdoval A, Jain M, Blair E et al (2024) Financial incentive programs and farm diversification with cover crops: assessing opportunities and challenges. Environ Res Lett 19:044063. https://doi.org/10.1088/1748-9326/ad35d8

Swinton SM, Rector N, Robertson GP et al (2015) Farmer decisions about adopting environmentally beneficial practices. In: Hamilton SK, Doll JE, Robertson GP (eds) The ecology of agricultural landscapes: long-term research on the path to sustainability. Oxford University Press, New York, pp 340–359

Thamo T, Pannell DJ (2016) Challenges in developing effective policy for soil carbon sequestration: perspectives on additionality, leakage, and permanence. Clim Pol 16:973–992. https://doi.org/10.1080/14693062.2015.1075372

Trencher G, Nick S, Carlson J, Johnson M (2024) Demand for low-quality offsets by major companies undermines climate integrity of the voluntary carbon market. Nat Commun 15:6863. https://doi.org/10.1038/s41467-024-51151-w

UNFCCC (1997) Kyoto protocol to the United Nations framework convention on climate change. In: Kyoto Climate Change Conference—December 1997. Kyoto
UNFCCC (2002) A guide to the climate change convention and its protocol. UNFCCC
UNFCCC (2016) The Paris agreement. United Nations Framework Convention on Climate Change (UNFCCC)
Ureta JC, Motallebi M, Vassalos M et al (2022) Estimating residents' WTP for ecosystem services improvement in a payments for ecosystem services (PES) program: a choice experiment approach. Ecol Econ 201:107561. https://doi.org/10.1016/j.ecolecon.2022.107561
Uri ND, Atwood JD, Sanabria J (1999) The environmental benefits and costs of conservation tillage. Environ Geol 38:111–125. https://doi.org/10.1007/s002540050407
USDA (2023) Report to congress: a general assessment of the role of agriculture and forestry in U.S. Carbon Markets
van Kooten GC, Zanello R (2023) Carbon offsets and agriculture: options, obstacles, and opinions. Canad J Agric Econ/Revue canadienne d'agroeconomie 71:375–391. https://doi.org/10.1111/cjag.12340
VCMI (2025) Voluntary Carbon Markets Integrity Initiative. https://vcmintegrity.org/. Accessed 9 Jan 2025
Wang T, Jin H, Clay D et al (2024) Carbon supply elasticity and determinants of farmer carbon farming decisions. Appl Econ Perspect Policy 46:1190–1213. https://doi.org/10.1002/aepp.13442
Ward RD (2020) Carbon sequestration and storage in Norwegian Arctic coastal wetlands: impacts of climate change. Sci Total Environ 748:141343. https://doi.org/10.1016/j.scitotenv.2020.141343
Wettestad J (2011) EU emissions trading: achievements and challenges. In: Toward a common European Union energy policy. Palgrave Macmillan US, New York, pp 87–111
White AC, Faulkner JW, Conner DS et al (2022) "How can you put a price on the environment?" farmer perspectives on stewardship and payment for ecosystem services. J Soil Water Conserv 00041:00041. https://doi.org/10.2489/jswc.2022.00041
Wongpiyabovorn O, Plastina A, Crespi JM (2023) Challenges to voluntary Ag carbon markets. Appl Econ Perspect Policy 45:1154–1167. https://doi.org/10.1002/aepp.13254
Wunder S, Engel S, Pagiola S (2008) Taking stock: a comparative analysis of payments for environmental services programs in developed and developing countries. Ecol Econ 65:834–852. https://doi.org/10.1016/j.ecolecon.2008.03.010
Xia Q, Li L, Dong J, Zhang B (2021) Reduction effect and mechanism analysis of carbon trading policy on carbon emissions from land use. Sustainability 13:9558. https://doi.org/10.3390/su13179558
Zhu Y, Chen J (2024) How do we motivate farmers to adopt low-carbon production? Analysis of extrinsic incentives' internalization. J Environ Psychol 94:102186. https://doi.org/10.1016/j.jenvp.2023.102186

Chapter 9
The Role of Technology in Carbon Farming

Abstract Technological innovation is increasingly central to enhancing carbon (C) farming systems, addressing both agricultural productivity and climate mitigation. This chapter examines the role of precision agriculture, smart sensors, remote sensing, artificial intelligence, Internet of Things (IoT), and blockchain in improving soil C sequestration through optimized resource management, real-time monitoring, and transparent C accounting. The integration of these technologies supports higher C capture rates and economic incentives via C credit markets. Key challenges—including sensor calibration, data integration, infrastructure gaps, scalability, and regulatory constraints—are analyzed alongside potential solutions. Framed by two hypotheses, the chapter demonstrates how advanced technologies can enhance C sequestration efficiency and drive the adoption of C farming practices, contributing to both environmental resilience and economic viability.

9.1 Introduction

The role of technology in carbon (C) farming is increasingly pivotal in addressing the challenges of rising food demand and climate change. As the global population grows, with food demand expected to increase by 70% by 2050 (FAO 2009), the need for innovative solutions to improve agricultural productivity while reducing C emissions becomes urgent (Lal 2016). Advanced technologies, such as precision agriculture (PA), smart farming, and artificial intelligence (AI), offer significant potential to optimize farming practices, enhance crop yields, and improve resource use efficiency. These technologies facilitate C sequestration (CS) by enabling better monitoring of soil health, optimizing irrigation, and reducing the C footprint of agricultural activities through innovations such as GPS equipment, variable rate fertilizer applicators, and precision irrigation systems (Hedley 2015). Furthermore, innovations like the Internet of Things (IoT) and robotic farming allow for real-time data collection, which helps farmers make informed decisions that enhance sustainability. Table 9.1 describes several key technological terms in agriculture that highlight innovations aimed at improving efficiency, sustainability, and productivity. By

N. Loria, R. Lal, *Carbon Farming*, https://doi.org/10.1007/978-3-032-00842-8_9

Table 9.1 Technological terms in agriculture

Term	Description	References
Precision agriculture	Uses technology to optimize farm inputs based on site-specific crop and soil conditions, aiming to improve efficiency and productivity by addressing field variability.	Gassner et al. (2013)
Digital farming	Integrates a network of machines, sensors, and data to optimize farming systems, enabling farmers to access real-time information and manage operations efficiently through connected devices like smartphones and tablets.	Tran et al. (2020)
Internet of things (IoT)	Integrates technologies like sensors, radio-frequency identification (RFID), and mobile devices to enable real-time monitoring and optimization of farming processes via internet connectivity	Abbasi et al. (2019); Salunkhe et al. (2023)
Artificial intelligence (AI)	Mimics human cognition through machine learning, robotics, and data analytics to enhance farming processes, support better decision-making, and improve productivity and sustainability.	Javaid et al. (2023)
Smart agriculture	Uses advanced technologies like IoT, sensors, and big data to enhance farm management efficiency, lower production costs, and reduce environmental impact.	Zhai et al. (2023)
Robotic farming	Leverages advanced technologies like sensors, IoT, and AI to enhance agricultural efficiency, reduce land damage, and improve crop health by automating tasks like navigation and crop monitoring.	Zhai et al. (2023)

reducing emissions and addressing financial and technical challenges, particularly in developing countries, technology in C farming can drive both environmental and economic benefits, promoting a more sustainable and resilient agricultural future (Calvin et al. 2023). This chapter aims to explore the role of advanced technologies in enhancing CS within farming systems, assessing their impact on both the environmental and economic dimensions of C farming. Specifically, it will examine how technologies such as PA, AI, remote sensing (RS), blockchain, smart sensors, and the IoT can improve C capture, promote sustainable agricultural practices (SAPs), and offer economic incentives for farmers. Additionally, the chapter will identify challenges related to the adoption of these technologies and propose potential solutions to overcome these obstacles.

Two key hypotheses guide the focus of this chapter. Hypothesis 1 suggests that the integration of advanced technologies, such as PA, AI, and RS, significantly enhances the effectiveness of CS in farming systems. This leads to higher rates of C capture and improves agricultural sustainability. Hypothesis 2, on the other hand, posits that the adoption of technology-driven C farming practices, supported by tools like blockchain for C credit tracking and soil sensors for monitoring, provides economic incentives for farmers. This, in turn, accelerates the widespread implementation of sustainable farming techniques. These objectives and the underlying hypotheses aim to highlight how technological advancements can contribute to a more sustainable and economically viable agricultural future by enhancing C farming practices.

9.2 Technologies Enabling Soil Carbon Sequestration

9.2.1 Precision Agriculture

Modern agriculture consumes significant energy through chemicals, fertilizers, and machinery, leading to significant energy consumption and resource inefficiency, often harming the environment (Sharma et al. 2019). The adoption of PA can address these challenges by optimizing energy use and resource management through the recognition of natural variability on farms. By tailoring treatments to suit spatial and temporal factors, PA shifts from large-scale uniform practices to more individualized crop and animal management, improving efficiency and reducing resource overuse that can degrade soil structure and C storage capacity (Blackmore 2016) and adversely impact the environment such as emission of greenhouse gases (GHGs) (Balafoutis et al. 2017) and eutrophication of water (Saleem et al. 2023). Technologies like soil sensors, GPS-guided machinery, RS, and data analytics enable farmers to apply only the necessary amount of water, fertilizers, and pest control, minimizing waste and promoting sustainable practices (Khose et al. 2023). This approach helps reduce nutrient runoff, lower GHG emissions, and increase soil organic C (SOC) sequestration (Jarecki and Lal 2003). The Fig. 9.1 illustrates the step-by-step process of PA in optimizing resource use and ensuring sustainable practices.

PA can be categorized into map-based and sensor-based approaches, each offering unique benefits for farm management. Map-based PA utilizes spatial data from technologies like GPS, satellite imagery, and geostatistics to create detailed maps of field variability, helping farmers make informed decisions on irrigation, fertilization, and crop management (Uddin et al. 2020). For instance, the use of electrical conductivity (EC) mapping has been shown to identify areas with higher organic matter (OM) and better soil structure, allowing for more targeted and effective management practices (Adamchuk and Rossel 2010). Sensor-based PA, on the other hand, involves real-time data collection from various sensors such as soil moisture, temperature, and nutrient sensors, enabling adaptive management by providing continuous monitoring of crop and soil conditions (Laveglia et al. 2024). Technologies

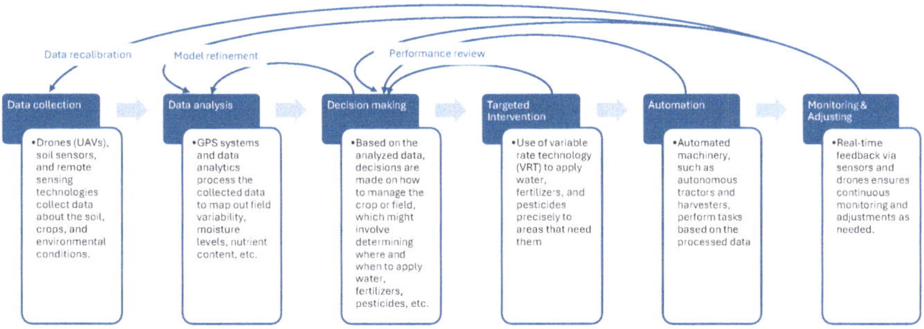

Fig. 9.1 Precision agriculture process flow

like wireless soil sensors, integrated with cloud computing and machine learning (ML) algorithms, enable precise nutrient management by providing personalized fertilizer recommendations and continuous monitoring, facilitating informed, adaptive decision-making in PA (Banu et al. 2024). Additionally, technologies such as Light Detection and Ranging (LiDAR), drones, and autonomous vehicles, enable precise crop management by offering detailed insights into crop yield forecasting, biomass mapping, and navigation (Farhan et al. 2024). Together, these approaches help improve efficiency, reduce resource waste, and promote SAPs through data-driven, tailored interventions.

Studies emphasize the role of PA in managing SOC variability through precise, site-specific interventions, which enhance CS and optimize land use (Uddin et al. 2020; Schimmelpfennig and Lowenberg-DeBoer 2021). When combined with conservation agriculture (CA) practices, PA technologies can reduce SOC losses by up to 63% and CO_2 emissions by up to 56% compared to conventional tillage, demonstrating the potential of PA to mitigate/adapt climate change and promote sustainable C-farming practices (Cillis et al. 2018). The integration of ML and AI into PA, such as the automation of all-terrain vehicles (ATVs), further enhances efficiency by optimizing planting, harvesting, and spraying, improving accuracy and minimizing resource consumption (Padhiary et al. 2024). Additionally, innovations like autonomous spraying systems utilizing LiDAR technology ensure precise input application, contributing to resource conservation and reducing environmental impacts (Danton et al. 2020).

9.2.2 *Smart Sensors*

Smart sensors are essential for optimizing C farming by enabling precise, real-time monitoring of soil health, crop growth, water usage, and GHG emissions (Adamchuk et al. 2004; Farooq et al. 2019; Yin et al. 2021; Rajak et al. 2023; Schmidinger et al. 2024; Thilakarathne et al. 2025). Key sensor types, including mechanical, electrochemical, and optical sensors, measure properties such as soil moisture content and potential temperature regime, pH, nutrient stock and fluxes, and GHG emissions, which help improve irrigation, soil health, and fertilizer use efficiency while reducing emissions (Table 9.2). Sensors provide a fast, cost-effective alternative to traditional soil analysis, enabling site-specific monitoring for PA, enhancing agronomic recommendations, and supporting CS efforts (Viscarra Rossel and Bouma 2016). These sensors help reduce input waste and the C footprint, supporting precision farming for better environmental outcomes (Rajak et al. 2023; Otieno et al. 2024).

Soil moisture sensors are essential in C farming as they help optimize irrigation, conserve water, and improve crop yield by monitoring key properties such as dielectric permittivity, matric potential, and volumetric moisture content (Zhang et al. 2024). Various types of sensors, including capacitive, electro-magnetic induction (EMI), ultrasonic, and optical, provide valuable data to tailor irrigation schedules and enhance water-use efficiency or WUE (Kodešová et al. 2011; Bobrov et al.

Table 9.2 Smart sensors in carbon farming

Category	Sensor type	Principle soil property measured	Use case in carbon farming	Limitations	Reference
Mechanical Sensors	Soil Strength Profile Sensor (SSPS)	Soil compaction at multiple depths (up to 50 cm)	Helps identify compacted zones, optimizing tillage and reducing soil disturbance, enhancing soil organic carbon (SOC) retention	Limited accuracy in highly heterogeneous soils	Sudduth et al. (2008)
	Instrumented Disc Coulter with Depth-Measuring Sensors (Rotary Potentiometer & Ultrasonic Proximity Sensor)	Soil Mechanical Resistance (Soil Strength, Cone Index, Penetration Energy)	Helps assess soil compaction and mechanical resistance, enabling precision tillage to minimize soil disturbance.	Affected by soil moisture variations and tractor sinkage, impacting accuracy	Hemmat et al. (2008)
	On-Line Soil Strength Sensor (Thin Blade + Transducer)	Soil strength variations (Draft force, Vertical force, and Moment)	Regulate tillage parameters in real-time, reducing excessive soil disturbance and preserving SOC	Accuracy influenced by soil moisture content; limited resolution in highly heterogeneous fields	Sirjacobs et al. (2002)
Electromagnetic Sensors	Printed circuit board (PCB)-based electromagnetic sensor with inductance and capacitance	SOC content (affecting electrical conductivity, dielectric permittivity, and magnetic permeability)	Estimates SOC, improving soil quality, optimizing crop yields, and enhancing water retention	Impedance measurement error ~ 5%	Afridi et al. (2023)

(continued)

Table 9.2 (contined)

Category	Sensor type	Principle soil property measured	Use case in carbon farming	Limitations	Reference
Optical/ Elemental Sensors	Vis-NIR spectroscopy (on-the-go, proximal)	SOC content	Enables real-time monitoring of SOC content, supporting carbon farming efforts, optimizing fertilization.	Requires model-building for accurate SOC prediction	Reyes and Ließ (2022)
	Portable X-ray fluorescence (PXRF) spectrometry, Visible Near Infrared Diffuse Reflectance Spectroscopy (VisNIR DRS)	Total Carbon (TC), Total Nitrogen (TN) in organic soil horizons	Rapid in-field measurement of organic soil properties, aiding in carbon sequestration and soil productivity	Primarily used for mineral soils, so performance on organic soils may need further validation	Cardelli et al. (2017)
	Non-dispersive Infrared (NDIR) CO_2 Sensor	Soil CO_2 Concentration	Monitors soil respiration, carbon sequestration, and greenhouse gas emissions tracking	Calibration dependence, restricted gas diffusion in high moisture, slower response time in automated systems	Chirinda et al. (2014)
	Fiber Optical pH Optodes	Soil pH	Monitors pH levels, influencing microbial activity, nutrient availability, and carbon cycling in precision agriculture	Limited pH range; requires calibration and investigation at varying water contents	Merl et al. (2022)

Electrochemical sensors	Resistive micro-sensor with interdigitated electrodes (IDE)	Soil moisture content	Helps optimize irrigation and improve soil health, enhancing carbon sequestration	Performance influenced by soil type, requires regular calibration for varying conditions	Kalita et al. (2016)
	Ion-selective electrodes (ISEs)	Soil pH, available potassium, nitrate-nitrogen, sodium	Enables precise nutrient management, supporting sustainable agricultural practices	Performance varies by soil property; nitrate-nitrogen and sodium accuracy is lower	Adamchuk et al. (2005)
	All-Solid-State Soil pH Sensor (Antimony Electrode & Solid Reference Electrode)	Soil pH levels, acidity/alkalinity	In-situ pH monitoring to optimize soil conditions for carbon sequestration and microbial activity	Limited pH range (3–8); may not suit extreme soil conditions	Nair et al. (2022)
	Electrochemical impedance spectroscopy (EIS), cyclic voltammetry, chromatocoulometry	Soil carbonate content	Supports carbon sequestration monitoring, improves soil fertility in carbon farming systems	Limited by in-field calibration challenges and sensor sensitivity to heterogeneous soil environments	Dhamu et al. (2023)
	Molecularly Imprinted Sensor (MIS)	Tryptophol concentration (biomarker for mycorrhizal activity)	Monitors soil mycorrhizal fungi activity, aiding in carbon sequestration and soil health	Interference from similar molecules affects accuracy; requires precise calibration	Ruiz-Gonzalez et al. (2024)
	Ion-Selective Electrode (ISE) with Electrochemical Impedance Spectroscopy (EIS)	Soil nitrate (NO_3^-) concentration	Optimizes nitrogen application, enables farmers to make informed decisions on fertilizer use, reducing environmental impact	Performance dependent on soil texture; limited to nitrate range of 5–512 ppm; requires recalibration	Eldeeb et al. (2023a)
	8-in-1 Soil Nutrients Monitoring Sensor	Temperature, moisture, salinity, EC, pH, nitrogen, phosphorus, potassium	Supports precision agriculture, optimizing resource use, improving soil health, and minimizing environmental impact.	Limited by sensor accuracy, continuous maintenance, and soil heterogeneity	Shahab et al. (2025)

(continued)

Table 9.2 (contined)

Category	Sensor type	Principle soil property measured	Use case in carbon farming	Limitations	Reference
Environmental Sensors	GHG Sensors Nanostructure-based Sensors (Pd-doped SnO2, In2O3, WO3, Au-La2O3-doped SnO2)	CO_2, CH_4, NO_2 emissions	Monitors greenhouse gas emissions, assessing carbon flux and improving mitigation strategies	Difficulty distinguishing gases, performance degradation, challenges detecting low concentrations in complex environments	Bassous et al. (2024); Rastgou et al. (2024)
	RFID-based temperature sensor	Soil temperature	Monitors temperature in solarization treatments, optimizing heat exposure for soil-borne pathogens and weeds control	Lower reliability in clay soils at high moisture, depth affects readings, and weather variability impacts effectiveness	Luvisi et al. (2016)
	FPGA-based photosynthesis smart sensor	Photosynthetic rate, chlorophyll fluorescence, CO_2 assimilation	Assesses plant carbon uptake, optimizes crop growth, and carbon sequestration efficiency.	Environmental variability, complexity in signal processing, limited crop applicability	Millan-Almaraz et al. (2013)

2019; Yin et al. 2021). These sensors help maintain soil conditions conducive to CS, supporting SAPs and reducing environmental impact. For instance, a study on wheat productivity found that irrigation scheduling based on soil moisture indicators (SMI) resulted in similar yields to the recommended irrigation water/cumulative USDA class A pan evaporation (IW/Pan-e) = 0.9 but with greater WUE (Kumar and Kahlon 2022). Additionally, ML models integrating data from EMI sensors and gamma-ray spectrometers have proven effective in assessing the variability of SOC and soil moisture content at the field scale (Rentschler et al. 2020). The combination of moisture sensor-based irrigation with organic amendments, such as farmyard manure (FYM), further enhances sustainability and efficient water management (Kumar and Kahlon 2022). Moreover, Fares et al. (2016) highlighted the influence of soil OM (SOM) content on the accuracy of soil moisture sensors, emphasizing the need for calibration adjustments to ensure precise performance across varying soil conditions.

Soil temperature, influenced by factors like climate, topography, vegetation, and soil type, and surface condition (presence or absence of crop residues) and tillage methods plays a critical role in crop planting, growth, nutrient absorption, moisture retention, and various soil physical processes (Liu et al. 2011). Temperature sensors track fluctuations that impact SOM decomposition and microbial activity (Arnold et al. 1999; Qi et al. 2016), ultimately affecting SOC composition and C stocks, particularly through the loss of cold-preferring bacteria (Li et al. 2024a). Temperature increases soil respiration across soil types, with properties influencing the response, providing key insights into climate feedback and GHG gas targets (Dacal et al. 2022). Various sensor types, such as thermocouples, resistance temperature detectors, thermistors, and semiconductor-based sensors, help measure soil temperature regime (Yin et al. 2021). Studies like those by Liu et al. (2020) highlighted the significant role of temperature in soil humification and C stability, offering valuable insights for integrating temperature and moisture sensors in C farming. Additionally, research by Futagawa et al. (2013) highlighted how multimodal sensors, integrated with temperature and EC sensors, can track nutrient concentration and temperature variations at different soil depths, providing real-time data on soil-quality and its functions. These findings demonstrate the importance of temperature sensors in enhancing soil monitoring and management, optimizing CS, and improving C farming strategies.

Soil pH is a critical factor in regulating soil C dynamics, influencing the stability, decomposition, and sequestration of both SOC and soil inorganic C (SIC) (Jin and Wang 2018). pH regulation affects microbial activity and nutrient turnover, with higher pH levels promoting OM decomposition and nitrogen (N) release (Kemmitt et al. 2006). To optimize C farming practices, real-time, high-resolution soil pH monitoring is crucial, and recent advancements in soil pH sensors have made this process more efficient. Mobile pH sensors and low-cost resistivity sensors enable precise, real-time monitoring of soil pH and moisture content, supporting better soil management and CS (Staggenborg et al. 2007; Tan et al. 2019). Electrochemical sensors, such as those using titanium nitride (TiN) and composite coatings, offer high sensitivity and reliability for continuous pH measurement, helping optimize nutrient availability and crop productivity (Eldeeb et al. 2023b; Mamilla et al. 2025). Furthermore, biodegradable wireless sensors made from degradable materials provide an eco-friendly, cost-effective solution for wide-area pH monitoring, contributing to sustainable agriculture and improved soil health management (Sakabe et al. 2024). These innovations in soil pH sensor technology are essential for enhancing C farming, as it allows for precise adjustments to soil management practices.

Soil nutrient sensors play a vital role in C farming by providing real-time, precise information about soil nutrient content, which is crucial for optimizing soil health and enhancing CS. Traditional methods of soil nutrient analysis are labor-intensive and time-consuming, making them impractical for large-scale soil testing (Liu et al. 2024). In contrast, modern soil nutrient sensors, including optical and electrochemical sensors, offer more efficient, cost-effective solutions. For instance, optical sensors that detect N, phosphorus (P), and potassium (K) through light absorption have shown promising results in providing accurate nutrient measurements for soil management (Yokota et al. 2007; Masrie et al. 2018). Portable NPK sensors, such as those developed for irrigated and rainfed paddy soils in Java, Indonesia, further enhance PA by offering real-time nutrient data and enabling better fertilizer optimization (Sulaeman et al. 2024). Additionally, novel approaches combining electronic sensing with ML have demonstrated the potential to predict soil nutrient content accurately, with models achieving high performance (Liu et al. 2024). Capillary electrophoresis-based sensors also offer rapid, efficient detection of essential nutrients, providing results comparable to laboratory tests (O'Flaherty et al. 2000). A study in Embu County, Kenya, observed that combining a portable X-ray fluorescence (PXRF) sensor with a soil reflectance sensor provided reliable, in situ soil assessments, demonstrating their potential to replace conventional laboratory methods (Piikki et al. 2016). Integrating these advanced soil sensors into PA enhances soil management practices, leading to improved CS and sustainable farming.

GHG sensors are critical in C farming, offering real-time monitoring of GHG like CO_2, methane (CH_4), and nitrous oxide (N_2O) in soil, which helps farmers assess the environmental impact of their agricultural practices. These sensors utilize advanced technologies, including nanomaterials, micromechanical units, infrared, electrochemical, optical, and calorimetric methods, to enhance detection sensitivity, selectivity, and stability (Saxena and Shukla 2023; Bassous et al. 2024; Rath et al. 2024). They operate by detecting changes in physical properties like capacitance, EC, or temperature, which occur in response to specific gases (Saxena and Shukla 2023). For example, Lockhart et al. (2023) used non-dispersive infrared (NDIR) CO_2 sensors to measure soil CO_2 concentrations in no-till (NT) and organic farming systems. Their findings revealed significantly higher CO_2 efflux rates in organic farming, suggesting that organic practices may enhance soil CS compared to conventional methods. Similarly, Heger et al. (2020) introduced the K33SOIL, a low-cost soil CO_2 probe with high accuracy and minimal deviation in laboratory and field experiments, proving valuable for long-term soil and plant studies. Lazik et al. (2019) developed a line sensor for CO_2 measurement that accounted for environmental factors like soil temperature and air pressure without requiring separate calibration, ensuring precise measurements in heterogeneous soils. Additionally, Bandara et al. (2021) designed a cost-effective NDIR device for N_2O monitoring using a submerged silicone diffusion cell, achieving high accuracy ($R^2 > 0.9995$) and a wide measuring range (1–2000 ppm). Furthermore, semiconductor sensors are increasingly being used for real-time ammonia monitoring in agriculture due to their compact size, low cost, and quick response (Wang et al. 2024). These sensors can help optimize fertilization and irrigation practices, contributing to more sustainable farming.

9.2.3 Remote Sensing

RS refers to the technology that captures information about objects or areas from a distance, typically via satellites, aircraft, or UAVs, without direct contact with the surface (Pouladi et al. 2023). This technology has revolutionized the way data is collected and analyzed in various fields, particularly in environmental monitoring and land management (Congalton 2010). By offering detailed, non-destructive, and large-scale data on soil properties and environmental conditions, this technology plays a pivotal role in enhancing C capture, improving farming efficiency, and promoting sustainability (Abdulraheem et al. 2023). Through the integration of advanced technologies such as digital soil mapping (DSM), ML, and image fusion, RS has proven essential in enhancing the accuracy and stability of SOC concentration and stock predictions, optimizing CS and supporting SAPs (Lamichhane et al. 2019; Li et al. 2024b).

9.2.3.1 Enhancing Carbon Capture

RS methods have been used extensively to accurately map SOC content, as highlighted by Chen et al. (2005), who demonstrated the feasibility of using RS images to predict SOC content in agricultural fields. By establishing strong relationships between SOC and image reflectance, they achieved high regression values ($R^2 = 0.89$–0.95), ensuring that RS can enhance C capture efforts through precise monitoring. Further, the study by Jonard et al. (2018) emphasized the importance of organic soils in global warming due to their substantial C storage capacity. By using space-borne L-band brightness temperature observations, RS improves the understanding of organic soils' hydrological properties, contributing to better C storage and soil moisture management (Jonard et al. 2018). Additionally, RS supports the mapping of SOC content in tidal wetlands using ML models, as demonstrated by Sharma et al. (2022). This innovative approach to predicting both surface and subsurface SOC enhances the ability to monitor and manage these critical ecosystems, where traditional methods are often challenging to apply. These advancements ensure that SOC content in diverse ecosystems is accurately assessed, thus enhancing C capture initiatives.

9.2.3.2 Improving Farming Efficiency

RS also plays an integral role in improving farming efficiency by enabling large-scale, accurate predictions of soil properties such as SOC and moisture content, and texture (Abdulraheem et al. 2023). Through spectral reflectance analysis, thermal infrared imaging, and radar imaging, RS provides non-invasive and cost-effective assessments that enhance agricultural decision-making (Karmakar et al. 2024). For instance, Guo et al. (2021) highlighted how Sentinel 2 data improved SOC predictions in low-relief agricultural areas, overcoming traditional mapping challenges and offering insights that enable more precise management of soil health. This ability to predict soil characteristics efficiently leads to optimized land use, helping farmers implement practices that conserve resources and improve crop yields. Moreover, the integration of RS data with topographic and soil variables has proven effective in predicting SOC distribution across diverse landscapes, as seen in the study by Sodango et al. (2021). By incorporating soil moisture, soil surface temperature, and vegetation indices, RS supports C farming practices by improving soil health and guiding agricultural interventions. These advancements contribute to more efficient farming operations by ensuring that soil management practices are tailored to the specific needs of the land.

9.2.3.3 Promoting Sustainability

Sustainability in agriculture is further enhanced by using RS to monitor and predict changes in soil properties over time, supporting the long-term health of ecosystems. RS enables better soil management by providing real-time data on soil degradation, moisture levels, and C stocks, which are essential for mitigating climate change and improving soil fertility. In the study by Xu et al. (2023), the fusion of multispectral and panchromatic data from Sentinel 2 and Landsat 8 improved SOC prediction accuracy, thus facilitating better soil management and promoting sustainability. The study's findings demonstrate how image fusion enhances the performance of spatial prediction models, which are crucial for ensuring the responsible management of SOC content and stock.

RS also supports more accurate and higher-resolution mapping of SOC content, which is critical for sustainable land use planning. Aitkenhead and Coull (2020) utilized digital soil mapping, legacy survey data, and RS to estimate SOC stocks across Scotland, generating high-resolution maps that aid in soil management and C farming efforts. These maps provide policymakers and land managers with valuable information for making informed decisions on landscape management, supporting the sustainable development of agricultural and peatland areas.

9.2.4 Artificial Intelligence

AI technology is increasingly playing a pivotal role in advancing C farming and climate change mitigation efforts. One prominent application is in the prediction of SOC content, which directly impacts CS in agricultural lands. Bernardini et al. (2024) highlighted how ML algorithms, especially Random Forest and Support Vector Machines, outperformed traditional process-based models in prediction of SOC content, particularly when large datasets were available. This capability allows for more efficient soil management, aiding farmers in identifying areas of their land where SOC contents are suboptimal and taking corrective actions. Moreover, Tziolas et al. (2024) demonstrated how deep learning models, such as Convolutional Neural Networks (CNN), paired with RS data from Copernicus Sentinel-2 satellites, can predict SOC levels at a large scale. By utilizing environmental data in conjunction with satellite imagery, their hybrid approach significantly improved the accuracy of SOC predictions, thereby enhancing soil management strategies. Mallik et al. (2022) also contributed to the integration of AI with RS, showing how ML can improve the accuracy of SOC predictions by using a combination of Sentinel-2 Band 6 and terrain data. The integration of hybrid models such as artificial neural network- ordinary kriging (ANN-OK) and advanced geostatistical-based empirical Bayesian Kriging regression (EBKR) further highlights how AI methods can surpass traditional approaches, making them a valuable tool for C farming. In plant breeding, Harfouche et al. (2019) demonstrated how AI-driven techniques are transforming breeding programs by enabling the selection of climate-resilient crops, crucial for mitigating climate-related risks and enhancing CS capabilities.

Another significant area where AI is making strides is in the optimization of nutrient and water management. Ennaji et al. (2023) reviewed how AI-based systems are being applied to precision nutrient management in agriculture. By using large datasets, AI models can predict the required amounts of fertilizers, reducing both excess application and associated environmental harm, such as nutrient runoff, which contributes to CO_2 emissions. These AI-based systems enable more sustainable farming by ensuring that fertilizers are applied only when necessary and in the right amounts, improving both yields and environmental outcomes. In irrigation management, Preite and Vignali (2024) introduced an AI-based three-layer system that predicts irrigation needs with high accuracy, achieving nearly 99% accuracy in reducing false predictions. By incorporating multi-layer perceptron, support vector machines, and k-neighbors classifiers, the system helps optimize irrigation, reducing water and energy usage by 27% and 57%, respectively. González Perea et al. (2018) further explored AI's role in irrigation optimization by combining neural networks, fuzzy logic, and genetic algorithms to predict daily irrigation depths, demonstrating the model's high accuracy in forecasting irrigation needs for crops like rice, maize, and tomatoes. Similarly, Goap et al. (2018) proposed an IoT-based smart irrigation system that uses sensors to monitor soil moisture, temperature, and other variables to predict irrigation needs. The system uses cloud-based AI algorithms to process real-time data and autonomously control irrigation, optimizing water use while minimizing resource waste. These AI advancements in nutrient and irrigation management not only increase efficiency but also significantly reduce agriculture's environmental impact, contributing to the global effort to reduce GHG emissions and ensure sustainable food production.

The integration of AI and nanoparticle-assisted technologies has proven highly effective in improving pesticide management within C farming. By leveraging AI for rapid data analysis and nanoparticle sensors for precise pesticide detection, this approach offers a more efficient means of managing pesticide use, ultimately minimizing its adverse effects on soil and water quality (Banerjee et al. 2024). Additionally, Trivedi et al. (2024) used AI to assess agroforestry land use systems by classifying various crops and trees using U-Net CNN architecture on high-resolution images. This study highlighted the potential of AI to map agroforestry practices, which are crucial for enhancing biodiversity and CS. The integration of AI-driven models, like the ImpelERO model developed by Afshar et al. (2016), showcases the potential of neural networks and decision trees in predicting soil erosion, a key factor in soil health. By assessing soil vulnerability, soil loss, and erosion risk, this AI model helps identify areas at risk of degradation, allowing farmers to implement conservation practices that protect the soil and improve its C storage capacity. The integration of AI with blockchain technology is emerging as a powerful solution in the realm of C farming, ensuring greater transparency and security in C credit transactions. Kim and Huh (2020) demonstrated how AI-driven sensors and satellite imagery can monitor CS in agricultural lands while blockchain technology secures C credit transactions. This combination enables accurate verification of CS and ensures that C credits are traded transparently and tamper-proof, making it easier for farmers to participate in C markets.

Overall, the integration of AI across various domains—SOC prediction, water and nutrient management, precise fertilizer, agroforestry and the transparent verification of C credits—demonstrates its immense potential in advancing C farming. Figure 9.2 illustrates the 10-step process for implementing AI-driven C farming, outlining key stages such as identifying high CS sites, forecasting climate risks, and optimizing soil health to ensure effective and sustainable practices. Through these technologies, farmers and researchers can optimize practices, reduce C footprints, and contribute to climate change mitigation.

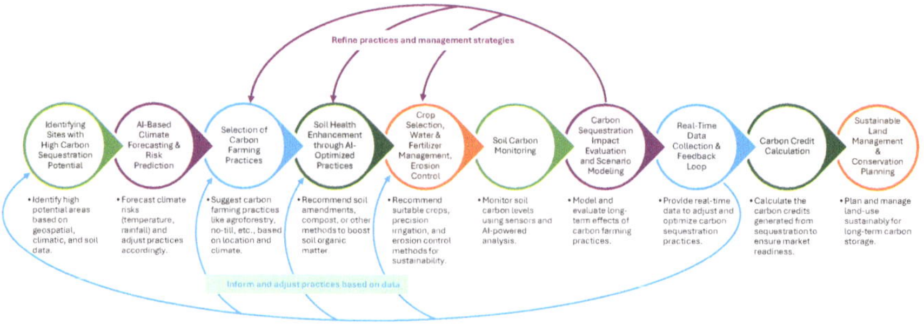

Fig. 9.2 AI-driven carbon farming process

9.2.5 *Internet of Things*

The IoT in agriculture integrates crops, livestock, environmental factors, and farming equipment with digital technology through sensors and communication networks (Xu et al. 2022). This interconnected system enables real-time data collection, exchange, and analysis to optimize farming practices, improve efficiency, and enhance decision-making (Shahab et al. 2025). While still in the experimental stage (FuBing 2012), advancements in sensors, communication protocols, cloud platforms, and automation have accelerated IoT adoption in agriculture (Chamara et al. 2022).

9.2.5.1 IoT Architecture for Carbon Farming

IoT systems in C farming operate through a layered architecture, ensuring efficient data collection, transmission, and analysis (Fig. 9.3). Environmental sensors, including soil moisture, temperature, gas sensors, and weather stations, collect real-time data, which is transmitted via LoRaWAN, cellular networks, or Wi-Fi to cloud-based platforms (Farooq et al. 2019). ML models then process these data to generate actionable insights on farming practices like crop rotation, NT farming, and irrigation management, aiding farmers in optimizing CS (Sarkar and Somenahally 2023). However, challenges such as limited real-time nutrient sensing, system heterogeneity, and power constraints remain. Edge computing, drone-based monitoring, and automated analysis offer promising solutions, enhancing IoT's role in soil health improvement and C storage (Chamara et al. 2022). An edge-fog-cloud architecture can reduce energy consumption by 36%, CO_2 emissions by 43%, and network congestion by 86% compared to traditional cloud-based models, improving real-time weather, soil moisture content, and irrigation management (Alharbi and Aldossary 2021).

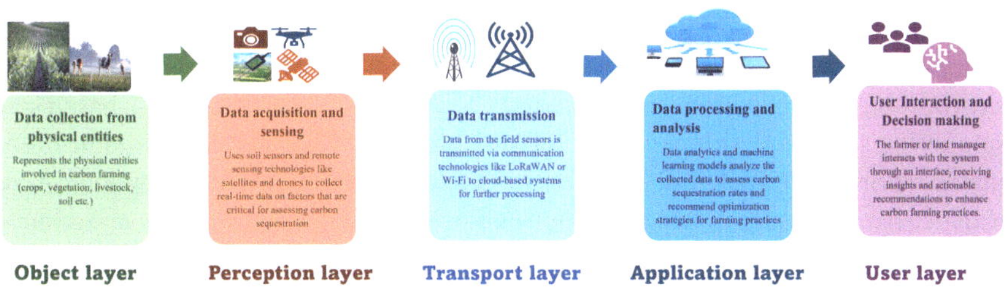

Fig. 9.3 Five-layer IoT architecture for carbon farming (Tzounis et al. 2017; Xu et al. 2022)

9.2.5.2 IoT for Soil Organic Carbon Monitoring

IoT enables real-time monitoring of SOC content by tracking changes in soil properties across different growing periods. For example, IoT sensors deployed in a wheat demonstration base in Shandong, China, monitored SOC content variations, allowing researchers to simulate SOC content changes over time and assess the impact of straw return on soil health (Wu et al. 2017). Going a step further, the Cognitive Soil Digital Twin integrates IoT sensor data, RS, and field measurements to track the physical, chemical, and biological properties of soil (Tsakiridis et al. 2023). This framework enables real-time SOC content monitoring, scenario analysis, and visualization of soil-C interactions, empowering farmers, researchers, and policymakers to identify and adopt sustainable land management (SLM) strategies.

9.2.5.3 IoT for Crop Residue Cover Management

Crop residue cover (CRC) is critical for CS, enhancing SOM content and reducing risks of accelerated soil erosion. Traditional residue measurement methods are time-intensive and often impractical for large-scale farming. IoT-based RS and ground-based sensors enhance CRC monitoring by accurately mapping surface residue cover and minimizing errors. ML and Pedotransfer Functions (PTFs) have been employed to predict soil properties and residue cover, filling data gaps for soil-C modeling (Zheng et al. 2012). Additionally, Convolutional Neural Networks (CNNs) applied to Landsat images effectively detected CRC patterns and their impact on soil fertility and CS (Kazemi Garajeh et al. 2023). By using the Normalized Difference Vegetation Index (NDVI) to assess vegetation density over time, the study confirmed that CRC positively impacts both crop yields and soil quality, further reinforcing the importance of IoT in sustainable C farming.

9.2.5.4 IoT for Real-Time Soil Health Monitoring

IoT-based real-time soil health assessment enhances precision farming and CS. A Smart Agro Device, integrating IoT with ML, measures key soil parameters such as NPK, pH, moisture and SOC content, optimizing fertilization, cost efficiency, and C storage potential (Babu et al. 2024). This IoT-enabled system enhances precision fertilization, cost reduction, and improved CS potential. Similarly, an IoT-based soil monitoring system, developed using STM32 boards and Bluetooth communication, enables farmers to track soil pH, moisture, and temperature via smartphones (Na et al. 2016). Ongoing enhancements, such as 6LoWPAN networking and cloud-based data storage, will further support C farming by optimizing soil conditions for long-term C storage.

9.2.5.5 IoT in Precision Agriculture and Fertilizer Management

IoT-powered PA contributes significantly to CS. A cost-effective IoT-based soil monitoring system, using Bluetooth communication to transmit real-time soil data, enables farmers to optimize irrigation and pesticide application, reducing resource wastage and enhancing SOC retention (Kim et al. 2020). Additionally, IoT-based fertilizer management systems analyze soil nutrient reserves and dynamics, calculate optimal fertilizer quantities, and automate application, preventing excessive fertilizer use and improving nutrient-use efficiency (G et al. 2020).

9.2.5.6 AI-IoT Integration for Carbon Farming

Advancements in AI and IoT integration further strengthen CS efforts. The Multi-Context Fusion Network (MCFN) enhances crop disease detection by integrating IoT and deep learning, achieving 97.5% accuracy in classifying 77 crop diseases (Zhao et al. 2020). By improving crop health and biomass production, IoT-driven disease detection indirectly supports CS by reducing crop losses and optimizing land use. Furthermore, IoT-GIS integration enables real-time soil health monitoring, utilizing sensors and cloud databases to track moisture, pH, temperature, and nutrients (Sivakumar et al. 2023). GIS mapping provides a spatial overview of SOC trends, assisting farmers, researchers, and policymakers in implementing climate-smart farming strategies.

9.2.6 Blockchain Technology

Blockchain is a decentralized, distributed database technology that is rapidly gaining traction in agriculture, offering enhanced transparency, security, and efficiency in managing C credits and climate-smart practices (Parra-López et al. 2024). Unlike traditional centralized systems, where a single entity controls the database, blockchain operates across a network of computers, ensuring that no one party can alter or erase transaction records (Sarmah 2018). Once a transaction is verified, it is added to an immutable chain, forming a transparent and secure history of actions. This feature is especially valuable in C farming, where tracking the generation, verification, and trading of C credits is critical for ensuring accountability and credibility (Singh and Sharma 2023). Blockchain provides farmers with a means to securely generate C credits by verifying and documenting their sustainable practices, such as organic farming or conservation tillage, and these credits can be traded in a transparent marketplace. A notable example of blockchain's application in C farming is the Blockchain-NFTs (non-fungible tokens) Trading System for Agricultural C Credits implemented in rural Uttar Pradesh, India (Khanna and Maheshwari 2024). This system incentivizes farmers to adopt eco-friendly farming methods by allowing them to generate C credits, which are then verified and tokenized as NFTs on a blockchain platform. This approach not only provides farmers with an additional income stream but also opens access to global markets for C credit trading, facilitating participation in the global C offset market.

Blockchain also integrates well with other technologies such as IoT, as demonstrated by the system proposed by Dey and Shekhawat (2021). They integrated blockchain with IoT networks to create an efficient, secure platform for data collection, validation, and management in e-agriculture. This combination addresses common challenges like data duplication, rights management, and enhances decision-making within agricultural value chains. Furthermore, platforms like "FarMarketplace" (Leduc et al. 2021) leverage blockchain to enable secure, direct trade between farmers and third-party stakeholders, ensuring that climate-friendly practices are verified, and C credits are securely exchanged via smart contracts. These smart contracts automate transactions, reducing costs and complexities involved in the C credit market. Key stakeholders in blockchain-based C farming can be grouped into three categories based on their roles and involvement (van Wassenaer et al. 2021a, b), as illustrated in Fig. 9.4. Each group plays a crucial role in generating, verifying, and trading C credits within a secure, transparent, and decentralized ecosystem.

9.2.6.1 Key Benefits in Carbon Farming

Ensuring Transparency and Trust One of the main challenges in the C credit market is ensuring transparency and trust between all stakeholders, such as farmers, C offset buyers, and verification bodies (Delacote et al. 2024). Blockchain provides a transparent and immutable record of every transaction, from the generation of C credits to their eventual sale (Patel et al. 2020). Each C credit transaction is recorded on a blockchain, which can be publicly verified by anyone, ensuring that the credits are legitimate, correctly measured, and not double-counted. For example, once a farmer implements a C farming practice (like crop residue retention or NT farming),

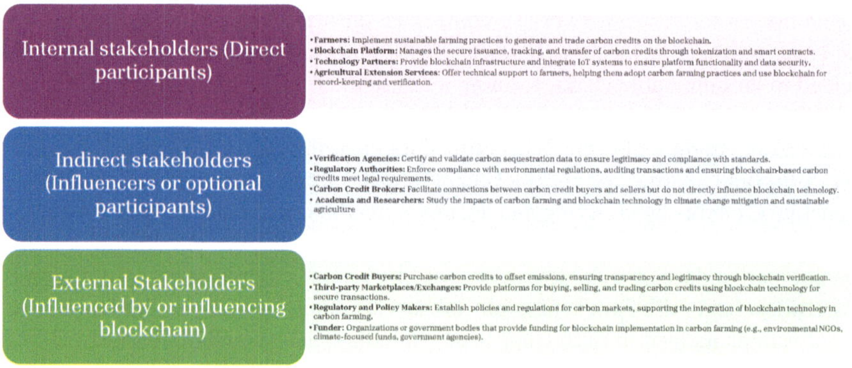

Fig. 9.4 Stakeholders in Blockchain Technology for carbon farming

the resulting C credits can be registered on a blockchain, and each credit is given a unique digital identifier (Patel et al. 2020). This creates a traceable history of the C credit's origin and its journey in the market, preventing fraud or the creation of fake credits.

Facilitating Secure and Efficient Transactions Blockchain enables secure and instantaneous transactions between C offset buyers and sellers, bypassing intermediaries and reducing transaction costs (Sadawi et al. 2021). Traditional C credit markets often involve several middlemen, such as brokers and certification bodies, which can slow down transactions and introduce inefficiencies (Andrew 2021). With blockchain, farmers can directly sell their C credits to buyers, streamlining the process and ensuring that funds reach farmers more quickly and efficiently (Khanna and Maheshwari 2024). Additionally, smart contracts, a key feature of blockchain, can be used to automate these transactions. These self-executing contracts ensure that when certain predefined conditions are met (e.g., the verification of CS by a third party), the transaction is automatically executed (Sadawi et al. 2021). This reduces the risk of human error and the possibility of disputes, creating a more seamless and trustworthy transaction process.

Improved Verification and Monitoring Verification is essential in C farming to ensure genuine CS, with third-party organizations often handling the process. Blockchain technology can improve verification by providing a transparent and easily accessible record of all the data related to a C farming project, such as soil health metrics, crop yields, and CS rates (van Wassenaer et al. 2021a). By integrating AI and IoT with blockchain, farmers can securely store and share data with auditors, making the verification process faster, more affordable, and more reliable (Chen et al. 2023; Pham et al. 2024; Hung 2024). As demonstrated by Rehman et al. (2023), blockchain, combined with IoT, enhances food security by improving real-time monitoring, traceability, and sustainability. Similarly, blockchain in C farming can also ensure the integrity of C credits, streamline resource management, and optimize trading through decentralized markets and smart contracts. Just as blockchain enhances transparency and traceability in agricultural supply chains by recording planting dates, fertilizer usage, and harvest times, it can also verify sustainable practices in C farming (Chen et al. 2023).

Enhancing Market Accessibility Blockchain can democratize access to the C credit market for small-scale farmers and landowners (Hung 2024). Traditionally, entering the C credit market has been difficult for smaller producers due to high entry costs and complex processes. Blockchain allows for the creation of tokenized C credits, which can be divided into smaller, tradable units, making it easier for small-scale farmers to participate (Burgoyne and Olexiuk 2025). This opens up the C farming market to a wider range of agricultural producers and landowners who may not have had the resources or infrastructure to engage with traditional C credit systems (Hung 2024).

9.2.6.2 Challenges

Despite its potential, blockchain technology may face several challenges in C farming, including concerns about data privacy and security, as sensitive agricultural data may need to be kept confidential (Kaur et al. 2022). The high energy consumption of some blockchain networks contradicts C reduction goals (Sedlmeir et al. 2020; Babaei et al. 2025), while integration with existing systems and ensuring interoperability can be complex (Gupta et al. 2022). Adoption barriers exist due to the lack of technical expertise among farmers, and regulatory and legal uncertainty surrounding C credits could hinder progress (Hung 2024). Scalability issues and the high cost of implementation may deter smaller stakeholders (Xiong et al. 2020), while the lack of standardized C credit measurement protocols could undermine the reliability of blockchain-based solutions (Dupla et al. 2024). These challenges require coordinated efforts to address technological, regulatory, and practical barriers to blockchain adoption in C farming (Tripoli and Schmidhuber 2018).

9.3 Technology Challenges in Carbon Farming

While significant progress has been made in developing advanced technologies, several challenges persist, limiting the widespread adoption and effectiveness of C farming practices. These challenges can range from sensor limitations to data management issues and scalability concerns, all of which hinder the full potential of C farming technologies.

9.3.1 Sensor Accuracy and Calibration

One of the primary technological challenges in C farming is ensuring the accuracy of sensors used to monitor soil conditions such as moisture content, pH, and SOC content. Sensors can struggle with accuracy, particularly in heterogeneous soils where variations in soil properties (e.g., texture, moisture, and SOC) affect sensor readings. For example, a comparison between the Stenon FarmLab handheld sensor and the Vis-NIR on-line multi-sensor platform for SOC measurements showed a low correlation ($R^2 = 0.29$), indicating significant discrepancies that hinder reliable monitoring of CS across different soil types (Steiger et al. 2025). Furthermore, frequent calibration is often necessary to maintain sensor accuracy, adding operational complexity and cost (Sudduth et al. 2008; Kalita et al. 2016).

9.3.2 Data Integration and Management

C farming technologies generate large volumes of data from a range of sensors, environmental monitoring systems, and platforms. Integrating and managing these data in a coherent and actionable way presents significant challenges. The lack of standardization in sensor outputs, coupled with the complexity of combining data from different sources (e.g., soil, weather, crop, and environmental data), makes it difficult for farmers to make real-time decisions (Ishaq et al. 2013). Moreover, analyzing and interpreting these data often require advanced computational methods, such as ML and AI, which may not be accessible or practical for all farmers (Rajak et al. 2023; Ahmed and Shakoor 2025). The sheer volume of data can also lead to data overload, complicating efforts to derive useful insights for effective C farming management (Wan et al. 2007).

9.3.3 High Initial Costs and Maintenance

The upfront investment required for advanced C farming technologies, including sensors, drones, autonomous vehicles, and automated monitoring systems, remains a significant barrier, especially for small-scale or resource-limited farmers (Choruma et al. 2024; Aarif et al. 2025). These technologies can be prohibitively expensive, and the ongoing maintenance, calibration, and replacement of sensors further adds to the operational costs (Sharp 2024; Guebsi et al. 2024). For example, drones and sensors may require periodic servicing to ensure that these tools provide accurate data over time. Additionally, the energy consumption associated with certain technologies, such as artificial lighting in plant factories, raises concerns about the C footprint of these systems, which conflicts with the overall goal of C neutrality (Villa-Henriksen et al. 2020; Santiteerakul et al. 2020). These financial and operational burdens limit the adoption of C farming technologies, particularly in developing regions.

9.3.4 Connectivity and Infrastructure

Poor connectivity and lack of adequate infrastructure pose significant challenges for the deployment of C farming technologies, particularly in remote areas where such technologies are often most needed (Choruma et al. 2024). Many C farming tools require reliable internet access for real-time monitoring (Padhiary and Mishra 2020), data transmission, and cloud-based storage, but in rural regions with limited connectivity, these technologies may be ineffective or difficult to use (Otieno et al. 2024). Without robust infrastructure to support these systems, farmers may struggle to implement or fully benefit from C farming technologies.

9.3.5 Scalability

While C farming technologies have been successful in small-scale or controlled environments, upscaling these technologies to larger agricultural operations or across diverse geographic regions remains challenging (Padhiary et al. 2024). Variability in soil properties, local climate conditions, and farm management practices requires customized solutions that may be difficult to implement on a broader scale (Snapp 2022). Technologies that work well on one type of farm may not be suitable for another, complicating efforts to monitor and manage CS at a larger, regional level (Viscarra Rossel and Bouma 2016). Additionally, there are limitations in applying RS tools, such as drones and LiDAR, effectively across different types of crops and farming systems (Farhan et al. 2024).

9.3.6 Data Privacy and Security

As C farming technologies rely more on interconnected sensor networks and farm management systems, data privacy and security become growing concerns (Rajak et al. 2023). Farmers may be hesitant to share sensitive agricultural data, such as crop yields, soil health, and management practices, due to fears of misuse or unauthorized access. This reluctance could slow the adoption of C farming technologies, particularly in regions where data privacy regulations are not well-defined. Additionally, cybersecurity threats could compromise sensitive data, leading to potential economic losses or privacy violations, further discouraging farmers from adopting these technologies (Sontowski et al. 2020; Sharp 2024; Farhan et al. 2024).

9.3.7 Regulatory and Policy Barriers

The lack of clear, standardized policies on C measurement, verification, and the integration of C farming technologies into C credit markets is a significant barrier to the widespread adoption of these technologies (Cariappa et al. 2024). Policymakers have yet to fully adapt regulatory frameworks to accommodate emerging technologies in C farming, such as soil sensors and RS tools (Guebsi et al. 2024). Without clear and supportive policies, the economic incentives for farmers to adopt C farming practices may be insufficient. Furthermore, the uncertainty around C credits and the lack of robust verification mechanisms can undermine confidence in the effectiveness of these technologies (Schmidinger et al. 2024).

9.3.8 Socio-Economic Barriers

Social networks and farmers' attitudes toward technology play a crucial role in the adoption of C farming technologies. Farmers with smaller operations or limited financial resources are less likely to invest in precision farming technologies compared to those with larger, more profitable farms (Villa-Henriksen et al. 2020). Age and access to social networks also play significant roles, as older farmers and those without strong social connections may be less inclined to embrace new technologies (Blasch et al. 2022). Furthermore, digital literacy is a crucial factor, as farmers lacking the necessary skills or training may struggle to effectively use advanced technologies, limiting their adoption (Choruma et al. 2024). These socio-economic challenges highlight the need for targeted support, such as financial incentives, educational programs, and enhanced extension services, to encourage broader uptake of C farming technologies, especially in resource-constrained settings like Zambia (Abdulai et al. 2021; Rajak et al. 2023).

9.4 Way Forward

To advance the role of technology in C farming, future priorities must focus on developing robust policy frameworks and ensuring equitable access to advanced technologies. Governments need to establish clear and standardized regulations for C credits, data ownership, and the verification of CS practices to ensure transparency and credibility within C markets (Bispo et al. 2017; Hansen et al. 2022). A global consensus on C data exchange and market access is essential to streamline efforts across regions and provide farmers with a consistent platform for participation (Kharel et al. 2022). Additionally, promoting the development and accessibility of innovative technologies such as RS, drones, AI, and soil sensors will be crucial for improving the efficiency and accuracy of C farming practices. Ensuring that these technologies are both accessible and affordable, especially for smallholder farmers who often face resource and infrastructure limitations, is essential (John et al. 2023). To facilitate widespread and inclusive adoption, tailored solutions that address the unique needs and challenges of diverse farming systems must be developed, ensuring that all farmers, regardless of scale or location, can effectively integrate these technologies into their operations.

Beyond technological development, financial support, education, and collaborative partnerships are key to scaling up C farming practices. Governments and financial institutions must provide incentives such as subsidies, grants, and tax benefits to make the adoption of C farming economically viable for farmers (Khan et al. 2025). These financial incentives can be coupled with payment for ecosystem services or market-based incentives, offering farmers a direct economic return for their C sequestration efforts, which will motivate them to adopt SAPs and embrace technological advancements (Antle and Stoorvogel 2009). Education and training programs will be crucial for empowering farmers with the digital literacy and technological skills required to manage and implement C farming technologies effectively (Yuan and Sun 2024). These programs should focus on enhancing farmers' understanding of the technologies and practices involved in CS, thereby building their capacity to integrate these innovations into their day-to-day operations. Moreover, fostering public-private partnerships will be essential for driving innovation and providing farmers with the support they need to scale up C farming. Collaboration among governments, technology providers, agricultural businesses, and environmental organizations can leverage resources and expertise to create scalable, region-specific solutions that benefit both the environment and farming communities (Hansen et al. 2022; Ahmed et al. 2024). Investment in ongoing research and development will also be crucial for optimizing existing technologies and creating new, more efficient solutions for C farming (Ahmed et al. 2024). Ultimately, by combining regulatory clarity, technological innovation, financial incentives, education, and collaborative partnerships, the full potential of technology in C farming can be unlocked to combat climate change and promote global environmental sustainability.

9.5 Conclusion

The integration of advanced technologies like PA, AI, RS, blockchain, smart sensors, and the IoT plays a crucial role in enhancing the effectiveness of CS in farming systems. Hypothesis 1, which suggests that these technologies significantly improve CS, holds true as they enable more precise and efficient management of soil health and crop growth. PA, AI, and RS facilitate real-time monitoring and data analysis, helping farmers optimize inputs, manage C stocks, and adopt sustainable practices that promote higher rates of C capture. RS, for example, provides valuable insights into soil moisture regime, SOC content, and environmental conditions, which can guide better soil management and improve CS potential. Additionally, smart sensors and AI algorithms offer advanced capabilities in measuring and modeling SOC dynamics, thus contributing to more accurate and scalable CS strategies that align with sustainability goals. The IoT, by connecting sensors and devices across the farm, allows for seamless data collection and analysis, enabling farmers to make informed decisions in real-time. However, challenges such as sensor accuracy, data integration, connectivity limitations, and regulatory uncertainties must be addressed to fully leverage these technological advancements.

Hypothesis 2 emphasizes the economic incentives driven by technology, which is a key factor in accelerating the adoption of C farming techniques. The use of blockchain for tracking C credits, IoT-based monitoring systems, and the integration of soil sensors for monitoring soil health can help farmers gain tangible benefits, such as verifiable C credits and access to C markets. Blockchain technology ensures transparency and accountability, making it easier to track and verify C storage, thus providing a reliable basis for farmers to participate in C credit schemes.

IoT enhances this by providing continuous, real-time data from connected devices that enable more efficient resource use and improved farming practices. However, challenges like data privacy, high initial costs, sensor calibration, and limited connectivity still pose barriers to widespread adoption. Addressing these challenges requires focused efforts to enhance infrastructure, policy support, and financial incentives, especially for small landholder or resource-limited farmers. Together, these technologies and incentives have the potential to foster a more sustainable and economically viable agricultural system, benefiting both the environment and farmers in the long run.

References

Aarif KOM, Alam A, Hotak Y (2025) Smart sensor technologies shaping the future of precision agriculture: recent advances and future outlooks. J Sens 2025. https://doi.org/10.1155/js/2460098

Abbasi M, Yaghmaee MH, Rahnama F (2019) Internet of things in agriculture: a survey. In: 2019 3rd International Conference on Internet of Things and Applications (IoT). IEEE, pp 1–12

Abdulai AN, Abdul-Rahaman A, Issahaku G (2021) Adoption and diffusion of conservation agriculture technology in Zambia: the role of social and institutional networks. Environ Econ Policy Stud 23:761–780. https://doi.org/10.1007/s10018-020-00298-z

Abdulraheem MI, Zhang W, Li S et al (2023) Advancement of remote sensing for soil measurements and applications: a comprehensive review. Sustainability 15:15444. https://doi.org/10.3390/su152115444

Adamchuk VI, Rossel RAV (2010) Development of on-the-go proximal soil sensor systems. In: Proximal soil sensing. Springer, Dordrecht, pp 15–28

Adamchuk VI, Hummel JW, Morgan MT, Upadhyaya SK (2004) On-the-go soil sensors for precision agriculture. Comput Electron Agric 44:71–91. https://doi.org/10.1016/j.compag.2004.03.002

Adamchuk VI, Lund ED, Sethuramasamyraja B et al (2005) Direct measurement of soil chemical properties on-the-go using ion-selective electrodes. Comput Electron Agric 48:272–294. https://doi.org/10.1016/j.compag.2005.05.001

Afridi WAK, Kumar N, Jayasundera K et al (2023) Design an electromagnetic sensor to measure the organic carbon in soil and its validation with standard Walkley–Black method. IEEE Sens Lett 7:1–4. https://doi.org/10.1109/LSENS.2023.3328591

Afshar E, Yarnia M, Bagherzadeh A et al (2016) The effects of cropping systems on soil erosion risks and crop productivity using ImpelERO model and GIS in northeast of Iran. Model Earth Syst Environ 2:164. https://doi.org/10.1007/s40808-016-0142-6

Ahmed N, Shakoor N (2025) Advancing agriculture through IoT, big data, and AI: a review of smart technologies enabling sustainability. Smart Agricultural Technology 100848:100848. https://doi.org/10.1016/j.atech.2025.100848

Ahmed B, Shabbir H, Naqvi SR, Peng L (2024) Smart agriculture: current state, opportunities, and challenges. IEEE Access 12:144456–144478. https://doi.org/10.1109/ACCESS.2024.3471647

Aitkenhead M, Coull M (2020) Mapping soil profile depth, bulk density and carbon stock in Scotland using remote sensing and spatial covariates. Eur J Soil Sci 71:553–567. https://doi.org/10.1111/ejss.12916

Alharbi HA, Aldossary M (2021) Energy-efficient edge-fog-cloud architecture for IoT-based smart agriculture environment. IEEE Access 9:110480–110492. https://doi.org/10.1109/ACCESS.2021.3101397

Andrew H (2021) Agricultural carbon markets: how might they work?

Antle JM, Stoorvogel JJ (2009) Payments for ecosystem services, poverty and sustainability: the case of agricultural soil carbon sequestration. In: Payment for environmental Services in Agricultural Landscapes. Springer, New York, pp 133–161

Arnold SS, Fernandez IJ, Rustad LE, Zibilske LM (1999) Microbial response of an acid forest soil to experimental soil warming. Biol Fertil Soils 30:239–244. https://doi.org/10.1007/s003740050614

Babaei A, Tirkolaee EB, Boz E (2025) Optimizing energy consumption for blockchain adoption through renewable energy sources. Renew Energy 238:121936. https://doi.org/10.1016/j.renene.2024.121936

Babu S, Madhusudanan S, Sathiyanarayanan M et al (2024) Soil mapping for farming productivity: internet of things (IoT) based sustainable agriculture. Microsyst Technol 31:679–694. https://doi.org/10.1007/s00542-024-05608-z

Balafoutis A, Beck B, Fountas S et al (2017) Precision agriculture technologies positively contributing to GHG emissions mitigation, farm productivity and economics. Sustainability 9:1339. https://doi.org/10.3390/su9081339

Bandara KMTS, Sakai K, Nakandakari T, Yuge K (2021) A low-cost NDIR-based N2O gas detection device for agricultural soils: assembly, calibration model validation, and laboratory testing. Sensors 21:1189. https://doi.org/10.3390/s21041189

Banerjee D, Adhikary S, Bhattacharya S et al (2024) Breaking boundaries: artificial intelligence for pesticide detection and eco-friendly degradation. Environ Res 241:117601. https://doi.org/10.1016/j.envres.2023.117601

Banu MS, Abuthahir AAM, Mohandas R, et al (2024) Cloud based intelligent nutrient management system for precision agriculture using CNN. In: 2024 2nd International Conference on Sustainable Computing and Smart Systems (ICSCSS). IEEE, pp 971–976

Bassous NJ, Rodriguez AC, Leal CIL et al (2024) Significance of various sensing mechanisms for detecting local and atmospheric greenhouse gases: a review. Adv Sensor Res 3. https://doi.org/10.1002/adsr.202300094

Bernardini LG, Rosinger C, Bodner G et al (2024) Learning vs. understanding: when does artificial intelligence outperform process-based modeling in soil organic carbon prediction? New Biotechnol 81:20–31. https://doi.org/10.1016/j.nbt.2024.03.001

Bispo A, Andersen L, Angers DA et al (2017) Accounting for carbon stocks in soils and measuring GHGs emission fluxes from soils: do we have the necessary standards? Front Environ Sci 5. https://doi.org/10.3389/fenvs.2017.00041

Blackmore S (2016) Towards robotic agriculture. In: Valasek J, Thomasson JA (eds) Autonomous air and ground sensing systems for agricultural optimization and phenotyping. p 986603

Blasch J, van der Kroon B, van Beukering P et al (2022) Farmer preferences for adopting precision farming technologies: a case study from Italy. Eur Rev Agric Econ 49:33–81. https://doi.org/10.1093/erae/jbaa031

Bobrov PP, Belyaeva TA, Kroshka ES, Rodionova OV (2019) Soil moisture measurement by the dielectric method. Eurasian Soil Sci 52:822–833. https://doi.org/10.1134/S106422931905003X

Burgoyne MT, Olexiuk (2025) Tokenized carbon credits: how blockchain is revolutionizing carbon markets. In: Osler

Calvin K, Dasgupta D, Krinner G, et al (2023) IPCC, 2023: Climate Change 2023: Synthesis Report, Summary for Policymakers. Contribution of Working Groups I, II and III to the Sixth Assessment Report of the Intergovernmental Panel on Climate Change [Core Writing Team, H. Lee and J. Romero (eds.)]. IPCC, Geneva

Cardelli V, Weindorf DC, Chakraborty S et al (2017) Non-saturated soil organic horizon characterization via advanced proximal sensors. Geoderma 288:130–142. https://doi.org/10.1016/j.geoderma.2016.10.036

Cariappa AAG, Konath NC, Sapkota TB, Krishna VV (2024) Evaluating the potential and eligibility of conservation agriculture practices for carbon credits. Sci Rep 14:9193. https://doi.org/10.1038/s41598-024-59262-6

Chamara N, Islam MD, Bai G, Frank et al (2022) Ag-IoT for crop and environment monitoring: past, present, and future. Agric Syst 203:103497. https://doi.org/10.1016/j.agsy.2022.103497
Chen F, Kissel DE, West LT et al (2005) Mapping surface soil organic carbon for crop fields with remote sensing. J Soil Water Conserv 60:51–57
Chen H-Y, Sharma K, Sharma C, Sharma S (2023) Integrating explainable artificial intelligence and blockchain to smart agriculture: research prospects for decision making and improved security. Smart Agricul Technol 6:100350. https://doi.org/10.1016/j.atech.2023.100350
Chirinda N, Plauborg F, Heckrath G et al (2014) Carbon dioxide in arable soil profiles: a comparison of automated and manual measuring systems. Commun Soil Sci Plant Anal 45:1278–1291. https://doi.org/10.1080/00103624.2014.884107
Choruma DJ, Dirwai TL, Mutenje MJ et al (2024) Digitalisation in agriculture: a scoping review of technologies in practice, challenges, and opportunities for smallholder farmers in sub-saharan africa. J Agric Food Res 18:101286. https://doi.org/10.1016/j.jafr.2024.101286
Cillis D, Maestrini B, Pezzuolo A et al (2018) Modeling soil organic carbon and carbon dioxide emissions in different tillage systems supported by precision agriculture technologies under current climatic conditions. Soil Tillage Res 183:51–59. https://doi.org/10.1016/j.still.2018.06.001
Congalton RG (2010) Remote Sensing: An Overview. GIsci Remote Sens 47:443–459. https://doi.org/10.2747/1548-1603.47.4.443
Dacal M, Delgado-Baquerizo M, Barquero J et al (2022) Temperature increases soil respiration across ecosystem types and soil development, but soil properties determine the magnitude of this effect. Ecosystems 25:184–198. https://doi.org/10.1007/s10021-021-00648-2
Danton A, Roux J-C, Dance B, et al (2020) Development of a spraying robot for precision agriculture: An edge following approach. In: 2020 IEEE Conference on Control Technology and Applications (CCTA). IEEE, pp 267–272
Delacote P, L'Horty T, Kontoleon A et al (2024) Strong transparency required for carbon credit mechanisms. Nat Sustain 7:706–713. https://doi.org/10.1038/s41893-024-01310-0
Dey K, Shekhawat U (2021) Blockchain for sustainable e-agriculture: literature review, architecture for data management, and implications. J Clean Prod 316:128254. https://doi.org/10.1016/j.jclepro.2021.128254
Dhamu VN, Muthukumar S, Prasad S (2023) E-SCAN: electrochemical scanning of carbonates, an in situ approach for screening and quantifying inorganic carbon in soil. J Agric Food Chem 71:15954–15962. https://doi.org/10.1021/acs.jafc.3c02948
Dupla X, Bonvin E, Deluz C et al (2024) Are soil carbon credits empty promises? Shortcomings of current soil carbon quantification methodologies and improvement avenues. Soil Use Manag 40. https://doi.org/10.1111/sum.13092
Eldeeb MA, Dhamu VN, Paul A et al (2023a) Electrochemical soil nitrate sensor for in situ real-time monitoring. Micromachines (Basel) 14:1314. https://doi.org/10.3390/mi14071314
Eldeeb MA, Dhamu VN, Paul A et al (2023b) Espial: electrochemical soil pH sensor for in situ real-time monitoring. Micromachines (Basel) 14:2188. https://doi.org/10.3390/mi14122188
Ennaji O, Vergütz L, El Allali A (2023) Machine learning in nutrient management: a review. Artif Intell Agric 9:1–11. https://doi.org/10.1016/j.aiia.2023.06.001
FAO (2009) How to feed the world in 2050, Rome
Fares A, Awal R, Bayabil H (2016) Soil water content sensor response to organic matter content under laboratory conditions. Sensors 16:1239. https://doi.org/10.3390/s16081239
Farhan SM, Yin J, Chen Z, Memon MS (2024) A comprehensive review of LiDAR applications in crop management for precision agriculture. Sensors 24:5409. https://doi.org/10.3390/s24165409
Farooq MS, Riaz S, Abid A et al (2019) A survey on the role of IoT in agriculture for the implementation of smart farming. IEEE Access 7:156237–156271. https://doi.org/10.1109/ACCESS.2019.2949703
FuBing (2012) Research on the agriculture intelligent system based on IOT. In: 2012 International Conference on Image Analysis and Signal Processing. IEEE, pp 1–4

Futagawa M, Ban Y, Kawashima K, Sawada K (2013) On-site monitoring of soil condition for precision agriculture by using multimodal microchip integrated with EC and temperature sensors. In: In: 2013 Transducers & Eurosensors XXVII: The 17th International Conference on Solid-State Sensors, Actuators and Microsystems (Transducers & Eurosensors XXVII). IEEE, pp 112–115

G L, C R, P G (2020) An automated low cost IoT based fertilizer intimation system for smart agriculture. Sustainable Computing: Informatics and Systems 28:100300. https://doi.org/10.1016/j.suscom.2019.01.002

Gassner A, Coe R, Sinclair F (2013) Improving food security through increasing the precision of agricultural development. In: Oliver M, Bishop T, Marchant B (eds) Precision agriculture for sustainability and environmental protection, 1st edn. Routledge, Taylor & Francis Group, pp 34–57

Goap A, Sharma D, Shukla AK, Rama Krishna C (2018) An IoT based smart irrigation management system using machine learning and open source technologies. Comput Electron Agric 155:41–49. https://doi.org/10.1016/j.compag.2018.09.040

González Perea R, Camacho Poyato E, Montesinos P, Rodríguez Díaz JA (2018) Prediction of applied irrigation depths at farm level using artificial intelligence techniques. Agric Water Manag 206:229–240. https://doi.org/10.1016/j.agwat.2018.05.019

Guebsi R, Mami S, Chokmani K (2024) Drones in precision agriculture: a comprehensive review of applications, technologies, and challenges. Drones 8:686. https://doi.org/10.3390/drones8110686

Guo L, Fu P, Shi T et al (2021) Exploring influence factors in mapping soil organic carbon on low-relief agricultural lands using time series of remote sensing data. Soil Tillage Res 210:104982. https://doi.org/10.1016/j.still.2021.104982

Gupta M, Sharma S, Sakshi, Sharma C (2022) Security and privacy issues in Blockchained IoT. In: Vyas S, Shukla VK, Gupta S, Prasad A (eds) Blockchain technology, 1st edn. CRC Press, Boca Raton, pp 27–56

Hansen BD, Leonard E, Mitchell MC et al (2022) Current status of and future opportunities for digital agriculture in Australia. Crop Pasture Sci 74:524–537. https://doi.org/10.1071/CP21594

Harfouche AL, Jacobson DA, Kainer D et al (2019) Accelerating climate resilient plant breeding by applying next-generation artificial intelligence. Trends Biotechnol 37:1217–1235. https://doi.org/10.1016/j.tibtech.2019.05.007

Hedley C (2015) The role of precision agriculture for improved nutrient management on farms. J Sci Food Agric 95:12–19. https://doi.org/10.1002/jsfa.6734

Heger A, Kleinschmidt V, Gröngröft A et al (2020) Application of a low-cost NDIR sensor module for continuous measurements of *in situ* soil CO_2 concentration. J Plant Nutr Soil Sci 183:557–561. https://doi.org/10.1002/jpln.201900493

Hemmat A, Adamchuk VI, Jasa P (2008) Use of an instrumented disc coulter for mapping soil mechanical resistance. Soil Tillage Res 98:150–163. https://doi.org/10.1016/j.still.2007.11.003

Hung AH-C (2024) The curious case of blockchain in rural China: unravelling power, profit, and surveillance. Big Data Soc 11. https://doi.org/10.1177/20539517241259674

Ishaq I, Carels D, Teklemariam G et al (2013) IETF standardization in the field of the internet of things (IoT): a survey. J Sens Actuator Netw 2:235–287. https://doi.org/10.3390/jsan2020235

Jarecki MK, Lal R (2003) Crop management for soil carbon sequestration. CRC Crit Rev Plant Sci 22:471–502

Javaid M, Haleem A, Khan IH, Suman R (2023) Understanding the potential applications of artificial intelligence in agriculture sector. Adv Agrochem 2:15–30. https://doi.org/10.1016/j.aac.2022.10.001

Jin SF, Wang HS (2018) Relationship between soil pH and soil carbon in China's carbonate soils. Fresenius Environ Bull 27:605–611

John D, Hussin N, Shahibi MS et al (2023) A systematic review on the factors governing precision agriculture adoption among small-scale farmers. Outlook Agric 52:469–485. https://doi.org/10.1177/00307270231205640

Jonard F, Bircher S, Demontoux F et al (2018) Passive L-band microwave remote sensing of organic soil surface layers: a tower-based experiment. Remote Sens 10:304. https://doi.org/10.3390/rs10020304

Kalita H, Palaparthy VS, Baghini MS, Aslam M (2016) Graphene quantum dot soil moisture sensor. Sens Actuators B Chem 233:582–590. https://doi.org/10.1016/j.snb.2016.04.131

Karmakar P, Teng SW, Murshed M et al (2024) Crop monitoring by multimodal remote sensing: a review. Remote Sens Appl 33:101093. https://doi.org/10.1016/j.rsase.2023.101093

Kaur J, Hazrati Fard SM, Amiri-Zarandi M, Dara R (2022) Protecting farmers' data privacy and confidentiality: recommendations and considerations. Front Sustain Food Syst 6. https://doi.org/10.3389/fsufs.2022.903230

Kazemi Garajeh M, Hassangholizadeh K, Bakhshi Lomer AR et al (2023) Monitoring the impacts of crop residue cover on agricultural productivity and soil chemical and physical characteristics. Sci Rep 13:15054. https://doi.org/10.1038/s41598-023-42367-9

Kemmitt S, Wright D, Goulding K, Jones D (2006) pH regulation of carbon and nitrogen dynamics in two agricultural soils. Soil Biol Biochem 38:898–911. https://doi.org/10.1016/j.soilbio.2005.08.006

Khan A, Qu X, Gong Z (2025) Decision mechanism of farmers' low-carbon agricultural technology adoption: an evolutionary game theory approach. Empirica 52:297–331. https://doi.org/10.1007/s10663-025-09645-3

Khanna A, Maheshwari P (2024) Blockchain-powered NFTs: a paradigm shift in carbon credit transactions for traceability, transparency, and accountability. In: Papadaki M, Themistocleous M, Al Marri K, Al Zarouni M (eds) Information systems. Springer, Cham, pp 75–87

Kharel TP, Ashworth AJ, Owens PR (2022) Linking and sharing technology: partnerships for data innovations for management of agricultural big data. Data (Basel) 7:12. https://doi.org/10.3390/data7020012

Khose SB, Dhokale KB, Shekhar S (2023) The role of precision farming in sustainable agriculture: advancements and impacts. Agric Food 5:115–119

Kim S-K, Huh J-H (2020) Blockchain of carbon trading for UN sustainable development goals. Sustainability 12:4021. https://doi.org/10.3390/su12104021

Kim W-S, Lee W-S, Kim Y-J (2020) A review of the applications of the internet of things (IoT) for agricultural automation. J Biosyst Eng 45:385–400. https://doi.org/10.1007/s42853-020-00078-3

Kodešová R, Kodeš V, Mráz A (2011) Comparison of two sensors ECH2O EC-5 and SM200 for measuring soil water content. Soil and Water Research 6:102–110. https://doi.org/10.17221/6/2011-SWR

Kumar T, Kahlon MS (2022) Sensor-based irrigation scheduling, soil physical properties and wheat productivity under different organic amendments. Commun Soil Sci Plant Anal 53:2297–2307. https://doi.org/10.1080/00103624.2022.2071439

Lal R (2016) Feeding 11 billion on 0.5 billion hectare of area under cereal crops. Food Energy Secur 5:239–251. https://doi.org/10.1002/fes3.99

Lamichhane S, Kumar L, Wilson B (2019) Digital soil mapping algorithms and covariates for soil organic carbon mapping and their implications: a review. Geoderma 352:395–413. https://doi.org/10.1016/j.geoderma.2019.05.031

Laveglia S, Altieri G, Genovese F et al (2024) Advances in sustainable crop management: integrating precision agriculture and proximal sensing. AgriEngineering 6:3084–3120. https://doi.org/10.3390/agriengineering6030177

Lazik D, Vetterlein D, Salas SK et al (2019) New sensor technology for field-scale quantification of carbon dioxide in soil. Vadose Zone J 18:1–14. https://doi.org/10.2136/vzj2019.01.0007

Leduc G, Kubler S, Georges J-P (2021) Innovative blockchain-based farming marketplace and smart contract performance evaluation. J Clean Prod 306:127055. https://doi.org/10.1016/j.jclepro.2021.127055

Li S, Delgado-Baquerizo M, Ding J et al (2024a) Intrinsic microbial temperature sensitivity and soil organic carbon decomposition in response to climate change. Glob Chang Biol 30:e17395. https://doi.org/10.1111/gcb.17395
Li X, Li Z, Qiu H et al (2024b) Multi-scale spatial and spectral feature fusion for soil carbon content prediction based on hyperspectral images. Ecol Indic 160:111843. https://doi.org/10.1016/j.ecolind.2024.111843
Liu C, Ren W, Zhang B, Lv C (2011) The application of soil temperature measurement by LM35 temperature sensors. In: Proceedings of 2011 International Conference on Electronic & Mechanical Engineering and Information Technology. IEEE, pp 1825–1828
Liu Q, Tang J, He C, Chan CCK (2020) Use of temperature sensors in testing soil humus content in saline wetland in response to freeze–thaw cycles. Sens Mater 32:3355. https://doi.org/10.18494/SAM.2020.2930
Liu S, Chen X, Xia X et al (2024) Electronic sensing combined with machine learning models for predicting soil nutrient content. Comput Electron Agric 221:108947. https://doi.org/10.1016/j.compag.2024.108947
Lockhart SRA, Keller CK, Evans RD et al (2023) Soil CO2 in organic and no-till agroecosystems. Agric Ecosyst Environ 349:108442. https://doi.org/10.1016/j.agee.2023.108442
Luvisi A, Panattoni A, Materazzi A (2016) RFID temperature sensors for monitoring soil solarization with biodegradable films. Comput Electron Agric 123:135–141. https://doi.org/10.1016/j.compag.2016.02.023
Mallik S, Bhowmik T, Mishra U, Paul N (2022) Mapping and prediction of soil organic carbon by an advanced geostatistical technique using remote sensing and terrain data. Geocarto Int 37:2198–2214. https://doi.org/10.1080/10106049.2020.1815864
Mamilla S, Zahid Yousuf S, Mohan Avulapati M et al (2025) A low-cost, reusable all-solid-state TiN-based EGFET soil pH sensor. IEEE Sensors J 25:4184–4191. https://doi.org/10.1109/JSEN.2024.3514735
Masrie M, Rosli AZM, Sam R et al (2018) Integrated optical sensor for NPK nutrient of soil detection. In: 2018 IEEE 5th International Conference on Smart Instrumentation, Measurement and Application (ICSIMA). IEEE, pp 1–4
Merl T, Rasmussen MR, Koch LR et al (2022) Measuring soil pH at in situ like conditions using optical pH sensors (pH-optodes). Soil Biol Biochem 175:108862. https://doi.org/10.1016/j.soilbio.2022.108862
Millan-Almaraz JR, Torres-Pacheco I, Duarte-Galvan C et al (2013) FPGA-based wireless smart sensor for real-time photosynthesis monitoring. Comput Electron Agric 95:58–69. https://doi.org/10.1016/j.compag.2013.04.009
Na A, Isaac W, Varshney S, Khan E (2016) An IoT based system for remote monitoring of soil characteristics. In: 2016 International Conference on Information Technology (InCITe)—The Next Generation IT Summit on the Theme—Internet of Things: Connect your Worlds. IEEE, pp 316–320
Nair N, V AA, Joseph J (2022) An in-situ soil pH sensor with solid electrodes. IEEE Sens Lett 6:1–4. https://doi.org/10.1109/LSENS.2022.3194200
O'Flaherty BD, Barry EF, Cholli AL (2000) A rapid soil nutrient sensor device based on capillary zone electrophoresis. J Environ Sci Health A 35:189–201. https://doi.org/10.1080/10934520009376963
Otieno F, Nalakurthi S-RNV, Raji M et al (2024) Farmer's attitudes towards GHG emissions and adoption to low-cost sensor-driven smart farming for mitigation: the case of Ireland tillage and horticultural farmers. Smart Agricultural Technology 9:100622. https://doi.org/10.1016/j.atech.2024.100622
Padhiary GG, Mishra SL (2020) Agricultural sensors: a step towards smart agriculture. Just Agriculture 1:
Padhiary M, Saha D, Kumar R et al (2024) Enhancing precision agriculture: a comprehensive review of machine learning and AI vision applications in all-terrain vehicle for farm automation. Smart Agric Technol 8:100483. https://doi.org/10.1016/j.atech.2024.100483

Parra-López C, Ben Abdallah S, Garcia-Garcia G et al (2024) Integrating digital technologies in agriculture for climate change adaptation and mitigation: state of the art and future perspectives. Comput Electron Agric 226:109412. https://doi.org/10.1016/j.compag.2024.109412

Patel D, Britto B, Sharma S, et al (2020) Carbon credits on Blockchain. In: 2020 International Conference on Innovative Trends in Information Technology (ICITIIT). IEEE, pp 1–5

Pham HC, Truong HQ, Thi Binh AD, Yap TL (2024) Enhancing agriculture resilience of smallholder farming to climate change through Blockchain technology adoption. In: Encyclopedia of Sustainable Technologies. Elsevier, pp 876–884

Piikki K, Söderström M, Eriksson J et al (2016) Performance evaluation of proximal sensors for soil assessment in smallholder farms in Embu County, Kenya. Sensors 16:1950. https://doi.org/10.3390/s16111950

Pouladi N, Gholizadeh A, Khosravi V, Borůvka L (2023) Digital mapping of soil organic carbon using remote sensing data: a systematic review. Catena (Amst) 232:107409. https://doi.org/10.1016/j.catena.2023.107409

Preite L, Vignali G (2024) Artificial intelligence to optimize water consumption in agriculture: a predictive algorithm-based irrigation management system. Comput Electron Agric 223:109126. https://doi.org/10.1016/j.compag.2024.109126

Qi R, Li J, Lin Z et al (2016) Temperature effects on soil organic carbon, soil labile organic carbon fractions, and soil enzyme activities under long-term fertilization regimes. Appl Soil Ecol 102:36–45. https://doi.org/10.1016/j.apsoil.2016.02.004

Rajak P, Ganguly A, Adhikary S, Bhattacharya S (2023) Internet of things and smart sensors in agriculture: scopes and challenges. J Agric Food Res 14:100776. https://doi.org/10.1016/j.jafr.2023.100776

Rastgou M, He Y, Jiang Q (2024) An analytical comparison of the performance of various sensing materials and mechanisms for efficient detection capability of greenhouse gas emissions. Engineering. https://doi.org/10.1016/j.eng.2024.11.008

Rath RJ, Talebian S, Giaretta J et al (2024) Organic-based chemiresistive sensors for detection of water-soluble gases: strategies and roadmap for enhancing sensing performance. Adv Funct Mater 35. https://doi.org/10.1002/adfm.202417323

Rehman KU, Andleeb S, Ashfaq M et al (2023) Blockchain-enabled smart agriculture: enhancing data-driven decision making and ensuring food security. J Clean Prod 427:138900. https://doi.org/10.1016/j.jclepro.2023.138900

Rentschler T, Werban U, Ahner M et al (2020) 3D mapping of soil organic carbon content and soil moisture with multiple geophysical sensors and machine learning. Vadose Zone J 19. https://doi.org/10.1002/vzj2.20062

Reyes J, Ließ M (2022) Can soil spectroscopy contribute to soil organic carbon monitoring on agricultural soils? EGUsphere

Ruiz-Gonzalez A, Kempson H, Haseloff J (2024) Development of a low-cost sensor system for accurate soil assessment and biological activity profiling. Micromachines (Basel) 15:1293. https://doi.org/10.3390/mi15111293

Sadawi AA, Madani B, Saboor S et al (2021) A comprehensive hierarchical blockchain system for carbon emission trading utilizing blockchain of things and smart contract. Technol Forecast Soc Change 173:121124. https://doi.org/10.1016/j.techfore.2021.121124

Sakabe K, Kan T, Onoe H (2024) Entirely biodegradable wireless pH sensor with Split-ring resonators for soil pH monitoring. Adv Mater Technol 9. https://doi.org/10.1002/admt.202400038

Saleem SR, Levison J, Haroon Z (2023) Environment: role of precision agriculture technologies. In: Precision Agriculture. Elsevier, pp 211–229

Salunkhe SS, Agarkar AA, Karyakarte M et al (2023) Integrated agriculture IoT based farm monitoring and management systems. Int J Nanotechnol 20:556–568. https://doi.org/10.1504/IJNT.2023.134014

Santiteerakul S, Sopadang A, Yaibuathet Tippayawong K, Tamvimol K (2020) The role of smart technology in sustainable agriculture: a case study of Wangree plant factory. Sustainability 12:4640. https://doi.org/10.3390/su12114640

Sarkar R, Somenahally A (2023) Machine learning soil-environmental impacts on agroecosystems for relating microbial biomass to soil carbon sequestration. Smart Agric Technol 4:100208. https://doi.org/10.1016/j.atech.2023.100208

Sarmah SS (2018) Understanding Blockchain technology. Comput Sci Eng 8:23–29

Saxena P, Shukla P (2023) A review on recent developments and advances in environmental gas sensors to monitor toxic gas pollutants. Environ Prog Sustain Energy 42. https://doi.org/10.1002/ep.14126

Schimmelpfennig D, Lowenberg-DeBoer J (2021) Precision agriculture adoption, farm size and soil variability. In: Precision agriculture '21. Brill | Wageningen Academic, pp 769–776

Schmidinger J, Barkov V, Tavakoli H et al (2024) Which and how many soil sensors are ideal to predict key soil properties: a case study with seven sensors. Geoderma 450:117017. https://doi.org/10.1016/j.geoderma.2024.117017

Sedlmeir J, Buhl HU, Fridgen G, Keller R (2020) The energy consumption of Blockchain technology: beyond myth. Bus Inf Syst Eng 62:599–608. https://doi.org/10.1007/s12599-020-00656-x

Shahab H, Naeem M, Iqbal M et al (2025) IoT-driven smart agricultural technology for real-time soil and crop optimization. Smart Agric Technol 10:100847. https://doi.org/10.1016/j.atech.2025.100847

Sharma A, Kumar V, Shahzad B et al (2019) Worldwide pesticide usage and its impacts on ecosystem. SN Appl Sci 1:1446. https://doi.org/10.1007/s42452-019-1485-1

Sharma R, Mishra DR, Levi MR, Sutter LA (2022) Remote sensing of surface and subsurface soil organic carbon in tidal wetlands: a review and ideas for future research. Remote Sens 14:2940. https://doi.org/10.3390/rs14122940

Sharp N (2024) 7 types of agricultural sensors driving the smart farming revolution. In: Escatec. https://www.escatec.com/blog/7-types-of-agricultural-sensors-driving-the-smart-farming-revolution. Accessed 23 Feb 2025

Singh V, Sharma SK (2023) Application of blockchain technology in shaping the future of food industry based on transparency and consumer trust. J Food Sci Technol 60:1237–1254. https://doi.org/10.1007/s13197-022-05360-0

Sirjacobs D, Hanquet B, Lebeau F, Destain MF (2002) On-line soil mechanical resistance mapping and correlation with soil physical properties for precision agriculture. Soil Tillage Res 64:231–242. https://doi.org/10.1016/S0167-1987(01)00266-5

Sivakumar VG, Baskar VV, Vadivel M, et al (2023) IoT and GIS integration for real-time monitoring of soil health and nutrient status. In: 2023 International Conference on Self Sustainable Artificial Intelligence Systems (ICSSAS). IEEE, pp 1265–1270

Snapp S (2022) Embracing variability in soils on smallholder farms: new tools and better science. Agric Syst 195:103310. https://doi.org/10.1016/j.agsy.2021.103310

Sodango T, Sha J, Li X et al (2021) Modeling the spatial dynamics of soil organic carbon using remotely-sensed predictors in Fuzhou City, China. Remote Sens (Basel) 13:1682. https://doi.org/10.3390/rs13091682

Sontowski S, Gupta M, Laya Chukkapalli SS, et al (2020) Cyber attacks on smart farming infrastructure. In: 2020 IEEE 6th International Conference on Collaboration and Internet Computing (CIC). IEEE, pp 135–143

Staggenborg SA, Carignano M, Haag L (2007) Predicting soil pH and buffer pH in situ with a real-time sensor. Agron J 99:854–861. https://doi.org/10.2134/agronj2006.0254

Steiger A, Qaswar M, Bill R et al (2025) Comparing the handheld Stenon FarmLab soil sensor with a Vis-NIR multi-sensor soil sensing platform. Smart Agric Technol 10:100717. https://doi.org/10.1016/j.atech.2024.100717

Sudduth KA, Chung S-O, Andrade-Sanchez P, Upadhyaya SK (2008) Field comparison of two prototype soil strength profile sensors. Comput Electron Agric 61:20–31. https://doi.org/10.1016/j.compag.2006.11.006

Sulaeman Y, Sutanto E, Kasno A et al (2024) Developing and testing a portable soil nutrient detector in irrigated and Rainfed Paddy soils from Java, Indonesia. Computers 13:209. https://doi.org/10.3390/computers13080209

Tan WY, Then YL, Lew YL, Tay FS (2019) Newly calibrated analytical models for soil moisture content and pH value by low-cost YL-69 hygrometer sensor. Measurement 134:166–178. https://doi.org/10.1016/j.measurement.2018.10.071

Thilakarathne NN, Abu Bakar MS, Abas PE, Yassin H (2025) Internet of things enabled smart agriculture: current status, latest advancements, challenges and countermeasures. Heliyon 11:e42136. https://doi.org/10.1016/j.heliyon.2025.e42136

Tran HT, Cao W, Reutemann M, et al (2020) GNSS Positioning and Navigation—A Foundational Element of Digital Farming. In: 2020 European Navigation Conference (ENC). IEEE, pp 1–8

Tripoli M, Schmidhuber J (2018) Emerging opportunities for the application of Blockchain in the agri-food industry, Rome and Geneva

Trivedi S, Vinod PV, Chandrasekaran B et al (2024) Geospatial assessment of agroforestry land use systems using very high-resolution satellite images and artificial intelligence. Agrofor Syst 98:2875–2895. https://doi.org/10.1007/s10457-024-01042-2

Tsakiridis NL, Samarinas N, Kalopesa E, Zalidis GC (2023) Cognitive soil digital twin for monitoring the soil ecosystem: a conceptual framework. Soil Syst 7:88. https://doi.org/10.3390/soilsystems7040088

Tziolas N, Tsakiridis N, Heiden U, van Wesemael B (2024) Soil organic carbon mapping utilizing convolutional neural networks and earth observation data, a case study in Bavaria state Germany. Geoderma 444:116867. https://doi.org/10.1016/j.geoderma.2024.116867

Tzounis A, Katsoulas N, Bartzanas T, Kittas C (2017) Internet of Things in agriculture, recent advances and future challenges. Biosyst Eng 164:31–48. https://doi.org/10.1016/j.biosystemseng.2017.09.007

Uddin MJ, Hooda PS, Mohiuddin ASM, Smith M (2020) Geospatial technologies in precision farming: a case study. Intell Environ, In, pp 133–142

van Wassenaer L, van Hilten M, van Ingen E, van Asseldonk M (2021a) Applying blockchain for climate action in agriculture: state of play and outlook. FAO and WUR, Rome/Wageningen

van Wassenaer L, Verdouw C, Wolfert S (2021b) What Blockchain are we talking about? An analytical framework for understanding Blockchain applications in agriculture and food. Front Blockchain 4. https://doi.org/10.3389/fbloc.2021.653128

Villa-Henriksen A, Edwards GTC, Pesonen LA et al (2020) Internet of things in arable farming: implementation, applications, challenges and potential. Biosyst Eng 191:60–84. https://doi.org/10.1016/j.biosystemseng.2019.12.013

Viscarra Rossel RA, Bouma J (2016) Soil sensing: a new paradigm for agriculture. Agric Syst 148:71–74. https://doi.org/10.1016/j.agsy.2016.07.001

Wan C-Y, Eisenman SB, Campbell AT, Crowcroft J (2007) Overload traffic management for sensor networks. ACM Trans Sens Netw 3:18. https://doi.org/10.1145/1281492.1281493

Wang W, Cao J, Zhang R et al (2024) Design strategies of semiconductor sensors toward ammonia monitoring in smart agriculture. J Environ Chem Eng 12:114380. https://doi.org/10.1016/j.jece.2024.114380

Wu Q, Liang Y, Li Y et al (2017) Factors acquisition and content estimation of farmland soil organic carbon based upon internet of things. Chin Geogr Sci 27:431–440. https://doi.org/10.1007/s11769-017-0875-9

Xiong H, Dalhaus T, Wang P, Huang J (2020) Blockchain technology for agriculture: applications and rationale. Front Blockchain 3. https://doi.org/10.3389/fbloc.2020.00007

Xu J, Gu B, Tian G (2022) Review of agricultural IoT technology. Artific Intell Agric 6:10–22. https://doi.org/10.1016/j.aiia.2022.01.001

Xu Y, Tan Y, Abd-Elrahman A et al (2023) Incorporation of fused remote sensing imagery to enhance soil organic carbon spatial prediction in an agricultural area in Yellow River Basin, China. Remote Sens (Basel) 15:2017. https://doi.org/10.3390/rs15082017

Yin H, Cao Y, Marelli B et al (2021) Soil sensors and plant wearables for smart and precision agriculture. Adv Mater 33:e2007764. https://doi.org/10.1002/adma.202007764

Yokota M, Okada T, Yamaguchi I (2007) An optical sensor for analysis of soil nutrients by using LED light sources. Meas Sci Technol 18:2197–2201. https://doi.org/10.1088/0957-0233/18/7/052

Yuan Y, Sun Y (2024) Practices, challenges, and future of digital transformation in smallholder agriculture: insights from a literature review. Agriculture 14:2193. https://doi.org/10.3390/agriculture14122193

Zhai C, Wang T, Xu Q, et al (2023) Research on the construction of intelligent agriculture based on agricultural robots. In: 2023 6th International Conference on Mechatronics, Robotics and Automation (ICMRA)(. IEEE, pp 71–75

Zhang X, Feng G, Sun X (2024) Advanced technologies of soil moisture monitoring in precision agriculture: a review. J Agric Food Res 18:101473. https://doi.org/10.1016/j.jafr.2024.101473

Zhao Y, Liu L, Xie C et al (2020) An effective automatic system deployed in agricultural internet of things using multi-context fusion network towards crop disease recognition in the wild. Appl Soft Comput 89:106128. https://doi.org/10.1016/j.asoc.2020.106128

Zheng B, Campbell JB, de Beurs KM (2012) Remote sensing of crop residue cover using multi-temporal Landsat imagery. Remote Sens Environ 117:177–183. https://doi.org/10.1016/j.rse.2011.09.016

Chapter 10
Global Perspectives on Carbon Farming

Abstract This chapter provides a comprehensive overview of carbon (C) farming as a globally relevant solution to the interconnected challenges of climate change, soil degradation, and food insecurity. It examines how countries across South Asia, Africa, Australia, Latin America, North America, and Europe are implementing practices such as conservation agriculture, agroforestry, and regenerative land management to enhance soil organic C sequestration and reduce greenhouse gas emissions. The chapter also reviews policy measures adopted in these regions to promote C farming, highlighting both enabling factors and key barriers to adoption, including financial incentives, policy design, monitoring challenges, and social and cultural dynamics. It emphasizes the importance of robust monitoring, reporting, and verification systems, inclusive payment models, and the integration of traditional knowledge into modern agricultural practices. Emerging research priorities such as soil inorganic C sequestration, the use of perennial vegetation, and integrated farming systems are also discussed. The chapter concludes that with supportive policies, strong institutional collaboration, and broad stakeholder engagement, C farming can be scaled as a climate resilient, socially inclusive, and economically viable pathway toward sustainable agricultural transformation.

10.1 Introduction

Climate change is one of the most pressing global challenges of the twenty-first century, drawing attention from governments, scientists, industries, and communities worldwide. Over the years, the debate surrounding climate change has evolved, shaped by diverse perspectives based on the varying knowledge, financial resources, and institutional capacities of different nations. These varying viewpoints often lead to different approaches in addressing climate change, particularly when it comes to greenhouse gas (GHG) emissions. Developed nations, having historically contributed the majority of global GHG emissions, remain among the largest current emitters (Matthews et al. 2014). In contrast, developing nations, while generally emitting fewer GHGs, face a unique set of challenges, with most emissions occurring within

N. Loria, R. Lal, *Carbon Farming*, https://doi.org/10.1007/978-3-032-00842-8_10

the Agriculture, Forestry, and Other Land Use (AFOLU) sector (Zeleke et al. 2016). These differences have complicated international climate negotiations and hindered the development of unified solutions.

The impacts of climate change have become increasingly evident, particularly in the agricultural sector, which is highly sensitive to shifts in temperature, rainfall, and extreme weather events. These disruptions have led to reduced productivity, more unpredictable cropping cycles, and growing food insecurity, especially in regions where agriculture is a key livelihood. Low-income and vulnerable populations are often the most affected, lacking the resources to adapt to climate-related shocks (Loria et al. 2015; Panwar et al. 2020).

In response, Carbon (C) farming has emerged as a promising solution, particularly for developing nations. It involves implementing land use practices that enhance or maintain C in natural sinks like soil and vegetation, aiming to reduce GHG emissions from agriculture (Leifeld 2023). Common practices include cover cropping, conservation agriculture (CA), agroforestry, grazing management, and diverse crop rotations (Vistarte et al. 2024). It offers a cost-effective means of both mitigating emissions and improving food security by integrating sustainable practices into agricultural systems. Where large-scale technological or financial interventions are not feasible, C farming provides a practical alternative. This chapter explores the global relevance of C farming, examining how different countries are adopting it based on their unique socio-economic and environmental contexts. It assesses the potential of C farming to reduce emissions, build agricultural resilience, and contribute to sustainable development, highlighting its importance as both a local and global climate solution.

10.2 Regional Perspectives on Carbon Farming

10.2.1 South Asia

C farming presents a promising pathway for addressing key agricultural challenges in South Asia, including food insecurity, soil degradation, and climate change (Jat et al. 2020). Agronomic practices, agroforestry, and livestock management offer effective climate change mitigation options with co-benefits for adaptation, but their widespread adoption is hindered by limited investment, cultural barriers, and policy gaps (Aryal et al. 2020). CA, particularly in the rice–wheat systems of the Indo-Gangetic Plains, is gaining momentum. India and Pakistan have adopted no-till practices on approximately 1.5 and 0.6 million hectares (M Ha), respectively, contributing to around 5 M Ha of no-till wheat across the region, although full CA adoption remains modest (Kassam et al. 2019). The slow uptake is largely attributed to resistance to change, challenges in crop residue management, and the impacts of erratic climate patterns. Initiatives like Promoting CA in Asia (PACA) and India's Crop Residue Management Program aim to address these barriers through awareness campaigns and technical support (Somasundaram et al. 2020).

Recent research mapping the evolution of climate-smart agriculture (CSA) in Bangladesh, India, Nepal, and Pakistan provides further insight into the socio-institutional dynamics affecting the adoption of sustainable agricultural practices (SAPs). The study identifies key drivers such as climate variability, the need for GHG mitigation, cost savings, and improved water management (Prutzer et al. 2023). However, it also highlights significant barriers, including limited financial resources, high initial costs, lack of access to context-specific knowledge, and persistent inequalities in power and access to resources. In India, public–private partnerships and alignment with national climate goals have supported CSA adoption (Agarwal et al. 2023), but smallholder exclusion, limited awareness, weak regulatory frameworks, low awareness, and insufficient access to finance and site-specific technologies remain significant challenges (Villalba et al. 2024; Vasavi et al. 2025). In Nepal, gender-sensitive project design and stronger institutional collaboration have the potential to improve CSA uptake; however, ongoing financial and structural barriers continue to limit women's access to resources and decision-making (Paudyal et al. 2019; Ghimire et al. 2022). In Bangladesh and Pakistan, adoption is closely tied to government support and the framing of interventions as tools for climate adaptation, but structural challenges like policy gaps and dependence on external technologies continue to limit impact (Prutzer et al. 2023).

Scaling up C farming in South Asia will require comprehensive policy backing, improved access to affordable technologies, and stronger regional collaboration (Miao et al. 2025). Promoting practices such as residue retention and enhancing soil organic C (SOC) storage is essential to securing long term soil productivity and food security. Encouraging adoption will depend on shifting farmer perceptions, improving technology transfer, and introducing targeted incentives, including C credits (Jat et al. 2023).

Insights from other parts of Asia provide valuable guidance. In China's Loess Plateau, a semi-arid region dominated by mixed crop and livestock systems, C farming strategies have demonstrated measurable success. A study by Tang et al. (2019) evaluated the impact of a C tax as an incentive for reducing agricultural emissions. Results showed a reduction in on farm emissions ranging from 16.6% to 33%, depending on the level of the tax, with relatively low marginal abatement costs. A major contributor to these reductions was the shift from livestock intensive systems to cropping systems, particularly through the replacement of oats and rapeseed with dry pea rotations. Dry peas contribute to lower emissions by fixing nitrogen (N) naturally, thereby reducing reliance on synthetic fertilizers. This case highlights how C pricing and land use incentives can guide farmers toward low emission practices without compromising productivity. The Loess Plateau experience serves as a model for other semi-arid and mixed farming regions in Asia, illustrating how policy and financial tools can drive the transition to sustainable agriculture. China's experiments with C markets and pricing mechanisms provide a useful reference for South Asian nations considering similar strategies. Beyond the Loess Plateau, national-level assessments suggest that CA and soil restoration practices across China's diverse landscapes have the potential to sequester between 105 and 198 teragrams (Tg) of SOC annually, and 7 to 138 Tg of soil inorganic C (SIC) (Lal

2002a). Cumulatively, this could amount to 11 Pg of soil CS by 2050, offsetting up to 20 percent of China's annual industrial CO_2 emissions.

A notable private sector example from India is Bayer's Rice C Program. By promoting Direct Seeded Rice (DSR) and alternate wetting and drying techniques across eleven states, the initiative reduces methane (CH_4) emissions from rice fields and supports farmers through expert advisory services. The program has generated up to 250 Gg of CO_2 equivalent (CO_2e) in certified credits, validated by Gold Standard and made available to climate focused companies (Bayer 2025). Alongside environmental gains, farmers benefit from reduced labor, lower water usage, and increased profitability. This example underscores how combining regenerative practices with market-based incentives can enhance agricultural resilience and sustainability across the region.

Together, these examples from South Asia and beyond demonstrate the importance of aligning local practices with global C mitigation goals. With the right policy instruments, technical support, and financial incentives, C farming can play a vital role in securing food systems and promoting climate resilient agriculture in the region.

10.2.2 Africa

C farming is gaining recognition in Africa as a critical approach for mitigating climate change while improving agricultural productivity and soil health (Schilling et al. 2023). In Southern Africa, however, studies have shown that the potential for CA to significantly increase SOC stocks is limited, largely due to low organic matter inputs (Cheesman et al. 2016). While conservation practices have proven beneficial in reducing GHG emissions, their impact on soil C accumulation remains modest, underscoring the challenge of enhancing soil C through current agricultural systems. Research in Zimbabwe indicated that communal grazing lands maintain relatively high levels of SOC, documenting the importance of conserving and improving the management of these landscapes (Nyawasha et al. 2025). Regenerative agricultural practices, such as agroforestry and improved fallow systems, show great promise for increasing SOC content and reducing GHG emissions (Vågen et al. 2005), yet adoption rates remain low in many parts of Sub-Saharan Africa (SSA). A significant barrier to scaling up C farming is the low awareness and adoption rates, although research indicates that farmers are increasingly open to practices like C farming if they receive proper incentives and support (Clarke et al. 2012).

In West Asia-North Africa (WANA), desertification, water scarcity, erratic rainfall, and unsustainable farming practices have contributed to severe soil degradation and low SOC (Lal 2002b). Integrated soil fertility management (ISFM), sustainable land practices, and water harvesting technologies are being promoted to restore soil health and sequester C, but more research and policy support are needed to scale these efforts (Diop et al. 2022). Furthermore, C farming holds potential for smallholder farmers in SSA, offering opportunities for increased productivity, climate change adaptation, and access to global C markets (Biramo 2018). A study in Northern Ghana showed that training programs on biochar and compost production led to improved soil quality, increased agricultural productivity, and poverty reduction (Okyere and Kornher 2023). Despite these positive outcomes, the financial expectations for C farming, especially regarding C credits, may be unrealistic, with farmers often anticipating returns that exceed actual earnings from SAPs (Clarke et al. 2012).

Overall, C farming presents a viable solution to enhance soil health, reduce emissions, and promote climate-resilient agriculture in SSA. However, achieving widespread adoption will require overcoming barriers such as low awareness, inadequate policy support, and financial challenges. Future efforts should focus on capacity building, research into compatible farming systems, and better integration of C credits into agricultural policies to promote a low-C economy across the continent (Hayo and Hasegawa 2024).

A noteworthy example of progress is the AgriCarbon initiative in Southern Africa, launched by the Anthesis Group. AgriCarbon is the first C farming program in Africa registered under Verra's VM0042 Agricultural Land Management methodology. The program promotes regenerative practices such as cover cropping, reduced tillage, and diversified crop rotations. These efforts not only enhance soil health but also convert farmlands into C sinks, helping to sequester CO_2 and improve climate resilience. Since its launch in South Africa in 2021, AgriCarbon has issued more than 43 Gg of C credits from 29 farmers operating across 18,000 ha. This initiative demonstrates the potential of C farming to generate new income streams for farmers while supporting broader sustainability goals (Von Riese 2025).

10.2.3 Australia

C farming has emerged as a critical strategy in Australia's efforts to mitigate climate change, supported by government incentives under the Emissions Reduction Fund (ERF). This initiative encourages landholders to implement practices that sequester C or reduce emissions, such as soil CS and vegetation management (Kragt et al. 2016). While the potential is significant, several challenges limit widespread adoption, particularly in soil C projects. A major concern is the credibility and permanence of soil C credits. For instance, although high sequestration rates were observed in 2023, they were largely attributed to exceptional rainfall rather than to SAPs (Mitchell et al. 2024). This trend underscores the need to strengthen the integrity of C credits by extending measurement periods to at least 5 years and setting science-based limits on expected SOC gains. This approach would ensure that credits reflect genuine, management-driven improvements rather than short-term climatic variability.

Farmer adoption of C farming practices is influenced by a mix of perceived benefits and uncertainties. A survey of dryland farmers in Western Australia found that while farmers valued co-benefits like improved soil quality and reduced erosion, concerns about policy stability, C price uncertainty, and potential impacts on productivity deterred many from participating (Dumbrell et al. 2016). Farmers tended to prefer practices such as stubble retention and no-till cropping, which offer clear agronomic advantages, but showed less interest in interventions like biochar application or tree planting, which were perceived as complex or less directly beneficial. To increase participation, policies must focus on tangible farm benefits, such as soil improvement and erosion control, rather than solely on financial incentives or C credits.

Regional differences also play a significant role in the willingness to invest in C offset projects. National surveys revealed that willingness to pay (WTP) for C offsets was lower in resource-dependent regions like Queensland and Western Australia. On average, respondents valued CO_2 reduction at AU$1.13 per Mg and native vegetation restoration at AU$19 per ha, indicating a stronger preference for co-benefits such as biodiversity and land rehabilitation over CS alone (Kragt et al. 2016).

Incorporating indigenous perspectives into C farming design is essential for inclusive and culturally appropriate outcomes. Sangha et al. (2024) emphasized that indigenous communities favor bottom-up approaches that align with their cultural values and social priorities in Payment for ESs (PES) schemes. Similarly, Pudasaini et al. (2024) noted that while high C prices offer strong incentives, complex program structures and barriers to entry such as limited knowledge, high upfront costs, and uncertain returns remain significant obstacles. Streamlining program procedures and increasing financial support will be critical to enabling broader participation.

Australia has also made considerable progress in integrating CA principles. Technologies such as no-till farming, precision agriculture, and mixed farming systems have led to improved productivity and sustainability outcomes, especially in Western Australia (Kirkegaard et al. 2014). However, improvements in SOC remain modest due to constraints such as sandy soils and low rainfall (Petersen and Hoyle 2016). Despite modest financial returns from SOC improvements, the long-term benefits to soil health and environmental resilience provide an economic incentive, particularly with stronger market signals like C pricing.

A successful example of C farming is the Swartz C Projects, managed by the Warnock family in Queensland. These projects span 2109 ha and have generated 6152 Australian C Credit Units (ACCUs) within 2 years (agriprove 2025). The Warnocks have implemented rotational grazing and invested in infrastructure, including electric fencing and additional watering points. These changes have not only enhanced soil C storage but also doubled their land's carrying capacity, highlighting the potential of grazing management to improve farm productivity while contributing to CS.

However, community concerns persist. In the Mulga Lands of Queensland, landholders were motivated by economic returns from C farming, but issues such as absentee landownership and rural depopulation raised concerns about the broader social impacts (Jassim et al. 2022). These concerns highlight the need for policies that account for local values and community impacts, ensuring C farming does not result in negative social outcomes.

To scale C farming effectively, a multi-objective approach is required, balancing social, economic, and environmental outcomes (Pham-Kieu et al. 2024). Policies should reflect regional priorities and community needs, highlighting co-benefits such as biodiversity, cultural heritage, and land restoration. Evidence suggests that C prices above AU$50 per ton CO_2-e are necessary to drive significant changes in land use practices and ensure financially viable outcomes.

Farmers' participation in PES and conservation programs is influenced by both financial and non-financial motivations. While many recognize the long-term value of ecosystem services (ESs), engagement is often limited by lack of awareness, unclear policy signals, and inconsistent government support (Page and Bellotti 2015). Future PES programs should be designed with clear, accessible information, flexible payment structures, and contract terms tailored to farmers' realities.

Australia's experience also highlights the promise of voluntary, informal landholder collaborations to overcome common barriers such as limited trusted advice and low institutional support. While formal joint ventures remain challenging due to the complexity of C markets and the focus on individual gains, cooperative models adapted to regional contexts may offer a practical way forward. As C prices rise and policy frameworks stabilize, innovative and collaborative approaches could become increasingly feasible (Baumber et al. 2024). Continued research and participatory governance will be essential to support these evolving models and ensure that C farming contributes meaningfully to Australia's climate, land, and community goals.

10.2.4 Brazil

As one of the world's leading agricultural exporters, Brazil increasingly views C farming as a strategic means to reconcile agricultural productivity with national climate goals. The country is the fifth-largest emitter of GHGs (Climate Trace 2024), primarily due to deforestation and unsustainable land use practices. Under its Nationally Determined Contributions (NDCs), Brazil has pledged to reduce emissions by 53% by 2030. In support of this goal, the government launched the Low C Agriculture Plan (ABC Plan) in 2010, which has since evolved into the ABC+ Plan (2020–2030). This policy framework promotes sustainable practices such as no-till farming, integrated crop-livestock-forestry (ICLF) systems, biological N fixation, and animal waste management (Hayashi 2024). By 2020, over 54 M Ha had adopted ABC technologies, contributing to the avoidance of approximately 194 Tg of CO_2 emissions.

Empirical studies provide strong support for the effectiveness of these strategies. de Oliveira Ferreira et al. (2021) found that CA systems in southern Brazil restored SOC to 97 percent of native vegetation levels at a depth of one meter, often outperforming experimental plots. Key contributors to this success included high biomass inputs, minimal soil disturbance, and strategic use of cover crops such as oilseed radish. Likewise, Torres et al. (2019) demonstrated that long-term no-till practices in the Cerrado biome increased SOC stock and improved soil structure over time. In another example, Bayer SA's regenerative agriculture initiative, involving 1900 farmers and covering 232,000 ha, created Brazil's largest on-farm C data platform. The initiative developed a predictive model for SOC management to support participation in global C markets (Cherubin et al. 2024).

Despite these advances, Brazil still faces challenges. A major barrier is the absence of a robust Monitoring, Reporting, and Verification (MRV) system, which is essential for accurately quantifying C benefits and unlocking access to C markets. Developing MRV systems for Brazil's diverse biomes and farming systems is technically complex. Additional barriers include farmer hesitance to share data, non-standardized emission factors, and fragmented governance systems (Perosa et al. 2023).

Land tenure insecurity presents another critical hurdle, particularly in the Amazon and Cerrado regions. Without formal land titles, many smallholders and Indigenous communities are excluded from rural credit programs and disincentivized from making long-term sustainability investments (Page and Bellotti 2015). Policy reforms aimed at land regularization, coupled with expanded access to PES, are crucial for ensuring equitable participation in C farming initiatives (Pagiola 2008). Although Brazil has a legal foundation for PES through its Forest Code and National PES Policy (Soterroni et al. 2018; Wienke and Mattos 2022), the framework remains underutilized. To fulfill its potential, it requires adequate funding, streamlined implementation pathways, and better integration with rural credit systems such as PRONAF (Ely et al. 2019). This could incentivize sustainable land management (SLM) at scale.

To scale C farming effectively, Brazil needs a national C farming strategy with clearly defined targets—such as hectares under C farming by 2030—aligned with the ABC+ Plan and National Bioeconomy Strategy (Demirkol 2024). Strengthening rural extension services is also crucial to provide smallholders and Indigenous communities with technical guidance (Bragança et al. 2022). The development of C credit aggregation hubs can help small farms pool resources, lower transaction costs, and access markets (Brammer and Bennett 2022).

Encouragingly, Brazil is already investing in such transitions. The National Program for the Conversion of Degraded Pastures (2023) aims to restore 40 M Ha using climate-smart methods. Meanwhile, the Ecological Transition Plan, launched at COP28, integrates sustainability across Brazil's broader development agenda (Hayashi 2024).

Financial alignment is critical to scaling C farming. Although initiatives like the ABC Program and Safra Plan provide subsidized loans for low C practices, only a small share of rural credit currently supports climate-smart agriculture (Hayashi 2024). Expanding access to green credit, linking finance more directly to sustainability outcomes, and improving MRV systems will be essential for mainstreaming C farming.

With its vast agricultural lands and rich biodiversity, Brazil holds one of the highest global potentials for nature-based climate solutions. By aligning policy, finance, and research through a coordinated national strategy, Brazil can not only protect its soils and ecosystems but also emerge as a global leader in sustainable agriculture and C farming.

10.2.5 United States of America

C farming has emerged as a vital strategy within the United States' broader agenda to address climate change and advance agricultural sustainability. Scientific modeling tools, policy instruments, and private sector initiatives are increasingly aligned to scale the climate benefits of regenerative land management. The Marginal Abatement Cost Curve (MACC) estimates that U.S. agriculture could reduce up to 120 Tg CO_2e at a C price of $100 per Mg, primarily through improvements in manure, tillage, and N management (Pape et al. 2016). Additionally, data from COMET-Planner suggests that scaling adoption of core conservation practices could further mitigate an additional 118.5 Tg CO_2e annually over the next decade (Moore et al. 2023). Long-term farm-level studies, such as those conducted on David Brandt's Ohio farm, demonstrate that combining no-till practices with multi-species cover crops significantly enhances SOC, N retention, microbial activity, and crop yields (Islam and Reeder 2014).

Rather than creating a separate national C market, the U.S. integrates C farming incentives into existing conservation and cost-share programs (USDA 2025). This embedded approach enables farmers to participate in climate mitigation efforts through familiar institutions while aligning agricultural practices with environmental goals. The U.S. supports these initiatives through PES programs funded under the Farm Bill (USDA 2024) and managed by the Natural Resources Conservation Service (NRCS) and the USDA's Farm Service Agency (FSA). Key programs like the Conservation Reserve Program (CRP), Conservation Stewardship Program (CSP), Environmental Quality Incentives Program (EQIP), and Regional Conservation Partnership Program (RCPP) provide financial support for regenerative practices such as reduced tillage, multi-species cover cropping, rotational grazing, and wetland restoration (Table 10.1). Originally aimed at addressing broader resource concerns like water quality, biodiversity, and habitat preservation, these programs have become essential tools for increasing SOC, reducing nitrous oxide emissions, and enhancing farm-level climate resilience. However, traditional PES contracts often rely on backward-looking penalties, which can discourage participation. Emerging research highlights the benefits of forward-looking PES models, which use projected environmental impacts to create more flexible and equitable contracts. These models improve contract completion rates and better target high-impact landscapes, thereby maximizing environmental outcomes per conservation dollar (Kim et al. 2024).

The United States employs two primary models to incentivize C farming: per-practice and per-output. The per-practice model, used by corporations like Archer Daniels Midland (ADM), Bayer, and PepsiCo, provides fixed payments—typically

Table 10.1 USDA pathways empowering farmers to advance carbon (C) farming (USDA 2025)

Program	Typical conservation activities	Benefit to producers	Contract term	Payment term
Conservation Reserve Program (CRP)	Riparian buffers, grassed waterways, preserving grasslands	Reduced agricultural input costs, minimize soil erosion, improved soil health & resilience	10–15 years	Annual rental payments based on soil productivity; up to 50% cost-share for establishing approved practice
Conservation Stewardship Program (CSP)	Cover cropping, no-till, crop-rotation, grazing management, conserving grassland	Annual stewardship payment, soil health improvement, cuts input costs, increase farm resilience to climate and market changes	5 years (renewable)	A minimum payment of USD $4000 per year covering both ongoing maintenance and any newly approved enhancements
Environmental Quality Incentives Program (EQIP)	Any agricultural activity addressing NRCS identified resource concerns	Reduces farm runoff and input costs, improves soil health, enhances C sequestration, reduces on-farm GHG emissions	Initial length of 5 years	Cost share reimbursements; advance payments possible for underserved producers
Regional Conservation Partnership Program (RCPP)	Land management, enteric methane reductions, enhanced watershed resilience, wetland restorations	Access to pooled federal-private funds for landscape C projects (e.g., bio-digester manure management)	5–10 years typical varies by Classic vs Alternative Funding Arrangements (AFA)	NRCS contracts (Classic) or partner reimbursements (AFA)
Agricultural Management Assistance (AMA)	Efficient irrigation upgrades, windbreak/tree plantings, soil erosion controls, integrated pest management, and organic diversification	Soil water conservation, C in trees/ hedgerows	1–10 years	Reimbursement covers 75% of the costs incurred (up to $50,000 annually), 100% of income foregone, and no payments for practices started before the contract unless waived

ranging from \$5 to \$25 per acre (\$12 to \$60 per Ha)—for implementing approved practices such as no-till farming or nutrient management (Plastina et al. 2024a). This model is favored for its simplicity and predictability, but it may overestimate environmental benefits and reward land that continues to emit more C than it sequesters. On the other hand, the per-output model, promoted by companies like Indigo Ag, Nori, and Truterra, compensates farmers based on the verified amount of C sequestered—usually between \$15 and \$35 per Mg of $CO_{2}e$ (Plastina et al. 2024b). Though more precise, the per-output model is burdened by complex verification procedures and slow adoption. A hybrid model that combines the accessibility of per-practice incentives with the scientific rigor of per-output verification is now gaining traction as a potential solution to scale participation while ensuring measurable impact.

Despite growing interest in C farming, participation in C markets by U.S. farmers remains relatively low. Several structural challenges persist, including the low price of C credits (\$10–\$20 per Mg of $CO_{2}e$ or one credit), opaque and often complex contracts, concerns over the long-term permanence of CS, and the lack of credit or recognition for early adopters (Thompson et al. 2021). Moreover, dominant C markets like California's cap-and-trade and the Regional GHG Initiative (RGGI) have not yet included agricultural soil C due to the absence of standardized verification protocols (Gramig and Widmar 2018). In this policy vacuum, the USDA's CRP has stepped up, retiring over 24 million acres (~10 M Ha) of environmentally sensitive land, preventing more than eight billion tons of soil erosion, and providing the largest federal CS benefits (Hanson and Schmidt 2017).

While scientific understanding of soil C dynamics is advancing rapidly, translating this knowledge into national policy remains a key challenge. International platforms like the Global Soil Partnership (GSP) and the Intergovernmental Technical Panel on Soils (ITPS) are working to bridge the gap between science and policy, aiming to align soil C strategies with the United Nations Sustainable Developments Goals (SDGs) (Montanarella 2017). Public perception is also a critical driver; recent studies indicate that increasing public consensus on climate change could not only reduce emissions but also yield up to \$4.75 trillion in welfare benefits through stronger policy support and market signals (El Armali and Rahimian 2024).

The U.S. approach to C farming is multifaceted, combining public conservation programs, private market initiatives, and scientific research. To fully realize its climate mitigation potential, key priorities include reforming PES structures to reward proactive land stewardship, standardizing verification protocols for SOC, expanding hybrid incentive models that balance simplicity with rigor, and recognizing early adopters and legacy practices. Equally important are efforts to enhance education and outreach, particularly in underserved farming communities. Holistic programs like Frontline to Farm exemplify how integrating education, community engagement, and regenerative agriculture can advance sustainability, economic opportunity, and social equity ((Fanatico et al. 2025). With stronger policy alignment, coordinated implementation, and improved market infrastructure, U.S. farmers are well positioned to lead the transition toward a more resilient and low C agricultural future, both nationally and globally.

10.2.6 Europe

The European Union (EU) views C farming as a central strategy to achieve its climate neutrality target by 2050, particularly in addressing residual emissions from agriculture and land use (Montanarella 2017). Through initiatives like the C Farming Initiative and the proposed C Removal Certification Framework (CRCF), and the EU promotes SLM practices that enhance SOC stocks while providing co-benefits for biodiversity, soil health, and nutrient efficiency (Günther et al. 2024). These efforts are deeply aligned with the European Green Deal, which sets out a roadmap for making the EU's economy sustainable and climate-neutral, including the 'Fit for 55' package and the Regulation on Land Use, Land-Use Change, and Forestry (LULUCF) as key pillars for reducing GHG emissions by at least 55% by 2030 (European Commission 2023). Events like the 2025 European C Farming Summit and partnerships such as Project Credible reflect Europe's broader efforts to align policy, finance, and practice in order to scale C farming across the continent (Climate KIC 2025). Recognized C farming practices include CA, cover cropping, agroforestry, livestock and manure management, nutrient management, peatland rewetting through paludiculture, application of organic fertilizers and biochar, and maintaining biodiversity-friendly features like flower strips (McDonald et al. 2021; Bumbiere et al. 2022; Panagos et al. 2022; Günther et al. 2024).

While the climate mitigation potential of these practices is widely acknowledged, the EU emphasizes that C removals should complement, not substitute, direct emissions reductions. This stems from concerns about the temporary nature of soil C storage, risks of overestimation, and double counting within C markets (Günther et al. 2024). Scientific uncertainties, particularly around MRV, as well as questions of permanence and additionality, further complicate scaling up C farming (McDonald et al. 2021).

Nevertheless, Europe supports a diverse suite of C farming solutions that balance climate mitigation with broader environmental and economic benefits. Key approaches include peatland rewetting, which curbs further C loss from drained soils (Panagos et al. 2022); agroforestry, where trees are integrated with crops and livestock to deliver long-term CS and biodiversity gains (Mosquera-Losada et al. 2018; Vanneste et al. 2025); and improved soil management through organic amendments and reduced tillage, particularly within organic farming systems that align with the EU's pesticide reduction goals (Krauss et al. 2022). Additionally, the integration of C farming with renewable energy—such as biogas and biomethane production supports both emissions reduction and energy security, advancing Europe's circular bioeconomy agenda (Bumbiere et al. 2022). A comparative assessment of C farming methods in Latvian conditions highlighted their varying potential for GHG sequestration and land applicability (Fig. 10.1).

Policy support is primarily delivered through the Common Agricultural Policy (CAP), which funds C farming under both Pillar I (direct payments linked to sustainability standards) and Pillar II (rural development co-funded by Member States) (Mosquera-Losada et al. 2018). However, the effectiveness of CAP varies by

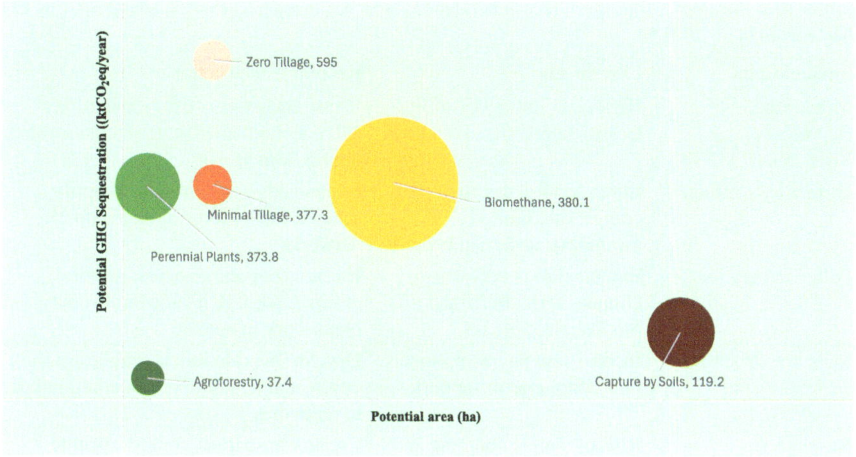

Fig. 10.1 The potential of carbon farming practices in Europe. Bubble size reflects the indicative budget investment (EUR) required for each practice. (Data for this figure are adapted from the study by Bumbiere et al. 2022)

country, depending on how national Strategic Plans are implemented (McDonald et al. 2021). Legal and administrative barriers—such as tree density restrictions in agroforestry and complex MRV requirements also slow adoption (Mosquera-Losada et al. 2018).

Europe's soil-related challenges, including widespread degradation, erosion, and C loss from mineral soils, have further underscored the urgency of advancing C farming (Panagos et al. 2022). The EU Soil Strategy for 2030 explicitly positions C farming as a critical solution to restoring soil health while delivering climate benefits. Research initiatives such as the Horizon Europe Soil Mission, European Joint Programme on Soil (Arias-Navarro et al. 2023), and EU funding schemes like LIFE and INTERREG (McDonald et al. 2021), are driving innovation in soil health monitoring, living labs, and farmer-led climate solutions. Together, these efforts are building the scientific, practical, and policy foundations to scale up C farming across Europe.

Despite uneven success so far due to financial, technical, and MRV challenges, the EU is steadily building an enabling framework to mainstream C farming (Table 10.2). Tools like the C Agroforestry Tool (CARAT) now help quantify sequestration in agroforestry systems, enabling evidence-based planning and incentivizing farmers through C credits (Vanneste et al. 2025).

To unlock C farming's full potential, Europe must invest in cost-effective MRV systems, streamline cross-sector policies, enhance farmer incentives, and implement safeguards to ensure environmental integrity and long-term C storage (McDonald et al. 2021; Günther et al. 2024). Coordinated public-private efforts, including certification schemes and C markets, are seen as essential for transitioning C farming from scattered pilots to mainstream practice across the continent (Bumbiere et al. 2022).

Table 10.2 Strategic gaps and recommendations for scaling up carbon farming in the EU (McDonald et al. 2021)

Strategic area	Current gap	Recommended action
Monitoring, Reporting & Verification (MRV)	High cost, low accessibility, lack of standardized methods	Invest in cost-effective, standardized MRV technologies (e.g., remote sensing, digital farm tools)
Farmer engagement	Low adoption due to complexity, limited awareness, uncertain benefits	Expand advisory services, simplify schemes, and offer training/support networks
Policy integration	Fragmentation between climate, agriculture, and biodiversity policies	Harmonize policies across sectors; ensure CAP, CRCF, and biodiversity plans work in tandem
Incentive structure	Inconsistent payment models, limited long-term support	Develop hybrid models (public-private), ensure stable, multi-year funding linked to outcomes
Baseline & Additionality Setting	Risk of double counting and rewarding non-additional practices	Create transparent, context-specific baselines and clear rules to avoid overlapping with existing CAP support
Equity across member states	Uneven capacity and uptake across EU regions	Tailor national CAP strategic plans to regional potential and support under-resourced areas

10.3 A Global Approach to Carbon Farming

The concept of C farming is increasingly gaining recognition worldwide, with many nations now establishing frameworks to support SAPs. These frameworks aim to improve soil productivity, restore soil health, and enhance SOC stocks—essential components in addressing the intertwined challenges of climate change and land degradation.

As the global climate continues to shift, agricultural systems face growing threats, including declining soil fertility (Mondal 2021), increased risk of crop failure, and food insecurity (Mirzabaev et al. 2023). C farming offers a promising path forward by not only mitigating GHG emissions but also by building agricultural resilience.

The successful adoption of C farming practices hinges on a range of social, economic, and institutional factors. One of the most critical factors influencing adoption is financial incentive. Financial incentives such as subsidies, C markets, or PES play a vital role in encouraging participation (Ingram et al. 2014). Both practice-based and result-based schemes that fairly compensate farmers for their efforts are essential for promoting widespread adoption (Raina et al. 2024). Direct payment programs, including Soil C Incentive Payments, help address market failures by rewarding farmers who implement C-sequestering LMP practices. To be effective, these programs should offer flexibility by allowing farmers to opt in or out and reduce risks for smallholders by minimizing opportunity costs and uncertainties related to long term benefits (Pascual et al. 2015).

In addition to incentives, perceived risk plays a major role. Farmers may be hesitant to shift from conventional practices due to concerns about short-term yield losses or uncertainty around long-term benefits (Dumbrell et al. 2016; Kragt et al. 2017). However, yield reductions in CA systems are typically not inherent to the practice itself but result from poor system adaptation—such as inadequate seeding equipment, insufficient residue cover, ineffective pest and weed control, and lack of operator experience (Derpsch et al. 2014). These challenges can lead to poor plant establishment and reduced productivity in the initial years. Without proper technical support and system-level adjustments, farmers may struggle to realize the full potential of CA. Evidence shows that adoption increases when farmers witness soil degradation firsthand and recognize improvements in productivity, labor efficiency, and input savings (Barrowclough et al. 2016). Therefore, effectively communicating the economic advantages, addressing yield concerns, and ensuring access to knowledge and technical assistance are essential for broad and sustained adoption. Informed and well-trained farmers are significantly more likely to adopt and maintain SAPs (Garbach et al. 2012). Without such support, even the best-designed programs may struggle to gain traction.

Landholding size, though often considered a determinant, does not consistently predict adoption behavior. While larger landowners may have more resources, smallholders often face higher barriers and risks, requiring targeted support (Loria et al. 2015). For example, studies from South Africa showed lower adoption rates of CA among larger farmers due to perceived incompatibility with their operational scale (Ntshangase et al. 2018). Hence, adoption strategies must go beyond land size and focus on inclusive approaches that ensure equitable participation across different farm types. This requires tailored support such as early-stage incentives, targeted training, enhanced extension services, and simplified eligibility criteria to effectively engage both smallholders and large-scale farmers (Schreefel et al. 2022; Okyere and Kornher 2023).

Regionally, C farming must be designed to address local degradation drivers such as soil erosion, desertification, deforestation, and biodiversity loss (Damian et al. 2021; Lal 2024; Moodley et al. 2024). Context-specific solutions—like forest-edge restoration, rangeland rehabilitation, or cover cropping—can align ecological restoration with income generation (Evans et al. 2015). A regionally nuanced approach that links local degradation drivers with viable C farming solutions can accelerate adoption and maximize impact. These localized interventions should be supported by strong monitoring and evaluation systems to measure outcomes and continuously refine practices.

From a scientific standpoint, soils represent one of the largest C sinks on the planet, storing around 1325 petagrams (Pg) of SOC in the top one meter, and nearly 3000 Pg when including deeper layers (Köchy et al. 2015). Agriculture, often viewed as a source of emissions, has the potential to contribute to climate solutions (Lal 2023a). The "4 per 1000" initiative illustrates this by proposing that an annual 0.4 percent increase in SOC stocks on managed lands could sequester 2 to 3 Pg of C annually, potentially offsetting up to 35% of global anthropogenic GHG emissions (Minasny et al. 2017). Achieving these goals demands coordinated, multi-level collaboration. Governments must enact enabling policies and fund pilot programs. Institutions such as NGOs, cooperatives, and extension services play a critical role in bridging policy and practice, while the research community advances methodologies for C measurement and site-specific implementation. Private sector involvement can introduce financing and innovation, while non-profits ensure social and environmental safeguards (Velten et al. 2021; Baumber et al. 2024; Miao et al. 2025). The private sector can bring innovation, technology, and financing; non-profits can advocate for social equity and environmental integrity; and governments can create the legislative environment needed to support all these efforts.

Investing in C farming today means investing in the future of agriculture, climate stability, and rural livelihoods. As countries seek to meet international climate targets such as those under the Paris Agreement, integrating C farming into national adaptation and mitigation strategies offers a scalable, science-based, and socially inclusive solution.

10.4 Emerging Research Priorities in Carbon Farming

While C farming has made commendable progress through research and pilot programs across various regions, there remain critical yet underrepresented areas that warrant greater attention. Much of the current focus has rightly been on developing cost-effective MRV systems (Perosa et al. 2023; Okoli and Birkenberg 2024), understanding SOC sequestration across diverse land types (Lal and Kimble 1997; Blanco-Canqui and Lal 2008; Patle et al. 2013; Blanco-Canqui 2022; Yu and Song 2023; Li and Tasnady 2023), and analyzing the agronomic and economic impacts of these practices (Antle et al. 2007; Dang et al. 2015). These foundational areas are essential for establishing credible C markets and supporting farmer participation. Equally important are the co-benefits of C farming, such as enhanced biodiversity, improved water retention, pest resilience, and the behavioral and institutional factors that influence farmer adoption (Kragt et al. 2016; Fleming et al. 2019; Hou et al. 2020; Gao et al. 2025).

A critical and rapidly evolving area is satellite-based monitoring, which offers scalable, real-time tools for tracking C fluxes, SOC dynamics, and land-use impacts. Multispectral sensors, Synthetic Aperture Radar (SAR), Unmanned Aerial Vehicles (UAVs), and AI-enabled platforms are being employed to support spatially explicit C assessments (Metrikaitytė Gudelė and Sužiedelytė Visockienė 2024; Conti et al. 2025). Integrating multi-source satellite data with field campaigns helps distinguish between natural and anthropogenic emissions and improves process-based C modeling (Wang et al. 2025). Model–data fusion approaches, such as combining satellite-derived Leaf Area Index (LAI) with C models, have shown promise in estimating C stocks in managed landscapes (Myrgiotis et al. 2020). Projects like Field Observatory Network (FiON) and INNO4CFIs are advancing real-time decision-support tools by integrating satellite imagery, weather data, and on-farm metrics (Nevalainen et al. 2022; Conti et al. 2025). This is especially valuable in smallholder contexts (Kearney et al. 2017) and challenging environments such as semi-arid zones and cloud covered regions (Hosseinpour-Zarnaq et al. 2024), where satellite data can overcome traditional monitoring limitations.

Recent studies have identified four key emerging research priorities in C farming: (1) linking soil microbiome diversity to CS and plant health, (2) understanding how farm management practices shape microbial communities and soil C dynamics, (3) valuing soil health for its combined climate and ecosystem benefits, and (4) strengthening science-policy-farmer networks to scale up data-driven, site-specific C farming solutions (Neher et al. 2022).

Despite growing interest, certain strategies remain relatively underexplored in mainstream C farming discourse. One such area is the SIC sequestration, particularly through enhanced weathering and mineral carbonation (Liang et al. 2024; Hasemer et al. 2024; Liao et al. 2025; Lei et al. 2025). A notable example is the Mati C project, which won the \$50 M XPRIZE C Removal award for its scalable approach using locally sourced rock powders to accelerate mineral weathering in soils and capture atmospheric CO_2 through stable carbonate formation (Mati 2025). Field studies have shown that mixing soils with fine concrete fragments can increase soil pH and bicarbonate (HCO_3^-) concentrations in leachates, indicating successful CO_2 capture through mineral reactions (Multer Hopkins et al. 2024). These materials not only promote durable C storage but also enhance soil chemistry and fertility. Further innovations such as microbe-enhanced silicate weathering and mineral-amended biochar offer dual benefits by improving nutrient availability and stabilizing C through microaggregation and mineral association (Zhu et al. 2017; Buss et al. 2022, 2024a, b). As research progresses, integrating these strategies into C farming systems could significantly expand the scope and permanence of soil-based C sequestration.

Another vital but less emphasized opportunity is the integration of perennial vegetation into agricultural systems. Perennials, including agroforestry species and perennial pastures, contribute to continuous C input, reduced erosion, improved water retention, and biodiversity support, all of which enhance long-term soil C accumulation (Pauli et al. 2018; Gross et al. 2022). Similarly, integrated farming systems that combine livestock, trees, and crops promote nutrient cycling, reduce reliance on external inputs, and lead to more resilient and low-C farming landscapes. In Benin, farms using cereal-legume-forage-livestock integration sequestered up to 158 kg C ha^{-1} $year^{-1}$ which is double that of pastoral systems (Idrissou et al. 2024). Addressing livestock emissions through region-specific emission intensity assessments and integrated environmental-economic planning is also crucial for supporting low-C livestock systems (Li and Xiao 2024).

To further enhance scalability and equity, C farming must advance MRV technologies that track both SOC and SIC more effectively (Ng et al. 2025; Liao et al. 2025), develop inclusive compensation models that support smallholder participation (Simone et al. 2017; Beka et al. 2022), and leverage traditional ecological knowledge for practice adaptation (Maharjan et al. 2025; Knorr and Augustin 2025). A critical research priority is designing inclusive PES models that balance farmer satisfaction, risk reduction, and economic returns. Incentive mechanisms for C farming are evolving, including result-based, action-based, and hybrid payment models (Raina et al. 2024). To build farmer confidence and encourage adoption particularly among smallholders, beginning farmers, and those already implementing C-enhancing practices action-based payments such as upfront payments or subsidies should be prioritized (Fanatico et al. 2025). These lower-risk models can ease participation barriers and provide immediate benefits. As farmers gain confidence and understanding, hybrid models that combine upfront payments with additional rewards based on verified mitigation outcomes can be introduced. Eventually, result-based payments can be promoted among farmers who are ready to independently engage with C markets, ensuring performance-based incentives without placing an undue burden on them. Further research must establish fair and equitable pricing models that compensate farmers appropriately for their sequestration efforts, while minimizing risks and transaction costs. Payments should reflect the growing societal value of C, which has risen from $30–35 per credit in 2014, to an estimated $45–55 per credit in 2024–2025 (Lal 2014, 2024).

The overarching goal should be to transition toward low-waste, high-efficiency agricultural systems based on conservation principles (Lal 2023b). While such systems may not deliver quick results, they are grounded, effective, and adaptable to local ecological and socio-economic contexts. Site-specific practices, tailored to soil types, climates, and community needs, can deliver genuine environmental benefits while sustaining farmer livelihoods (Santra et al. 2012). Ultimately, C farming must evolve into a holistic, site-specific, conservation-based approach that balances C goals with productivity, land restoration, and long-term farmer benefits.

10.5 From Science to Scalable Action

Translating the science of C farming into widespread, scalable action is both an opportunity and a necessity in the global response to climate change. It requires a systems approach that integrates robust science-policy interfaces, enforceable legislation, practical tools, and inclusive, farmer-centered implementation.

Global platforms like the Intergovernmental Panel on Climate Change (IPCC) have demonstrated success in converting complex science into actionable policy frameworks (Livingston and Rummukainen 2023). However, soil science has long been underrepresented in such discussions, limiting its influence on climate agendas. The establishment of the ITPS under the GSP marks a critical step forward, emphasizing threats such as erosion and contamination, and supporting the SDGs through Voluntary Guidelines on Sustainable Soil Management that aim to connect global governance with local action (Montanarella et al. 2016; Montanarella 2017).

Despite progress, challenges remain. Soil protection policies are often constrained by private land ownership complexities, national sovereignty concerns, and institutional fragmentation (Stankovics et al. 2020). To scale C farming, clear and enforceable legislation is essential. The proposed Healthy Soil Act, modeled on landmark U.S. laws like the Clean Air Act and Clean Water Act, aims to institutionalize soil health as a climate solution, establish performance standards, and incentivize sustainable soil management through C markets (Lal 2020, 2025). In parallel, a proposed C Farming Certification system would validate on-farm CS, improve access to financing, and integrate farmers into sustainable supply chains (Marks 2020).

Effective implementation must account for the spatial variability of soils and the diversity of ecological and socio-economic contexts (Santra et al. 2012). Practices such as CA, cover cropping, agroforestry, and grazing management show promise but require local adaptation to ensure relevance and success (Odebiri et al. 2021; Ogieriakhi and Woodward 2022). Integrating scientific research with farmers' local knowledge is essential for designing practical, acceptable solutions (Zhang et al. 2022; Maharjan et al. 2025; Knorr and Augustin 2025). The development of site-specific soil health indicators, informed by high-resolution mapping and machine learning, can help identify priority areas for C farming interventions (Hou et al. 2020).

Practical, field-based monitoring systems, such as those piloted in Argentina and Brazil, demonstrate how practical, locally adapted protocols can track soil C while guiding sustainable management decisions (Noellemeyer et al. 2024). However, short-term studies offer limited insight into long-term sequestration; thus, long-term trials are necessary to validate benefits and refine best practices.

Emerging technologies such as big data analytics, remote sensing, digital soil mapping, and precision agriculture tools offer transformative potential for tracking SOC at scale (Odebiri et al. 2021; Dibbern et al. 2024). Yet for these tools to be effective, they must be affordable, accessible, and farmer-friendly, ensuring adoption across diverse geographies and farm types. Industry and business, as highlighted by initiatives like 4 per 1000, also have a pivotal role through investment, innovation, and market-driven incentives for soil CS (Lal 2020).

Ultimately, scaling up C farming requires aligning science, policy, finance, and practice (Fig. 10.2). By integrating soil C metrics into policy frameworks, C credit systems, and farm-level innovations, C farming can move beyond isolated projects to become a global strategy that delivers measurable climate, ecological, and economic benefits. Achieving this transformation will depend on building strong connections between global climate goals and local farming realities, turning scientific insights into practical solutions and incentives that drive sustainable development and climate resilience worldwide.

Now is the time to act, not only to realize the scientific potential of soil CS but also to catalyze a just, inclusive, and sustainable transformation of agriculture.

10.6 Conclusion

C farming represents a transformative opportunity to simultaneously address the global crises of climate change, soil degradation, and food insecurity. With its capacity to sequester atmospheric CO_2, regenerate soil health, and enhance ESs, C

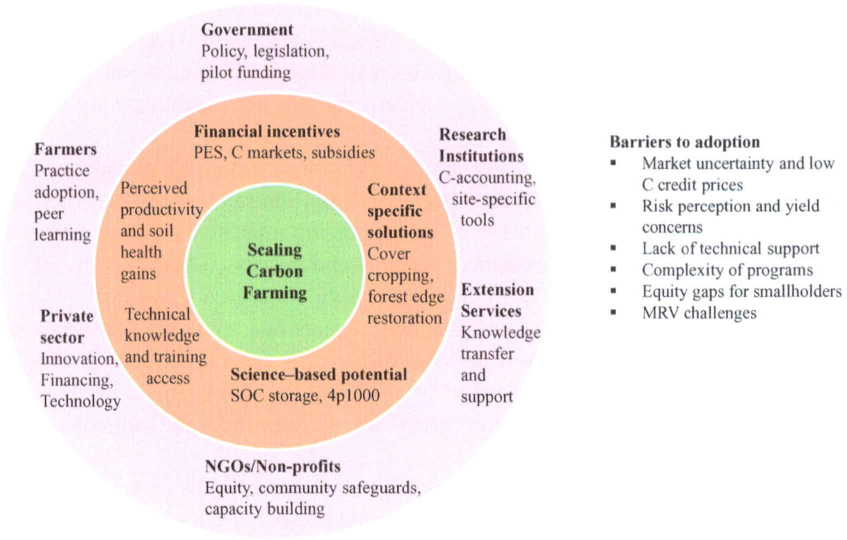

Fig. 10.2 A conceptual framework highlighting the enabling factors, stakeholder roles, and key barriers influencing the adoption and scalability of carbon farming

farming stands out as a nature-based solution that can contribute meaningfully to national and global climate targets. Across regions including the ABC Plus Plan in Brazil, Farm Bill programs in the United States, the Common Agricultural Policy in Europe, and emerging initiatives in South Asia and Africa, there is growing momentum toward integrating C farming into sustainable land management. However, the pace and scale of adoption remain uneven due to systemic challenges including limited financial incentives, weak monitoring frameworks, knowledge gaps, and socio-economic inequalities.

To overcome these barriers, it is essential to build strong science-policy interfaces, implement inclusive and adaptive payment models, and develop region-specific strategies that are responsive to local degradation drivers, farming realities, and land tenure conditions. Long-term field trials, accessible MRV systems, and the integration of farmer knowledge with cutting-edge technologies such as digital soil mapping, big data analytics, and AI-based/satellite-based decision support tools will be pivotal in scaling site-specific carbon farming solutions. Moreover, emerging frontiers like soil inorganic carbon sequestration, integration of perennial vegetation, and microbial-soil carbon interactions offer new avenues for research and innovation.

Multi-level collaboration is essential. Governments must provide enabling legislation, research institutions should continue to develop science-based metrics and models, the private sector can contribute to finance and innovation, and non-profit organizations can ensure that efforts are socially inclusive and environmentally sound. Legal frameworks such as the proposed Healthy Soil Act and C Farming Certification frameworks can institutionalize soil health within climate policies, ensure farmer participation in C markets, and build long-term resilience across agricultural landscapes.

In the long term, C farming has the potential to deliver multiple benefits such as climate mitigation, improved productivity, and rural livelihood support if implemented with equity, scientific rigor, and community engagement. Investing in C farming now means investing in a future where agriculture is not only productive and profitable, but also resilient, restorative, and climate friendly. The time to act is now, not only to realize the scientific potential of SOC sequestration but also to support an inclusive and sustainable transformation of agriculture for future generations.

References

Agarwal V, Malhotra S, Dagar V, Pavithra MR (2023) Coping with public-private partnership issues: a path forward to sustainable agriculture. Socio Econ Plan Sci 89:101703. https://doi.org/10.1016/j.seps.2023.101703

agriprove (2025) Swartz carbon project. https://agriprove.io/swartz-carbon-project. Accessed 7 Apr 2025

Antle JM, Stoorvogel JJ, Valdivia RO (2007) Assessing the economic impacts of agricultural carbon sequestration: terraces and agroforestry in the Peruvian Andes. Agric Ecosyst Environ 122:435–445. https://doi.org/10.1016/j.agee.2007.02.003

Arias-Navarro C, Panagos P, Jones A et al (2023) Forty years of soil research funded by the European Commission: trends and future. A systematic review of research projects. Eur J Soil Sci 74:e13423. https://doi.org/10.1111/ejss.13423

Aryal JP, Rahut DB, Sapkota TB et al (2020) Climate change mitigation options among farmers in South Asia. Environ Dev Sustain 22:3267–3289. https://doi.org/10.1007/s10668-019-00345-0

Barrowclough M, Stehouwer R, Alwang J et al (2016) Conservation agriculture on steep slopes in the Andes: promise and obstacles. J Soil Water Conserv 71:91–102. https://doi.org/10.2489/jswc.71.2.91

Baumber A, Cross R, Ampt P et al (2024) Soil-based carbon farming: opportunities for collaboration. J Rural Stud 108:103268. https://doi.org/10.1016/j.jrurstud.2024.103268

Bayer (2025) Bayer to issue first carbon credits from regenerative farming practices in India. Krishak Jagat. https://www.en.krishakjagat.org/crop-protection/bayer-to-issue-first-carbon-credits-from-regenerative-farming-practices-in-india/. Accessed 2 Apr 2025

Beka S, Burgess PJ, Corstanje R, Stoate C (2022) Spatial modelling approach and accounting method affects soil carbon estimates and derived farm-scale carbon payments. Sci Total Environ 827:154164. https://doi.org/10.1016/j.scitotenv.2022.154164

Biramo G (2018) The role of integrated nutrient management system for improving crop yield and enhancing soil fertility under small holder farmers in sub-Saharan Africa: a review article. Mod Concepts Dev Agron 2:221–229. https://doi.org/10.31031/MCDA.2018.02.000547

Blanco-Canqui H (2022) Cover crops and carbon sequestration: lessons from U.S. studies. Soil Sci Soc Am J 86:501–519. https://doi.org/10.1002/saj2.20378

Blanco-Canqui H, Lal R (2008) No-tillage and soil-profile carbon sequestration: an on-farm assessment. Soil Sci Soc Am J 72:693–701. https://doi.org/10.2136/sssaj2007.0233

Bragança A, Newton P, Cohn A et al (2022) Extension services can promote pasture restoration: evidence from Brazil's low carbon agriculture plan. Proc Natl Acad Sci 119:e2114913119. https://doi.org/10.1073/pnas.2114913119

Brammer TA, Bennett DE (2022) Arriving at a natural solution: bundling credits to access rangeland carbon markets. Rangelands 44:281–290. https://doi.org/10.1016/j.rala.2022.04.001

Bumbiere K, Diaz Sanchez FA, Pubule J, Blumberga D (2022) Development and assessment of carbon farming solutions. Environ Climate Technol 26:898–916. https://doi.org/10.2478/rtuect-2022-0068
Buss W, Wurzer C, Manning DAC et al (2022) Mineral-enriched biochar delivers enhanced nutrient recovery and carbon dioxide removal. Commun Earth Environ 3:67. https://doi.org/10.1038/s43247-022-00394-w
Buss W, Hasemer H, Ferguson S, Borevitz J (2024a) Stabilisation of soil organic matter with rock dust partially counteracted by plants. Glob Chang Biol 30:e17052. https://doi.org/10.1111/gcb.17052
Buss W, Hasemer H, Sokol NW et al (2024b) Applying minerals to soil to draw down atmospheric carbon dioxide through synergistic organic and inorganic pathways. Commun Earth Environ 5:602. https://doi.org/10.1038/s43247-024-01771-3
Cheesman S, Thierfelder C, Eash NS et al (2016) Soil carbon stocks in conservation agriculture systems of southern Africa. Soil Tillage Res 156:99–109. https://doi.org/10.1016/j.still.2015.09.018
Cherubin MR, Pinheiro Junior CR, Alves LA et al (2024) Carbon farming initiative: a national-scale public-private partnership to promote regenerative agriculture in Brazil. Exp Agric 60:e28. https://doi.org/10.1017/S0014479724000255
Clarke C, Shackleton S, Powell M (2012) Climate change perceptions, drought responses and views on carbon farming amongst commercial livestock and game farmers in the semiarid Great Fish River Valley, Eastern Cape province, South Africa. Afr J Range Forage Sci 29:13–23. https://doi.org/10.2989/10220119.2012.687041
Climate KIC (2025) Beyond sequestration: insights from the 2nd EU Carbon Farming Summit. https://www.climate-kic.org/news/beyond-sequestration-insights-from-the-2nd-eu-carbon-farming-summit/. Accessed 5 May 2025
Climate Trace (2024) Climate trace. Climate trace. https://www.climatetrace.org/inventory?sector=agriculture&country=USA&year_from=2024&year_to=2024&gas=co2e20. Accessed 1 Apr 2025
Conti N, Scatena L, Lombardi E et al (2025) Living hubs in the INNO4CFIs project. In: From ground to satellites to AI applications: integrating advanced technologies for carbon farming in agroforestry, pp 399–408
Damian JM, Durigan MR, Cherubin MR et al (2021) Deforestation and land use change mediate soil carbon changes in the eastern Brazilian Amazon. Reg Environ Chang 21:64. https://doi.org/10.1007/s10113-021-01796-w
Dang YP, Moody PW, Bell MJ et al (2015) Strategic tillage in no-till farming systems in Australia's northern grains-growing regions: II. Implications for agronomy, soil and environment. Soil Tillage Res 152:115–123. https://doi.org/10.1016/j.still.2014.12.013
de Oliveira Ferreira A, Amado TJC, Rice CW et al (2021) Comparing on-farm and long-term research experiments on soil carbon recovery by conservation agriculture in southern Brazil. Land Degrad Dev 32:3365–3376. https://doi.org/10.1002/ldr.4015
Demirkol S (2024) Brazil launches national bioeconomy strategy. Brazilian NR
Derpsch R, Franzluebbers AJ, Duiker SW et al (2014) Why do we need to standardize no-tillage research? Soil Tillage Res 137:16–22. https://doi.org/10.1016/j.still.2013.10.002
Dibbern T, Romani LAS, Massruhá SMFS (2024) Main drivers and barriers to the adoption of digital agriculture technologies. Smart Agric Technol 8:100459. https://doi.org/10.1016/j.atech.2024.100459
Diop M, Chirinda N, Beniaich A et al (2022) Soil and water conservation in Africa: state of play and potential role in tackling soil degradation and building soil health in agricultural lands. Sustainability 14:13425. https://doi.org/10.3390/su142013425
Dumbrell NP, Kragt ME, Gibson FL (2016) What carbon farming activities are farmers likely to adopt? A best–worst scaling survey. Land Use Policy 54:29–37. https://doi.org/10.1016/j.landusepol.2016.02.002

El Armali J, Rahimian M (2024) Public climate change agreement and GHG emissions in the US. Int Rev Appl Econ 38:498–504. https://doi.org/10.1080/02692171.2024.2316336
Ely RA, Parfitt R, Carraro A, Ribeiro FG (2019) Rural credit and the time allocation of agricultural households: the case of PRONAF in Brazil. Rev Dev Econ 23:1863–1890. https://doi.org/10.1111/rode.12606
European Commission (2023) The European green Deal – delivering the EU's 2030 climate targets. Publications Office of the European Union. https://data.europa.eu/doi/10.2775/783179. Accessed 19 May 2025
Evans MC, Carwardine J, Fensham RJ et al (2015) Carbon farming via assisted natural regeneration as a cost-effective mechanism for restoring biodiversity in agricultural landscapes. Environ Sci Pol 50:114–129. https://doi.org/10.1016/j.envsci.2015.02.003
Fanatico AC, Gibbard LG, Bramall SB (2025) Frontline to farm: sustainable farming training for military veterans and beginning farmers. Poult Sci 104:104443. https://doi.org/10.1016/j.psj.2024.104443
Fleming A, Stitzlein C, Jakku E, Fielke S (2019) Missed opportunity? Framing actions around co-benefits for carbon mitigation in Australian agriculture. Land Use Policy 85:230–238. https://doi.org/10.1016/j.landusepol.2019.03.050
Gao Y, Connor J, Summers D et al (2025) Carbon farming co-benefits optimization in south-eastern Australian rangeland. J Environ Manag 382:125315. https://doi.org/10.1016/j.jenvman.2025.125315
Garbach K, Lubell M, DeClerck FAJ (2012) Payment for ecosystem services: the roles of positive incentives and information sharing in stimulating adoption of silvopastoral conservation practices. Agric Ecosyst Environ 156:27–36. https://doi.org/10.1016/j.agee.2012.04.017
Ghimire R, Khatri-Chhetri A, Chhetri N (2022) Institutional innovations for Climate smart agriculture: assessment of Climate-Smart Village approach in Nepal. Front Sustain Food Syst 6:734319. https://doi.org/10.3389/fsufs.2022.734319
Gramig BM, Widmar NJO (2018) Farmer preferences for agricultural soil carbon sequestration schemes. Appl Econ Perspect Policy 40:502–521. https://doi.org/10.1093/aepp/ppx041
Gross CD, Bork EW, Carlyle CN, Chang SX (2022) Agroforestry perennials reduce nitrous oxide emissions and their live and dead trees increase ecosystem carbon storage. Glob Chang Biol 28:5956–5972. https://doi.org/10.1111/gcb.16322
Günther P, Garske B, Heyl K, Ekardt F (2024) Carbon farming, overestimated negative emissions and the limits to emissions trading in land-use governance: the EU carbon removal certification proposal. Environ Sci Eur 36:72. https://doi.org/10.1186/s12302-024-00892-y
Hanson KD, Schmidt JM (2017) US farm programs and the impacts on national and international soil security. In: Field DJ, Morgan CLS, McBratney AB (eds) Global soil security. Springer, Cham, pp 411–424
Hasemer H, Borevitz J, Buss W (2024) Measuring enhanced weathering: inorganic carbon-based approaches may be required to complement cation-based approaches. Front Clim 6:1352825. https://doi.org/10.3389/fclim.2024.1352825
Hayashi T (2024) Sustainable agriculture programs in Brazil – past present and future. https://apps.fas.usda.gov/newgainapi/api/Report/DownloadReportByFileName?fileName=Sustainable%20Agriculture%20Programs%20in%20Brazil-%20Past%20Present%20and%20Future_Brasilia_Brazil_BR2024-0006.pdf. Accessed 10 Apr 2025
Hayo L, Hasegawa H (2024) Enhancing emission reductions in South African agriculture: the crucial role of carbon credits in incentivizing climate-smart farming practices. Sustainable Futures 8:100260. https://doi.org/10.1016/j.sftr.2024.100260
Hosseinpour-Zarnaq M, Moshiri F, Jamshidi M et al (2024) Monitoring changes in soil organic carbon using satellite-based variables and machine learning algorithms in arid and semi-arid regions. Environ Earth Sci 83:582. https://doi.org/10.1007/s12665-024-11876-9
Hou D, Bolan NS, Tsang DCW et al (2020) Sustainable soil use and management: an interdisciplinary and systematic approach. Sci Total Environ 729:138961. https://doi.org/10.1016/j.scitotenv.2020.138961

Idrissou Y, Vall E, Blanfort V et al (2024) Integrated crop-livestock effects on soil carbon sequestration in Benin, West Africa. Heliyon 10:e28748. https://doi.org/10.1016/j.heliyon.2024.e28748
Ingram J, Mills J, Frelih-Larsen A et al (2014) Managing soil organic carbon: a farm perspective. EuroChoices 13:12–19. https://doi.org/10.1111/1746-692X.12057
Islam R, Reeder R (2014) No-till and conservation agriculture in the United States: an example from the David Brandt farm, Carroll, Ohio. Int Soil Water Conserv Res 2:97–107. https://doi.org/10.1016/S2095-6339(15)30017-4
Jassim D, Witt B, Evans MC (2022) Community perceptions of carbon farming: a case study of the semi-arid Mulga lands in Queensland, Australia. J Rural Stud 96:78–88. https://doi.org/10.1016/j.jrurstud.2022.10.010
Jat ML, Chakraborty D, Ladha JK et al (2020) Conservation agriculture for sustainable intensification in South Asia. Nat Sustain 3:336–343. https://doi.org/10.1038/s41893-020-0500-2
Jat ML, Gathala MK, Choudhary M et al (2023) Conservation agriculture for regenerating soil health and climate change mitigation in smallholder systems of South Asia. In: Sparks DL (ed) Advances in agronomy. Academic, pp 183–277
Kassam A, Friedrich T, Derpsch R (2019) Global spread of conservation agriculture. Int J Environ Stud 76:29–51. https://doi.org/10.1080/00207233.2018.1494927
Kearney SP, Coops NC, Chan KMA et al (2017) Predicting carbon benefits from climate-smart agriculture: high-resolution carbon mapping and uncertainty assessment in El Salvador. J Environ Manag 202:287–298. https://doi.org/10.1016/j.jenvman.2017.07.039
Kim Y, Lichtenberg E, Newburn DA (2024) Payments and penalties in ecosystem services programs. J Environ Econ Manag 126:102988. https://doi.org/10.1016/j.jeem.2024.102988
Kirkegaard JA, Conyers MK, Hunt JR et al (2014) Sense and nonsense in conservation agriculture: principles, pragmatism and productivity in Australian mixed farming systems. Agric Ecosyst Environ 187:133–145. https://doi.org/10.1016/j.agee.2013.08.011
Knorr D, Augustin MA (2025) Towards resilient food systems: interactions with indigenous knowledge. Trends Food Sci Technol 156:104875. https://doi.org/10.1016/j.tifs.2025.104875
Köchy M, Hiederer R, Freibauer A (2015) Global distribution of soil organic carbon—part 1: masses and frequency distributions of SOC stocks for the tropics, permafrost regions, wetlands, and the world. Soil 1:351–365. https://doi.org/10.5194/soil-1-351-2015
Kragt ME, Gibson FL, Maseyk F, Wilson KA (2016) Public willingness to pay for carbon farming and its co-benefits. Ecol Econ 126:125–131. https://doi.org/10.1016/j.ecolecon.2016.02.018
Kragt ME, Dumbrell NP, Blackmore L (2017) Motivations and barriers for Western Australian broad-acre farmers to adopt carbon farming. Environ Sci Pol 73:115–123. https://doi.org/10.1016/j.envsci.2017.04.009
Krauss M, Wiesmeier M, Don A et al (2022) Reduced tillage in organic farming affects soil organic carbon stocks in temperate Europe. Soil Tillage Res 216:105262. https://doi.org/10.1016/j.still.2021.105262
Lal R (2002a) Soil carbon sequestration in China through agricultural intensification, and restoration of degraded and desertified ecosystems. Land Degrad Dev 13:469–478. https://doi.org/10.1002/ldr.531
Lal R (2002b) Carbon sequestration in dryland ecosystems of West Asia and North Africa. Land Degrad Dev 13:45. https://doi.org/10.1002/ldr.477.abs
Lal R (2014) Societal value of soil carbon. J Soil Water Conserv 69:186A–192A. https://doi.org/10.2489/jswc.69.6.186A
Lal R (2020) The role of industry and the private sector in promoting the "4 per 1000" initiative and other negative emission technologies. Geoderma 378:114613. https://doi.org/10.1016/j.geoderma.2020.114613
Lal R (2023a) Carbon farming by recarbonization of agroecosystems. Pedosphere 33:676–679. https://doi.org/10.1016/j.pedsph.2023.07.024
Lal R (2023b) Farming systems to return land for nature: It's all about soil health and re-carbonization of the terrestrial biosphere. Farming System 1:100002. https://doi.org/10.1016/j.farsys.2023.100002

Lal R (2024) Carbon farming in global drylands. In: El-Beltagy A, Lal R, Malik K (eds) Climate change and sustainable agro-ecology in global Drylands. CABI, pp 56–76
Lal R (2025) Managing soil health in Africa (MASHA) by re-carbonization of AgroEcosystems. Egypt J Soil Sci 65:301–320. https://doi.org/10.21608/ejss.2024.334426.1913
Lal R, Kimble JM (1997) Conservation tillage for carbon sequestration. Nutr Cycl Agroecosyst 49:243–253. https://doi.org/10.1023/A:1009794514742
Lei K, Bucka FB, Teixeira PPC et al (2025) Balancing organic and inorganic carbon dynamics in enhanced rock weathering: implications for carbon sequestration. Glob Chang Biol 31:e70186. https://doi.org/10.1111/gcb.70186
Leifeld J (2023) Carbon farming: Climate change mitigation via non-permanent carbon sinks. J Environ Manag 339:117893. https://doi.org/10.1016/j.jenvman.2023.117893
Li S, Tasnady D (2023) Biochar for soil carbon sequestration: current knowledge, mechanisms, and future perspectives. C J Carbon Res 9:67. https://doi.org/10.3390/c9030067
Li M, Xiao H (2024) Spatiotemporal heterogeneity and its influencing factors: perspective of livestock carbon emission and emission intensity considering livestock farming and energy use. Clean Techn Environ Policy 27:2433–2450. https://doi.org/10.1007/s10098-024-03007-6
Liang J, Zhao Y, Chen L, Liu J (2024) Soil inorganic carbon storage and spatial distribution in irrigated farmland on the North China Plain. Geoderma 445:116887. https://doi.org/10.1016/j.geoderma.2024.116887
Liao Y, Deng L, Huang Y et al (2025) Inorganic carbon should be considered for carbon sequestration in agricultural soils. Glob Chang Biol 31:e70160. https://doi.org/10.1111/gcb.70160
Livingston JE, Rummukainen M (2023) Making policy-relevant knowledge in the IPCC special report on 1.5 degrees: an analysis of reviewer comments. Environ Sci Pol 147:305–314. https://doi.org/10.1016/j.envsci.2023.06.001
Loria N, Bhardwaj SK, Mahajan PK et al (2015) Vulnerability assessment of farming community to environmental changes in low hills of Himachal Pradesh in India. Int J Curr Microbiol Appl Sci 4:547–560
Maharjan KL, Gonzalvo CM, Baggo JC (2025) Balancing tradition and innovation: the role of environmental conservation agriculture in the sustainability of the Ifugao Rice terraces. Agriculture 15:246. https://doi.org/10.3390/agriculture15030246
Marks AB (2020) (Carbon) farming our way out of climate change, vol 97. University of Colorado Boulder, pp 498–555
Mati (2025) Mati carbon is the $50M grand prize winner. Mati Carbon. https://www.mati.earth/mati-carbon-is-the-grand-prize-winner-of-the-xprize-carbon-removal-competition/. Accessed 19 May 2025
Matthews HD, Graham TL, Keverian S et al (2014) National contributions to observed global warming. Environ Res Lett 9:014010. https://doi.org/10.1088/1748-9326/9/1/014010
McDonald H, Frelih-Larsen A, Lóránt A et al (2021) Carbon farming: Making agriculture fit for 2030. In: Study for the committee on Environment, Public Health and Food Safety (ENVI). Policy Department for Economic, Scientific and Quality of Life Policies, European Parliament, Luxembourg
Metrikaitytė Gudelė G, Sužiedelytė Visockienė J (2024) Remote sensing data application to monitor carbon farming practices. The international archives of the photogrammetry. Remote Sens Spat Inf Sci XLVIII-4/W9-2024:257–265. https://doi.org/10.5194/isprs-archives-XLVIII-4-W9-2024-257-2024
Miao S, Chen B, Jiang N (2025) Collaboration among governments, agribusinesses, and rural households for improving the effectiveness of conservation tillage technology adoption. Sci Rep 15:45. https://doi.org/10.1038/s41598-024-83827-0
Minasny B, Malone BP, McBratney AB et al (2017) Soil carbon 4 per mille. Geoderma 292:59–86. https://doi.org/10.1016/j.geoderma.2017.01.002
Mirzabaev A, Bezner Kerr R, Hasegawa T et al (2023) Severe climate change risks to food security and nutrition. Clim Risk Manag 39:100473. https://doi.org/10.1016/j.crm.2022.100473

Mitchell E, Takeda N, Grace L et al (2024) Making soil carbon credits work for climate change mitigation. Carbon Manag 15:2430780. https://doi.org/10.1080/17583004.2024.2430780
Mondal S (2021) Impact of climate change on soil fertility. In: Mishra DK, Varma A (eds) Climate change and the microbiome, sustenance of the ecosphere. Springer, Cham, pp 551–569
Montanarella L (2017) Translating soil science knowledge to public policy. In: Field DJ, Morgan CLS, McBratney AB (eds) Global soil security. Springer, Cham, pp 451–456
Montanarella L, Pennock DJ, McKenzie N et al (2016) World's soils are under threat. Soil 2:79–82. https://doi.org/10.5194/soil-2-79-2016
Moodley V, Phophi MM, Mafongoya PL (2024) Regenerative agriculture: a paradigm shift for food production in South Africa. Outlook Agric 53:336–350. https://doi.org/10.1177/00307270241292085
Moore JM, Manter DK, Bowman M et al (2023) A framework to estimate climate mitigation potential for US cropland using publicly available data. J Soil Water Conserv 78:193–206. https://doi.org/10.2489/jswc.2023.00132
Mosquera-Losada MR, Santiago-Freijanes JJ, Pisanelli A et al (2018) Agroforestry in the European common agricultural policy. Agrofor Syst 92:1117–1127. https://doi.org/10.1007/s10457-018-0251-5
Multer Hopkins B, Lal R, Lyons WB, Welch SA (2024) Carbon capture potential and environmental impact of concrete weathering in soil. Sci Total Environ 957:177692. https://doi.org/10.1016/j.scitotenv.2024.177692
Myrgiotis V, Blei E, Clement R et al (2020) A model-data fusion approach to analyse carbon dynamics in managed grasslands. Agric Syst 184:102907. https://doi.org/10.1016/j.agsy.2020.102907
Neher DA, Harris JM, Horner CE et al (2022) Resilient soils for resilient farms: an integrative approach to assess, promote, and value soil health for small- and medium-size farms. Phytobiomes J 6:201–206. https://doi.org/10.1094/PBIOMES-10-21-0060-P
Nevalainen O, Niemitalo O, Fer I et al (2022) Towards agricultural soil carbon monitoring, reporting, and verification through the Field Observatory Network (FiON). Geosci Instrum Methods Data Syst 11:93–109. https://doi.org/10.5194/gi-11-93-2022
Ng W, Padarian J, Dobarco MR et al (2025) Mapping the distribution and magnitude of soil inorganic and organic carbon stocks across Australia. Geoderma 456:117239. https://doi.org/10.1016/j.geoderma.2025.117239
Noellemeyer E, Álvarez L, Álvarez C et al (2024) From science to practice: the AGSUS protocol for monitoring and certification of sustainable soil management and carbon sequestration. Soil Tillage Res 241:106102. https://doi.org/10.1016/j.still.2024.106102
Ntshangase N, Muroyiwa B, Sibanda M (2018) Farmers' perceptions and factors influencing the adoption of no-till conservation agriculture by small-scale farmers in Zashuke, KwaZulu-Natal Province. Sustainability 10:555. https://doi.org/10.3390/su10020555
Nyawasha RW, Falconnier GN, Todoroff P et al (2025) Drivers of soil organic carbon stocks at village scale in a sub-humid region of Zimbabwe. Catena (Amst) 252:108843. https://doi.org/10.1016/j.catena.2025.108843
Odebiri O, Mutanga O, Odindi J et al (2021) Deep learning approaches in remote sensing of soil organic carbon: a review of utility, challenges, and prospects. Environ Monit Assess 193:802. https://doi.org/10.1007/s10661-021-09561-6
Ogieriakhi MO, Woodward RT (2022) Understanding why farmers adopt soil conservation tillage: a systematic review. Soil Secur 9:100077. https://doi.org/10.1016/j.soisec.2022.100077
Okoli AO, Birkenberg A (2024) Monitoring soil carbon in smallholder carbon projects: insights from Kenya. Clim Chang 177:143. https://doi.org/10.1007/s10584-024-03796-1
Okyere CY, Kornher L (2023) Carbon farming training and welfare: evidence from northern Ghana. Land Use Policy 134:106932. https://doi.org/10.1016/j.landusepol.2023.106932
Page G, Bellotti B (2015) Farmers value on-farm ecosystem services as important, but what are the impediments to participation in PES schemes? Sci Total Environ 515–516:12–19. https://doi.org/10.1016/j.scitotenv.2015.02.029

Pagiola S (2008) Payments for environmental services in Costa Rica. Ecol Econ 65:712–724. https://doi.org/10.1016/j.ecolecon.2007.07.033

Panagos P, Montanarella L, Barbero M et al (2022) Soil priorities in the European Union. Geoderma Reg 29:e00510. https://doi.org/10.1016/j.geodrs.2022.e00510

Panwar P, Bhatt VK, Pal S et al (2020) Vulnerability of agricultural households to climate change in the hill state of North Western Himalaya. Mausam 71:199–208

Pape D, Lewandrowski J, Steele R, et al (2016) Managing agricultural land for greenhouse gas mitigation within the United States. https://www.usda.gov/sites/default/files/documents/White_Paper_WEB71816.pdf. Accessed 20 Apr 2025

Pascual U, Termansen M, Abson DJ (2015) The economic value of soil carbon. In: Soil carbon: science, management and policy for multiple benefits. CABI, pp 179–187

Patle GT, Bandyopadhyay KK, Singh DK (2013) Impact of conservation agriculture and resource conservation technologies on carbon sequestration-a review. Indian J Agric Sci 83:3–13

Paudyal BR, Chanana N, Khatri-Chhetri A et al (2019) Gender integration in Climate change and agricultural policies: the case of Nepal. Front Sustain Food Syst 3:66. https://doi.org/10.3389/fsufs.2019.00066

Pauli N, Abbott LK, Rex R et al (2018) A farmer–scientist investigation of soil carbon sequestration potential in a chronosequence of perennial pastures. Land Degrad Dev 29:4301–4312. https://doi.org/10.1002/ldr.3184

Perosa B, Newton P, da Silva RFB (2023) A monitoring, reporting and verification system for low carbon agriculture: a case study from Brazil. Environ Sci Pol 140:286–296. https://doi.org/10.1016/j.envsci.2022.12.006

Petersen EH, Hoyle FC (2016) Estimating the economic value of soil organic carbon for grains cropping systems in Western Australia. Soil Res 54:383. https://doi.org/10.1071/SR15101

Pham-Kieu M, Ives S, Badgery W, Harrison MT (2024) Tensions between private and public benefit associated with carbon farming. Sustain Prod Consum 49:387–397. https://doi.org/10.1016/j.spc.2024.07.011

Plastina A, Jo H, Wongpiyabovorn O (2024a) The business case for carbon farming in the USA. Carbon Balance Manag 19:7. https://doi.org/10.1186/s13021-024-00253-5

Plastina A, Wongpiyabovorn O, Jo H (2024b) How to grow and sell carbon credits in US agriculture. In: Ag decision maker file A1–76. Iowa State University Extension and Outreach. https://www.extension.iastate.edu/agdm/crops/pdf/a1-76.pdf. Accessed 14 Apr 2025

Prutzer E, Patrick A, Ishtiaque A et al (2023) Climate-smart irrigation and responsible innovation in South Asia: a systematic mapping. Ambio 52:2009–2022. https://doi.org/10.1007/s13280-023-01895-4

Pudasaini K, Bhattarai T, Rolfe J (2024) Exploring the barriers to farmer participation in soil carbon projects under the Australian Carbon Credit Unit Scheme. Aust J Environ Manag 1–19:1–19. https://doi.org/10.1080/14486563.2024.2393246

Raina N, Zavalloni M, Viaggi D (2024) Incentive mechanisms of carbon farming contracts: a systematic mapping study. J Environ Manag 352:120126. https://doi.org/10.1016/j.jenvman.2024.120126

Sangha KK, Ahammad R, Russell-Smith J, Costanza R (2024) Payments for ecosystem services opportunities for emerging nature-based solutions: integrating indigenous perspectives from Australia. Ecosyst Serv 66:101600. https://doi.org/10.1016/j.ecoser.2024.101600

Santra P, Kumawat RN, Mertia RS et al (2012) Spatial variation of soil organic carbon stock in a typical agricultural farm of hot arid ecosystem of India. Curr Sci 102:1303–1309

Schilling F, Baumüller H, Ecuru J, Braun JV (2023) Carbon farming in Africa: opportunities and challenges for engaging smallholder farmers. Bonn, Germany

Schreefel L, van Zanten HHE, Groot JCJ et al (2022) Tailor-made solutions for regenerative agriculture in The Netherlands. Agric Syst 203:103518. https://doi.org/10.1016/j.agsy.2022.103518

Simone TE, Lambert DM, Cuvaca I, Eash NS (2017) Soil carbon sequestration, carbon markets, and conservation agriculture practices: a hypothetical examination in Mozambique. Int Soil Water Conserv Res 5:167–179. https://doi.org/10.1016/j.iswcr.2017.06.001

Somasundaram J, Sinha NK, Dalal RC et al (2020) No-till farming and conservation agriculture in South Asia – issues, challenges, prospects and benefits. CRC Crit Rev Plant Sci 39:236–279. https://doi.org/10.1080/07352689.2020.1782069

Soterroni AC, Mosnier A, Carvalho AXY et al (2018) Future environmental and agricultural impacts of Brazil's Forest code. Environ Res Lett 13:074021. https://doi.org/10.1088/1748-9326/aaccbb

Stankovics P, Montanarella L, Kassai P et al (2020) The interrelations of land ownership, soil protection and privileges of capital in the aspect of land take. Land Use Policy 99:105071. https://doi.org/10.1016/j.landusepol.2020.105071

Tang K, He C, Ma C, Wang D (2019) Does carbon farming provide a cost-effective option to mitigate GHG emissions? Evidence from China. Aust J Agric Resour Econ 63:575–592. https://doi.org/10.1111/1467-8489.12306

Thompson NM, Hughes MN, Nuworsu EKM, et al (2021) Opportunities and challenges associated with "carbon farming" for U.S. row-crop producers. Purdue University. https://ag.purdue.edu/commercialag/home/resource/2021/06/opportunities-and-challenges-associated-with-carbon-farming-for-u-s-row-crop-producers/. Accessed 15 Apr 2025

Torres JLR, Mazetto Júnior JC, Silva Júnior J et al (2019) Soil physical attributes and organic matter accumulation under no-tillage systems in the Cerrado. Soil Res 57:712. https://doi.org/10.1071/SR19047

USDA (2024) Farm bill 2024: themes in the proposed conservation titles. The National Agricultural Law Center. https://nationalaglawcenter.org/farm-bill-2024-themes-in-the-proposed-conservation-titles/#:~:text=The%20Conservation%20Title%20of%20the%20Farm%20Bill%2C%20also,ranchers%20address%20environmental%20resource%20concerns%20on%20private%20lands. Accessed 21 Apr 2025

USDA (2025) NRCS USDA programs and initiatives. https://www.nrcs.usda.gov/programs-initiatives. Accessed 22 Apr 2025

Vågen T-G, Lal R, Singh BR (2005) Soil carbon sequestration in sub-Saharan Africa: a review. Land Degrad Dev 16:53–71. https://doi.org/10.1002/ldr.644

Vanneste T, De Praetere L, Pardon P et al (2025) CARAT: an innovative tool for quantifying carbon sequestration in agroforestry systems. Agrofor Syst 99:68. https://doi.org/10.1007/s10457-025-01162-3

Vasavi S, Anandaraja N, Murugan PP et al (2025) Challenges and strategies of resource poor farmers in adoption of innovative farming technologies: a comprehensive review. Agric Syst 227:104355. https://doi.org/10.1016/j.agsy.2025.104355

Velten S, Jager NW, Newig J (2021) Success of collaboration for sustainable agriculture: a case study meta-analysis. Environ Dev Sustain 23:14619–14641. https://doi.org/10.1007/s10668-021-01261-y

Villalba R, Joshi G, Daum T, Venus TE (2024) Financing Climate-smart agriculture: a case study from the indo-Gangetic Plains. Mitig Adapt Strateg Glob Chang 29:33. https://doi.org/10.1007/s11027-024-10127-3

Vistarte L, Kubule A, Rozentale L, Pubule J (2024) Carbon farming: a systematic literature review on sustainable practices. Environm Clim Technol 28:895–909. https://doi.org/10.2478/rtuect-2024-0068

Von Riese WE (2025) Carbon credits for Africa's first regenerative farming program. https://esgreview.net/2025/03/26/carbon-credits-for-africas-first-regenerative-farming-program/. Accessed 2 Apr 2025

Wang T, Zhang Y, Yue C et al (2025) Progress and challenges in remotely sensed terrestrial carbon fluxes. Geo-spat Inf Sci 28:1–21. https://doi.org/10.1080/10095020.2024.2336599

Wienke FF, de Mattos RF (2022) National policy on payment for environmental services: a recent instrument for agroecological transition. Acad J Fac Law Recife 94:130. https://doi.org/10.51359/2448-2307.2022.251472

Yu H, Song W (2023) Research progress on the impact of land use change on soil carbon sequestration. Land (Basel) 12:213. https://doi.org/10.3390/land12010213

Zeleke A, Phung T, Tulyasuwan N et al (2016) Role of agriculture, forestry and other land use mitigation in INDCs and National Policy in Asia. https://winrock.org/wp-content/uploads/2016/05/AFOLU-LEDS-Working-Group-Techincal-paper-Role-of-AFOLU-mitigation-in-INDCs-and-national-policy-in-Asia-1.0-Feb-25-2016.pdf. Accessed 20 Apr 2025

Zhang T, Xiong W, Sapkota TB et al (2022) The optimization of conservation agriculture practices requires attention to location-specific performance: evidence from large scale gridded simulations across South Asia. Field Crop Res 282:108508. https://doi.org/10.1016/j.fcr.2022.108508

Zhu X, Chen B, Zhu L, Xing B (2017) Effects and mechanisms of biochar-microbe interactions in soil improvement and pollution remediation: a review. Environ Pollut 227:98–115. https://doi.org/10.1016/j.envpol.2017.04.032

The manufacturer's authorised representative in the EU is Springer Nature Customer Service Centre GmbH, Europaplatz 3, 69115 Heidelberg, Germany. If you have any concerns regarding our products, please contact ProductSafety@springernature.com

Printed and bound by CPI Group (UK) Ltd, Croydon, CR0 4YY
30/09/2026
02225564-0013